I0885428

Quantum Computing and Cryptography in Future Computers

Shyam R. Sihare
Dr. A.P.J. Abdul Kalam Government College, India

IGI Global
PUBLISHER of TIMELY KNOWLEDGE

Published in the United States of America by
IGI Global
Engineering Science Reference (an imprint of IGI Global)
701 E. Chocolate Avenue
Hershey PA, USA 17033
Tel: 717-533-8845
Fax: 717-533-8661
E-mail: cust@igi-global.com
Web site: http://www.igi-global.com

Library of Congress Cataloging-in-Publication Data

CIP Data Pending
ISBN: 978-1-7998-9522-0
eISBN: 978-1-7998-9524-4

British Cataloguing in Publication Data
A Cataloguing in Publication record for this book is available from the British Library.

All work contributed to this book is new, previously-unpublished material.
The views expressed in this book are those of the authors, but not necessarily of the publisher.

For electronic access to this publication, please contact: eresources@igi-global.com.

Advances in Systems Analysis, Software Engineering, and High Performance Computing (ASASEHPC) Book Series

Vijayan Sugumaran
Oakland University, Rochester, USA

ISSN:2327-3453
EISSN:2327-3461

MISSION

The theory and practice of computing applications and distributed systems has emerged as one of the key areas of research driving innovations in business, engineering, and science. The fields of software engineering, systems analysis, and high performance computing offer a wide range of applications and solutions in solving computational problems for any modern organization.

The **Advances in Systems Analysis, Software Engineering, and High Performance Computing (ASASEHPC) Book Series** brings together research in the areas of distributed computing, systems and software engineering, high performance computing, and service science. This collection of publications is useful for academics, researchers, and practitioners seeking the latest practices and knowledge in this field.

Coverage

- Engineering Environments
- Enterprise Information Systems
- Network Management
- Virtual Data Systems

Titles in this Series

For a list of additional titles in this series, please visit: www.igi-global.com/book-series

Revolutionizing Curricula Through Computational Thinking, Logic, and Problem Solving
Mathias Mbu Fonkam (Penn State University, USA) and Narasimha Rao Vajjhala (University of New York, Tirana, Albania)
Engineering Science Reference • copyright 2024 • 245pp • H/C (ISBN: 9798369319741) • US $245.00 (our price)

Harnessing High-Performance Computing and AI for Environmental Sustainability
Arshi Naim (King Khalid University, Saudi Arabia)
Engineering Science Reference • copyright 2024 • 401pp • H/C (ISBN: 9798369317945) • US $315.00 (our price)

Recent Trends and Future Direction for Data Analytics
Aparna Kumari (Nirma University, Ahmedabad, India)
Engineering Science Reference • copyright 2024 • 350pp • H/C (ISBN: 9798369336090) • US $345.00 (our price)

Advancing Software Engineering Through AI, Federated Learning, and Large Language Models
Avinash Kumar Sharma (Sharda University, India) Nitin Chanderwal (University of Cincinnati, USA) Amarjeet Prajapati (Jaypee Institute of Information Technology, India) Pancham Singh (Ajay Kumar Garg Engineering College, Ghaziabad, India) and Mrignainy Kansal (Ajay Kumar Garg Engineering College, Ghaziabad, India)
Engineering Science Reference • copyright 2024 • 354pp • H/C (ISBN: 9798369335024) • US $355.00 (our price)

Advancements, Applications, and Foundations of C++
Shams Al Ajrawi (Wiley Edge, USA & Alliant International University, USA) Charity Jennings (Wiley Edge, USA & University of Phoenix, USA) Paul Menefee (Wiley Edge, USA) Wathiq Mansoor (University of Dubai, UAE) and Mansoor Ahmed Alaali (Ahlia University, Bahrain)
Engineering Science Reference • copyright 2024 • 564pp • H/C (ISBN: 9798369320075) • US $295.00 (our price)

IGI Global
PUBLISHER of TIMELY KNOWLEDGE

701 East Chocolate Avenue, Hershey, PA 17033, USA
Tel: 717-533-8845 x100 • Fax: 717-533-8661
E-Mail: cust@igi-global.com • www.igi-global.com

Table of Contents

Detailed Table of Contents

In this chapter, quantum computing concepts are discussed in detail. Discussion about qubit and allied terms is done in brief. Light is also thrown on quantum ambiguity, key distribution, quantum entanglement, quantum simulation, digital quantum simulation, etc. Further sections explore applications of quantum in image processing, wireless sensor protocol, routing protocol, and cyber security. Bibliographic analysis of research trends in quantum computing in cyber security is also discussed in detail in this chapter.

Quantum computing plays a major role in modern computation. It can perform operations exponentially faster when compared to classical computation. It has applications in various areas like secure communication, drug design, artificial intelligence, cyber security, etc. Thus, the researchers and students are showing interest to perform experiments in quantum computing to design novel architectures. But to learn/understand quantum computing, one should have strong knowledge of its basics. Because quantum computing performs operations at the atomic level, so the learners need to understand basic concepts like Qubits, superposition, quantum gates, etc. This chapter gives a clear idea about the basic concepts of quantum computing like Qubits, superposition, entanglement, and quantum gates.

The author will discuss different quantum protocols in this chapter for guided media and open space communication. The author examines already developed quantum protocols for photons and electrons. Earlier developed quantum protocols are the basis of recently developed quantum protocols. The author makes a study on quantum protocols with the help of quantum mechanics features such as entanglement, superposition, uncertainty principle, and no cloning.

The aim of this chapter is to introduce the reader to various aspects of quantum entanglement. A detailed summary of the jargon of important mathematical notions, and concepts used in entanglement theory is provided. Various methods of entanglement generation are discussed followed by an introduction to the methods of detection and quantification of quantum entanglement in terms of various witnesses and measures. A brief account of some well-known applications of quantum entanglement is presented. Finally, a set of problems with solutions is provided, illustrating various important concepts discussed in the chapter.

This chapter discusses the significance of quantum cryptography in securing communication channels for the future. It highlights the challenges posed by quantum computing to traditional cryptographic systems and the potential solutions offered by quantum-resistant protocols. The chapter emphasizes the transition from classical to quantum-resistant cryptography, highlighting hybrid cryptosystems, algorithm agility, and standards development. It discusses the vulnerability of classical cryptographic systems to quantum algorithms, such as Shor's and Grover's algorithms. It also explains the concept of hybrid cryptosystems, which combine classical algorithms with post-quantum key exchange protocols.

Charles Bennett and Gilles Brassard, in 1984, proposed the first QKD protocol and called BB84. It is assumed the protocol shares a quantum key safely (between two parties). In 2000 it was implemented easily and showed a significant method for detecting an attacker that trying to get the shared key by utilizing the final key length. When the length exceeds a certain value, that is calculated before the transmitting the key, the majority of prior works agree with that, but this chapter showed a significant threat that affects the final key to same. There is no attacker at the middle. Moreover, this chapter analyzed the final key lengths and showed the harmed value of the final key length with the attacker effect. It also showed how often these values could be within the threshold. Furthermore, a solution was found to detect the attacker by using a machine learning technique. The results showed a promising accuracy to detect the attacker relying on the final key length.

The multifaceted landscape of quantum technologies, including their advances, challenges, and ethical implications, is explored in the chapter. It covers the diverse applications of quantum computing, collaborating across scientific domains, impacting society and the critical role of quantum literacy. The chapter explores different aspects of quantum technology, including interdisciplinary research, quantum communication networks, the relationship between quantum computing and artificial intelligence, and ethical considerations. It also highlights the societal responsibilities of the quantum community. Topics covered include ethical supply chains, media representation, disaster response, and international relations. The chapter advocates ethical governance, responsible innovation, and inclusive access to ensure harmonious integration into the fabric of our society as quantum technologies continue to develop.

Preface

It is with great pleasure and enthusiasm that I present to you this meticulously crafted reference book, *Quantum Computing and Quantum Cryptography in Future Computers*. As the editor of this comprehensive volume, it is my honor to guide you through the intricate realms of Quantum Computing and Quantum Cryptography, shedding light on their principles, applications, and profound impact on the future of computing.

In recent years, the fields of Quantum Computing and Quantum Cryptography have emerged as transformative forces, challenging the conventional paradigms of information processing and security. This book, born out of two years of dedicated research and exploration, aims to demystify these complex subjects and provide a solid foundation for understanding their nuances.

The primary objectives of this book are manifold. Firstly, it seeks to introduce our readers to the fundamental concepts of Quantum Computing, offering a clear and concise overview of topics such as Quantum Basics, Single Qubit and Multi Qubit Systems, Quantum Algorithms, and Quantum Entanglement. As we delve into the quantum realm, we explore the intricate dance of particles and the potential applications of Quantum Entanglement, laying the groundwork for a deeper understanding.

Beyond the fundamentals, the book delves into the practical applications of Quantum Computing, exploring its symbiotic relationship with fields such as security, Blockchain, Artificial Intelligence, and Machine Learning. The integration of Quantum Computing with these domains promises unprecedented advancements and capabilities, shaping the future of technology.

A significant portion of our exploration is dedicated to Quantum Cryptography, a vital companion in the quest for secure and unbreakable communication. We delve into the principles of Quantum Key Distribution and its protocols, showcasing their role in bolstering the security of our digital communications. Real-world case studies further illustrate the practical applications of Quantum Cryptography, offering valuable insights into its efficacy.

The impact of this book extends beyond the academic realm. It serves as a valuable resource for researchers, developers, engineers, and practitioners navigating the intricate landscapes of Quantum Computing, Quantum Cryptography, and their diverse applications. Whether you are a seasoned professional or a graduate/post-graduate student in computer science and engineering/information technology, this reference book is designed to be an invaluable companion in your journey.

As we stand on the precipice of a new era in computing, characterized by the marriage of quantum principles with practical applications, this book aims to be a beacon of knowledge. It is my sincere hope that the insights presented within these pages will inspire further research, innovation, and collaboration, paving the way for new opportunities and discoveries in the fascinating realm of Quantum Computing and Quantum Cryptography.

Chapter 1: "Quantum Computing: Throwing Light on Quantum Computing" provides a detailed exploration of Quantum Computing concepts. The authors, Shruti Aggarwal, Vishal Bharti, and Afroj Badhon, delve into essential elements such as qubits, quantum ambiguity, key distribution, and quantum entanglement. Additionally, the chapter explores the practical applications of quantum in image processing, Wireless Sensor Protocol, Routing Protocol, and Cyber Security. A bibliographic analysis of research trends in Quantum Computing in Cyber Security is also included, offering a comprehensive foundation for understanding the quantum landscape.

Chapter 2: "Fundamentals of Quantum Computation and Basic Quantum Gates," authored by Swathi Mummadi and Bhawana Rudra, aims to establish a clear understanding of the basic concepts essential for comprehending quantum computing. The chapter discusses crucial components like Qubits, Superposition, Quantum gates, and Entanglement. With an emphasis on the atomic-level operations of quantum computing, the authors lay the groundwork for researchers and students interested in experimenting with quantum computing and designing novel architectures.

Chapter 3: "Quantum Key Distribution Protocols" by Shyam Sihare provides an in-depth exploration of different quantum protocols for guided media and open space communication. Examining existing quantum protocols for photons and electrons, the author highlights how earlier developments serve as the foundation for recent advancements. The chapter conducts a study on quantum protocols utilizing features of quantum mechanics, such as entanglement, superposition, uncertainty principle, and no cloning.

Chapter 4: "Quantum Entanglement," written by Javid Naikoo, introduces readers to various aspects of quantum entanglement. The chapter covers mathematical notions, entanglement generation methods, detection, quantification, and applications. It concludes with problems and solutions that illustrate key concepts discussed, providing a comprehensive and practical understanding of this fundamental quantum phenomenon.

Chapter 5: "The Potential of Quantum Cryptography in Securing Future Communication Channels" by Shyam Sihare explores the significance of quantum cryptography in securing communication channels for the future. The chapter delves into the challenges posed by quantum computing to traditional cryptographic systems and discusses potential solutions offered by quantum-resistant protocols. The transition from classical to quantum-resistant cryptography, vulnerability analysis, and hybrid cryptosystems are highlighted, providing insights into the evolving landscape of secure communication.

Chapter 6: "Quantum Program: A Sequence of Quantum Circuits Using Qiskit" authored by Amlan Sengupta and Debotosh Bhattacharjee focuses on the rapid evolution of quantum theory and technology. The chapter explains the implementation of quantum algorithms using Qiskit, emphasizing the efficiency and speed enhancements offered by quantum computing. It provides a detailed explanation of quantum software, Qiskit, and its application in building and executing quantum circuits.

Chapter 7: "QKD Protocol for Securing the Communication With Real-Life Application Scenarios" by Hasan Al-Mohammed, Elias Yaacoub, and Khalid Abualsaud explores Quantum Key Distribution (QKD) techniques. The chapter discusses the principles of quantum security and cryptography, highlighting recent developments and potential applications in real-life scenarios such as healthcare and railway monitoring. The focus is on securing communication through well-known QKD protocols, BB84 and E91.

Chapter 8: "Machine Learning Techniques for Studying the Effect of The Final Key Lengths in the QKD Protocol on Detecting Attacks Problems and Solutions" by Hasan Al-Mohammed, Afnan Al-Ali, Elias Yaacoub, and Khalid Abualsaud delves into the analysis of QKD protocol's final key lengths. The chapter addresses threats to the final key length and proposes a machine learning technique to detect potential attackers. By analyzing the final key lengths, the authors present solutions and insights into maintaining secure communication in the presence of potential threats.

Chapter 9: "Embarking on Quantum Horizons" by Shyam Sihare explores the multifaceted landscape of quantum technologies. The chapter covers diverse applications, interdisciplinary research, quantum communication networks, the relationship between quantum computing and artificial intelligence, and ethical considerations. It advocates for ethical governance, responsible innovation, and inclusive access, ensuring the harmonious integration of quantum technologies into our society.

In conclusion, *Quantum Computing and Quantum Cryptography in Future Computers* stands as a testament to the profound advancements and transformative potential of quantum technologies. As the editor of this comprehensive volume, it has been a privilege to guide our readers through the intricate realms of Quantum Computing and Quantum Cryptography.

This meticulously crafted reference book has successfully achieved its primary objectives, offering a clear and concise overview of fundamental concepts in Quantum Computing, ranging from Quantum Basics to Quantum Entanglement. The exploration extends beyond the basics, delving into the practical applications of Quantum Computing and its symbiotic relationship with security, Blockchain, Artificial Intelligence, and Machine Learning.

A significant portion of our journey is dedicated to Quantum Cryptography, where we navigate through the principles of Quantum Key Distribution and its protocols. Real-world case studies provide tangible insights into the efficacy of Quantum Cryptography, further reinforcing its role in securing digital communications.

The impact of this book extends far beyond academia, serving as an invaluable resource for researchers, developers, engineers, and practitioners navigating the intricate landscapes of Quantum Computing and Quantum Cryptography. Whether you are a seasoned professional or a graduate/postgraduate student in computer science and engineering/information technology, this reference book is designed to be a reliable companion in your exploration of these cutting-edge fields.

As we stand on the precipice of a new era in computing, characterized by the fusion of quantum principles with practical applications, this book aims to be a beacon of knowledge. It is my sincere hope that the insights presented within these pages will not only demystify the complexities of Quantum Computing and Quantum Cryptography but also inspire further research, innovation, and collaboration. May this volume pave the way for new opportunities and discoveries in the fascinating realm of Quantum Computing and Quantum Cryptography, shaping the future of technology and ushering in a new era of secure and efficient computation.

Shyam R. Sihare

Dr. A.P.J. Abdul Kalam Government College, India

Chapter 1
Quantum Computing

Shruti Aggarwal
Chandigarh University, India

Vishal Bharti
Chandigarh University, India

Afroj Jahan Badhon
Chandigarh University, India

ABSTRACT

In this chapter, quantum computing concepts are discussed in detail. Discussion about qubit and allied terms is done in brief. Light is also thrown on quantum ambiguity, key distribution, quantum entanglement, quantum simulation, digital quantum simulation, etc. Further sections explore applications of quantum in image processing, wireless sensor protocol, routing protocol, and cyber security. Bibliographic analysis of research trends in quantum computing in cyber security is also discussed in detail in this chapter.

1. QUANTUM COMPUTING

Processed Data called information, is stored in computer in the form of Bits (Binary Digits) i.e., 0 and 1. Data storage in memory is related to magnetic energy. The binary bits are stored in memory by aligning the atomic magnetic dipoles permanently (Yusuf et al., 2015). Collection of many individual atoms with their magnetic moments align at certain location form a small bar magnet. Subatomic particles would behave strangely by not being localized to a specific position and not having specific momentum at a given instant of time.

DOI: 10.4018/978-1-7998-9522-0.ch001

1.1 Quantum Ambiguity

Atoms or subatomic particles behave like waves would substantiate their uncertainty of being not localized. The mathematical relation describing the wavelength λ_d of the matter waves obtained using a relation of energy to momentum: p=hv/c=h/ λ_d which implies de-Broglie wavelength λ_d = h/p.

Particle presence could only be described using probabilities, there exists a mathematical function describing particle waves called Wave Function, denoted by the Greek letter 'Ψ' (psi) (Agrawal et al., 2012) The wave function Ψ also carries the information of position or momenta, which could be obtained from $|\Psi|^2$, wherein the information would be in terms of probabilities. Finding the probabilities of the particles through the wave functions, and further running several simulations to reveal the probabilistic locations of electrons at different energy levels would lead you to get a clear understanding of different orbital structures.

At Quantum regime (atomic scale) individual atoms exist in multiple states as its electrons could exist in various available orbits (Wilhite, 2012). Quantum particles such as electrons, behaving like waves (spread around in space) substantiates their probabilistic existence at multiple locations.A given electron could exist at multiple locations, precisely getting position and momentum is not possible.

1.2 Qubit

Qubits can exist simultaneously as '0' and '1'. It encodes two complex numbers at once. A quantum computer promises to be immensely powerful because it can be in multiple states at once. Entanglement describes correlations between quantum systems that are much stronger than any classical correlations.

A given valency electron could be in outermost orbit or in the orbit within the outermost (Kumar Sharma et al., 2021). With suitable interacting field (laser), particular atomic levels could be selected for study. Excitation described in form of probabilities. At quantum regime, two level atom could be a potential device to store two data points, leading to Qubit. Excitation described in form of probabilities. At quantum regime, two level atom could be a potential device to store two data points, leading to Qubit. Apart from electron orbital locations, electron's spinning about their axis Imagine a physical process that emits two photons, one to the left and the other to the right, with the two photons having opposite polarizations – a photonic qubit in a singlet state (Li et al., 2021). The polarization of each of the photons is indeterminate until a measurement is done. At the instant the polarization for one photon is measured, the state of the other polarization becomes immediately fixed no matter how far away it is. could be chosen to store information.

1.3 Kochen-Specker Theorem (1967)

Non-contextuality: All measurable properties of a physical system do not depend on the context in which they are measured. But a non-contextual assignment of values to the observables is not possible in quantum world. Kochen-Specker studied the logical feature of the quantum theory in connection with the consistency of counterfactual propositions concerning the values of observables that are not jointly measurable.

1.4 Quantum Key Distribution (QKD)

The Future of Security Transmission of secure data typically relies on encryption and decryption "keys" generated by sophisticated algorithms and swapped between sender and receiver so encrypted data can be deciphered (Pemberton-Ross & Kay, 2011). These keys are generally considered secure, but their degree of security is highly dependent on how much computing power a third party has at its disposal.

1.5 Quantum Entanglement

Two or more than two particles can exist in entangled state. There is instantaneous information transfer from two photons X and Y, where X and Y can be physically separated from one another and any change of state in one photon can result in change in another photon (de Andrade et al., 2021). Superposition allows to simultaneously explore all possible solutions. Interference allows to engineer constructive interference towards desired results and destructive interference at undesired options. Entanglement allows instantaneous and secured information exchange (Yusuf et al., 2015). These methods help to understand the entanglement process to work in more than one state simultaneously.

1.6 Quantum Simulation

Quantum Simulation use principles of quantum physics to develop tools for solving challenging problems in basic sciences where quantum effects are prominent, in a range of fields, including physics, chemistry, biology and material sciences. The main objective of quantum simulator is to harness controllable quantum evolution in order to simulate other inaccessible complex quantum systems and to turn the exponential scaling of resources needed to simulate quantum systems on classical computers into a more favorable polynomial overhead on a quantum machine (Y. Zhao & Qiao, 2021). Quantum Simulation use principles of quantum physics to develop tools for solving challenging problems in basic sciences where quantum effects are prominent, in a range of fields, including physics, chemistry, biology and material sciences.

1.7 Digital Quantum Simulation

Decomposition of know Hamiltonian's of the complex quantum system into sum of local terms and approximating the combinations of discrete step-wise unitary operators with minimum errors is one of the standard ways for arriving at operations scheme for digital quantum simulators. These operations will be in the form is computational gates on qubits which will approximate the time evolution of arbitrary local Hamiltonians.

Analog Quantum Simulation represents an alternative approach that is not restricted to register of qubits and the dynamics are not necessarily build upon gates (Caleffi, 2017). A map is constructed that transfers the model of interest to the engineered dynamics of quantum simulators. This approach depends continuously on time and provides less flexibility due to their lack of universality.

Most quantum systems interact with the environment. Open quantum system Many studies have shown that noise can facilitate efficient transmission of energy in quantum system (M. Pant et al., 2019). The process is called Environment Assisted Quantum Transport (ENAQT). Noise presents a major challenge in building quantum simulators. Also, one has to engineer environmental effect to simulate open quantum system. One of the promising near-term applications of quantum simulators is to understand the dynamics in chemical systems.

2. INTRODUCTION OF IMAGE PROCESSING USING QUANTUM COMPUTING

The basic steps of image processing include image acquisition, restoration, enhancement, compression and segmentation. Image acquisition is the first process, followed by preprocessing. The goal of image restoration is to remove the noise in the image. Next is image enhancement, which will sharpen the image details of the human eye. Image compression is used to reduce the size of the image. Processing specific parts of the image is called segmentation. Image enhancement is used to improve the color quality and contrast in the image for visual interpretation. Image enhancement is second only to image restoration, because even after restoration, the image will still be treated as the original image. Unless enhanced, the image can no longer be used in any application. Enhancement technology is very important in image processing, because compared with the original image, the image perception and visual interpretation need to be increased. Regardless of the band, image enhancement plays a vital role in presenting details more clearly. These calculation algorithms are applied to multispectral images for improvement. The quantum-processed image is compared with the original image, and better results are obtained in terms of visual interpretation of the extracted information. Quantitatively evaluate the performance of the improved method and evaluate the applicability of the improved technology to different land types. Due to insufficient light produced globally and locally, the original image cannot be clearly understood. Depending on the application and characteristics of the problem, this defect should be resolved by improved technology.

2.1 Computational Image Processing

Computational image processing is a process of extracting, analyzing, and using the knowledge of images acquired from various detection platforms as shown in Figure 1.

Figure 1. Quantum Computing

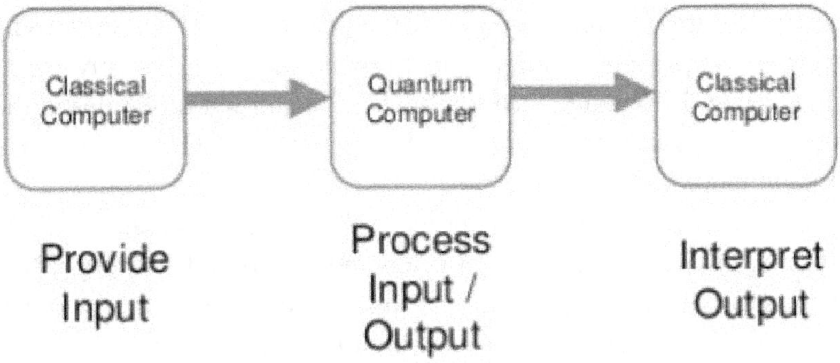

Computational image processing solves problems by using interdisciplinary technologies in advanced computing environments, including instrument design, phenomenon modeling, simulation, and experimentation. Computational image processing supports algorithm design and instrument experimentation.

2.2 Image Enhancement Computation Algorithm

Two novel computation enhancement methods are-

(i) Quantum computation-based quantum Fourier transform

(ii) Parallel computation-based modified data parallel algorithm

These computation enhancement algorithms are also suitable for multispectral satellite images. Both computational enhancement algorithms have achieved high fidelity and Earth Sciences have improved the display of real satellite images. The enhanced image is displayed along with the image histogram and statistical values to briefly present the enhancement process.

2.3 Need for the Quantum Computing in Image Processing

Research needs in order to facilitate image analysis and understanding of visual interpretation, the image must be enhanced from the original image to the enhanced image. In the enhancement technique, the method is trial and error, and you must highlight the appropriate features. The enhancement technology makes the original image interpretable and facilitates the perception of the image. There are several ways to enhance the technology, which can be done logically, requiring a lot of careful inspection and patience. It can also be said that enhancement technology is an art that allows images to increase survivability and highlights application-based features. Through histogram analysis of enhanced images, statistical features can highlight the features of digital images. In the past few years, researchers have been developing many proven techniques to enhance the image simply by stretching the digital image. Eliminate aliasing effects by using interpolation on the image. Therefore, it is necessary to develop an effective methodology from the traditional improvement methods.

This work uses advanced computer technology to process satellite image information through enhanced technology (Fitzsimons, 2017). Computing is generally defined as the activity of using and improving information technology and computer hardware and software. Here, quantum computing and parallel computing methods are used to enhance and extract features from multispectral satellite images.

2.3 Literature on Quantum Image Processing

A simulated image is a mathematical representation of a continuous range of values representing position and intensity. A digital image is a matrix of real (or) complex numbers represented by a finite number of digits. Images can be in the form of transparencies, slides, photos, or graphics, which are digitized and stored in computer memory as an array of binary numbers. The processing of two-dimensional images by digital computers is called digital image processing. Digital imaging has a wide range of applications, such as remote sensing, acoustic imaging, medical imaging, and computer vision (Shi & Qian, 2019). This chapter reviews the basics of image processing, remote sensing, and remote sensing instrument development. Focus on different methods of image enhancement technology. Research on image processing based on quantum computing and parallel computing. To perform digital image processing on remote sensing data, the image must be in digital format. Captured aerial photographs are in analog format. To convert an analog format to a digital format, it must be digitized. The images acquired by the satellite are in digital format.

2.4 Image Enhancement in Remote Sensing Data

In atmospheric science, image enhancement technology can reduce the impact of haze, fog, and turbulent weather on meteorological observations (Amer et al., 2020). Improved technology helps to detect the shape and structure of remote objects in environmental detection (L. Zhao et al., 2014). Satellite images must be restored and enhanced to eliminate noise. The analysis of the research and the use of the results aim to improve the visualization of images to provide different contextual information and to help achieve the end goal of designing effective training programs for clients (Pirker & Dür, 2019).

In synthetic aperture radar (SAR), image enhancement reveals subtle image details that might otherwise be overlooked (Jiang et al., 2012). Image enhancements can provide insight into the shape and contour of objects and provide important information for the human visual system. Image enhancement helps distinguish features by improving the perception of visual quality, and is used in many fields. Contrary to enhancement, land cover categories in satellite images are difficult to classify because overall enhancement increases the overall brightness of the image. Information about the earth is lost in the dark and bright areas of the satellite image. Therefore, for land cover areas, different technologies should be used for improvement. The technology is an artificial neural network, which interprets the output as the presence gradient and edge pixels.

Chen et al. (2008) used fuzzy set theory and a three-stage algorithm to solve the improvement problem. They are fuzzy using C Means' fuzzy algorithm. Then use the stretch model to stretch the model to build and defuzzify. By enhancing satellite images and better contrast images for visual interpretation and display, the image output performs better (Bernardino & Barbosa, 2009).

Use a probabilistic neural network (PNN) classifier to classify remote sensing images so that multispectral pixels represent specific categories (Mo & Xu, 2011). The complete PNN is implemented in multiple languages and platforms to measure the performance of computing technology. Generally, conversion from RGB to any color space will not change the hue. In the process of converting from one color space to another, a problem called Gramut is observed. The problem is that the values of the variables may not be in their respective ranges.

Naik and Murthy (2003) proposed a method in which hue-preserving color image enhancement is performed without color gamut issues. This is achieved by linear stretching, which is applied before the nonlinear transformation (Du & Swamy, 2016).

The simulation showed that the concept of quantum computing algorithm is much faster than the classical algorithm. Monte Carlo technology is used for quantum simulation and analysis (Alazzam & AlOmar, 2018).

Pang et al. (2006) designed an iterative quantum discrete cosine transformation algorithm for image compression. They found that the time complexity of the one-dimensional and two-dimensional Discrete Cosine Transform (DCT) is O ((N) 1/2) and O (N), respectively. They generated the famous Grover algorithm to solve complex unstructured search problems. Based on the two properties of DCT, the DCT quantum algorithm is designed. The first property is the DCT energy conservation transformation. The second characteristic is that the DCT coefficients are very close to zero; these coefficients are then discarded without severely affecting the quality of the reconstructed image. Image compression is based on 1DDCT quantum implementation. Quantum algorithms can do two different things. The first is the calculation of the DCT coefficients, which are performed at the same time. The second is the marking of the desired DCT coefficients, which is also done at the same time. They also compared the DCT quantum iteration and the Grover iteration (A. Srivastava & Chaudhury, 2021).

These algorithms are applied and studied, especially to enhance satellite images to improve visual interpretation. The first is quantum computing and the second is parallel computing. The enhancement is done by using a parallel computing algorithm on the same input image, which is used in quantum computing. Once these calculation algorithms have been applied, the improved images obtained from these two methods can be analyzed quantitatively.

2.5 Methodology of Image Processing by Quantum Computing

A new method of information processing is quantum computing. Existing traditional information processing methods are called classical information (S. Srivastava & Sahana, 2020). As shown in figure 2, all aspects of computer science, information theory, and quantum mechanics combine to form quantum computing. The suppression architecture of the Von Neumann computer has the computational complexity of the classical algorithm, which usually slows down the speed and causes the loss of information, so it is necessary to find a new method to obtain this information.

In classic computers, storage is done little by little (Rekaya et al., 2013). These bits are independent of each other. The connection of these independent bits is completed with software components.

Figure 2. Comparison of image processing by classical and quantum computers

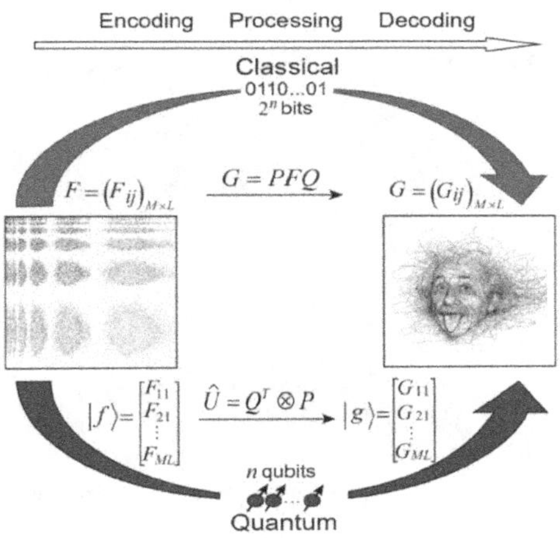

The bits in memory are not connected to each other, causing a loss of information. Each individual bit will represent certain attributes of the related image, namely spatial or light intensity. Image recovery is accomplished by extracting the binary data from hardware memory and using bit independent characteristics. Interconnection occurs in quantum computing. The computer in use today is called a classical computer. The calculation method used in classical calculations is the general Turing machine. In classical calculations, the `n` bit system forms a vector space of dimension` n`(Blum & López-Ibáñez, 2007). However, the "n" qubits of a quantum system result in a dimensional state space. The exponential growth of space will make quantum computing run faster, more powerful, and more efficient. Quantum computing algorithms are implemented in satellite images for feature extraction and enhancement. The transform used for the enhancement is the Quantum Fourier Transform (QFT). QFT can perform quantum mechanical calculations more efficiently to reduce computational complexity below O (I I I log I I I).

2.6 Applications and Future Scope

Quantum computing is used as like edge extraction, security and image denoising. Quantum technology can be used in various fields such as: Telecom, Aerospace, Transportation, healthcare, Government as shown in figure 3:

Figure 3. Application of quantum computing

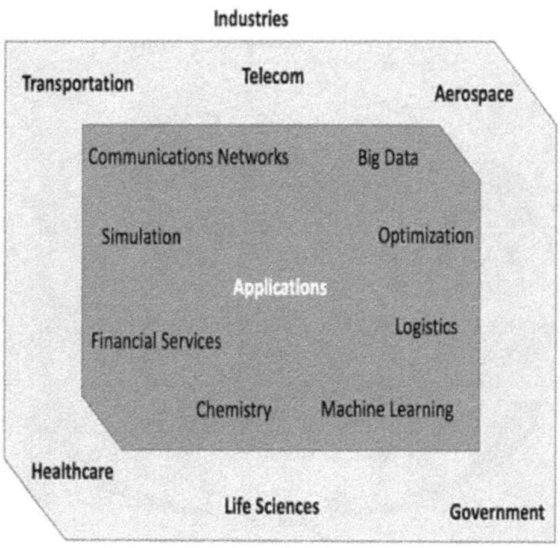

The Quantum Computing based Quantum Fourier Transform image enhancement works well for Low Resolution as well as the High-Resolution images. The Quantum Fourier Transform can also be extended to other areas of Satellite image processing.

3. INTRODUCTION OF WIRELESS SENSOR NETWORK OF QUANTUM COMPUTING

Quantum means the bundle of energy that in proportion to the frequency of energy radiation. Quantum Computing and quantum computers combined depend upon the basic fundamentals and principles of the quantum theory. At atomic or subatomic levels if the behaviors and the nature of matter and its energy explained that we called it as quantum theory.

Quantum computing basically the advanced version of classical computing with the combination of advanced capabilities of quantum theory and quantum physics. This Quantum computing is collection of some quantum states like superposition of elements or energy, entanglement to implement the computing, the devices that we used in this kind of computing called as quantum computers or quantum devices.

3.1 Quantum Computing in WSN

As we already know that in today's networking world security is very important, same in wireless sensor networks we need to make our sensor network reliable, strong, efficient and more secure. As the area or called as domain of sensor networks remain increase. So, to provide security to our wireless sensor networks we have quantum computing and known as quantum cryptography.

Wireless Sensor Network is a network of sensor nodes that are connected to each other and communicates with each other these nodes may receive or send the data, and for better communication and secure communication we require to select those nodes which require minimum energy consumption, loss minimum data, have data efficiency and reliability in the network sensor nodes (Bansal et al., n.d.). Quantum computing and quantum cryptography is technique by which you make your wireless network sensor nodes secure by providing number of protocols and also applying number of algorithms for minimum energy consumption of sensor nodes. The security of sensor networks depends upon the applications and also which attack we covered with.

Attacks in wireless sensors basically of two types:

a) External Attack: Attack from the outside the network boundaries or from the environment, like extra fake sensor node added to the network.
b) Internal Attack: Selection of cluster head within the network was wrong or some fake message passes with in network sensor nodes.

Figure 4. Wireless Sensor Network

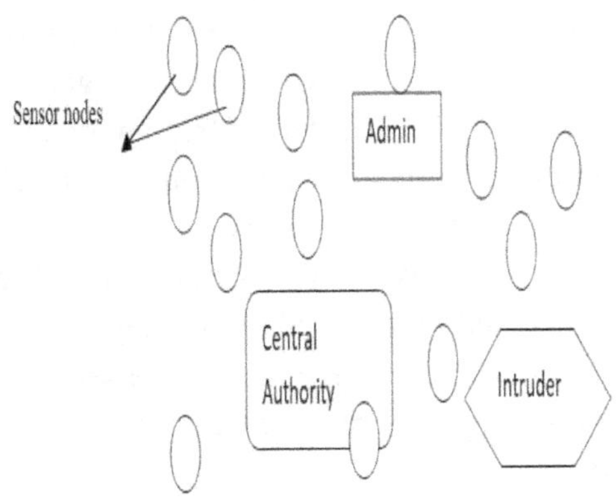

As above diagram of Wireless sensor network (Adalı & Institute of Electrical and Electronics Engineers, n.d.):

Admin: Take the queries related to the nodes also control to message passing to the whole network.

Central Authority (CA): That is given the security to the network.

Sensor nodes: Main parameter of network to check the communication through.

Intruder: Interact with the Environment for information gathering.

Our main Objective to make provide security to network for that we need to protect our network environment behavior from Intruder. For this purpose, we have following techniques, protocols and algorithms under the Quantum computing or quantum cryptography.

3.2 Key Management System Technique WSN

For the key management analysis in WSN, it depends upon three agreements:

- Having trusted Server
- Pre-Distribution environment
- Self-Enforcement learning

To achieve these above agreements in WSNs is difficult because all nodes in the sensor network have its own application and properties.

In WSN, number of KMP (key management Protocols) depended on Quantum Computing (QC), Following are some basic protocols:

Localized Encryption and Authentication Protocol (LEAP) (Agarwal et al., 2020):

LEAP type key management Quantum computing protocols used in large scale of WSN, having four types of keys such as pairwise shared keys, group keys, individual key the cluster keys either this technique or protocols is memory efficient protocol, but need high computation for the sensor nodes.

SPINS, Security protocol in Sensor Nodes, this is the protocols for data authentication, two respective nodes not established the authentication between each other. Base station selection is trusted on the basis of key management and authentication.

LEACH, Low Energy Adaptive Clustering Hierarchy one of the achieved protocols in WSN with minimum energy consumption, MP developed an improved or secure version of LEACH (Sardogan et al., 2018).

Sensor network or wireless sensor network having fixed sensor in the network, the base station sensor is used or selected as the key server having long lasting energy power among the sensors. The identity of the sensor defined by its power, communication and computing. Key management proved key with each sensor for security, communication and computing and sensor nodes have enough space to store these keys with them. We divide the key management into various parts like:

Key generation used to generate the new keys, key exchange used to communicate and transfer the data between the connected devices and nodes, key revolution for the invalidation and verification of the key and key agreement used to derived new key from the initial and original key. Further are some protocols that we used for the communication in WSN under the category of cryptography protocols and have scalable symmetric, asymmetric and lightweight symmetric are (Khan & Narvekar, 2020):

1. LAS (Lightweight Authentication Scheme), Used as very light and energy efficient protocols
2. IKDM (Improved Key Distribution Mechanism), this used as a scalable Symmetric Scheme protocol.
3. MSQPS (Modified Secured Query Processing Scheme) advanced of IKDM but used with cluster based Wireless Sensor networks
4. KRRP (Key Revolution and Renewal Protocol) one of the asymmetric schemes.

As above discussed, protocol with the symmetric and asymmetric cryptography, we learn that with both the symmetric and asymmetric type key management we can combine more protocols and algorithm via quantum computers to increase energy efficiency of the wireless sensor network nodes. Also, with the key management of symmetric or the asymmetric cryptography some problems we face like loss of data while key exchange, unauthorized user, attacks observed in this key managements so to overcome this we have advanced wireless sensor network like use hierarchy cluster routing protocols to ensure efficient sensor networks. Quantum Cryptography one of the best to use over the classical cryptography with symmetric and asymmetric key managements computing.

3.3 Quantum Cryptography in WSN

Classical computing has some weaknesses with key distribution also the algorithms that we used with this like RSA no longer trustworthy. Quantum computing as quantum cryptography here to advance the classical cryptography and also provide sensor security in sensor nodes of wireless sensor networks. Quantum theory effects are unavoidable, so the requirements of quantum computing are on peak (Ramentol et al., 2012).

In classical computing we use 2 bits form data 0 and 1.

In quantum computing data represents in Qubits.

Figure 5. Diagram of basic Qubit

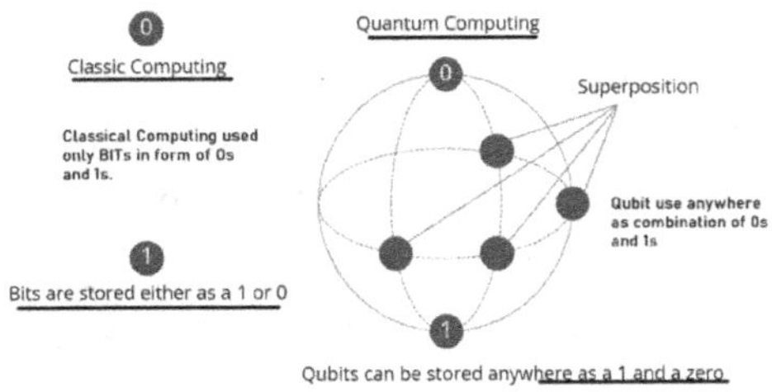

Quantum Cryptography have the following advantages over the classical computing:

1. Problems solved through the quantum data are much more time efficiency.
2. As Qubit compare to Bit data used in multiple manner any use anywhere and combination of bits.
3. Advanced property of Qubit make message more reliable to transfer and also check this message as a copy of desirable data.
4. States of Qubit that have superposition property is collapse and disappears in the qubit cloud that maintains the privacy of the share messages over quantum computing.

Classical bits only combination of 0 and 1, where the quantum bit called qubit are in states, that is the superposition of these 1s and 0s (D. Singh & Singh, 2020).

$1\alpha\beta + = \Psi$ is the combination of states where, Ψ denoted the superposition state.

In Quantum Computing the cryptographically system basically of two types:

Public Key Cryptography

Secret Key Cryptography

In first category the public key cryptography we used the pair of keys for encryption and decryption and we combine this to provide security in Quantum computing.

Secret Key Cryptography also called as symmetric cryptography, not pair only the same key common key we used both for encryption and decryption. For the secure and the reliable channel in key management symmetric key cryptography used as pre distribution security channel for communication (Awad & Khanna, 2015).

Number of algorithms that researched on quantum computing for wireless communication as discussed below table:

Table 1. Algorithms as quantum computing in wireless communication

Algorithms	Application Type
Deutsch Algorithm	Classification
Simon's Algorithm	Classification
Shor Algorithm	Factoring
Quantum Phase Estimation Algorithm	Order and finding factoring search
Grover's Algorithm	Search
Boyer-Brassard-Hoyer-Tapp Algorithm	Search
Quantum Computing Algorithm	Search
Quantum Heuristic Algorithm	Search
Quantum Genetic Algorithm	Search
Non-dominated Quantum Optimization	Search and solving nonlinear system problems
Quantum Mean Algorithm	Function Moment Finding
Quantum Weighted Sum Algorithms	Function Moment Finding

Further, more algorithms we have to make wireless communication secure with the wireless sensor networks (G & R., 2020). As we compare these algorithms, we observed that quantum computing is under specific complexity requirement, a gain performance technique. With the quantum computing and quantum cryptography a specific target reached at lower computing complexity

Challenges of Quantum Computing in Wireless Sensor Networks (L. L. Shen et al., 2007):

- To make our data communication more secure we advanced more algorithms in the quantum computing.
- As discussed above all symmetric and asymmetric cryptography on key management to add more protocol.
- Public and secret key cryptography is more used in quantum computing with the quantum cryptography for secure wireless communication.

4. INTRODUCTION ON CYBER SECURITY OF QUANTUM COMPUTING

Quantum Encryption is indispensable to guard our digital assets and substructure from attackers. If anyone think that they know everything about cybersecurity, this think is wrong. Cybersecurity is the defense of internet-related systems like as hardware, software and data from cyberthreats. Quantum computers will be accomplished of resolve difficulties that are more multifaceted for traditional computers to determine. This comprises resolving the algorithms behindhand encryption keys which are guard user information, data and the Internet's structure. This work concludes with an outline of quantum computing in cyber security (QCCS) and avenues for future research.

Quantum computing is the manipulation of cooperative possessions of quantum states, like as superposition and entanglement, to achieve the computation. (*Quantum Computing and Communications - Google Books*, n.d.)The devices which execute quantum computations are identified as quantum computers. Quantum computers accomplish intentions founded on the possibility of an item's condition beforehand the measurement - in preference to 1s or 0s - which identifies that they have the prospective to procedure exponentially additional information likened to conventional CPUs. (Steane, 1998)Quantum computers that can be utilized in taking large manufacturing data sets on effective failures and interpreting them to combinatoric trials that can be identified which part of a composite manufacturing procedure donated to incidents of invention failure. Quantum computation might revolution the world. It can change medication, breakdown encoding and reform transportations and artificial intelligence. Corporations similar IBM, Microsoft and Google are competing to develop dependable quantum computers. Linear Algebra is the basic math that allows quantum computing to perform its magic ("A Universal Two-Bit Gate for Quantum Computation," 1995). In quantum computing, everything is from the representation of qubits and gates to circuits' functionality describe using various forms of Linear Algebra. In a quantum computer, each qubit inspirations the other qubits round it, working together to reach as a result. Superposition and entanglement are the main that give quantum computers the ability to process more information in much faster. Around three minutes, the mechanism accomplished a scientifically considered design so multifaceted that it would take the world's greatest capable supercomputer named IBM's conference 10,000 years to complete the problem.

Cyber security has become a substance of worldwide issue now a day. Cyber security is important for every nation and every person. By now more than 50 nations have formally published their official stance on cybercrime. (Patil & Devane, 2019) The International Telecommunications Union (ITU) describes cyber security as likes: Cyber security is the gathering of gears, strategies, safety perceptions, safety protections, strategies, risk administration methods, movements, exercise etc. that can be used to defend the cyber situation and association and user's possessions. Association and user's possessions contain linked computation devices, staffs, substructure, submissions, facilities, broadcastings structures, and the entirety of communicated and/or deposited data in the cyber atmosphere. Cyber security attempts to certify the accomplishment and preservation of the safety belongings of the association and user's properties in contradiction of related security dangers in the cyber situation. (Von Solms & Van Niekerk, 2013)The over-all security purposes include the following:

1. Obtainability
2. Truthfulness, which may contain legitimacy and non-repudiation
3. Discretion

Cyber security is defending the structures or policies which are related to the cyberspace from the cyber-attacks or system muggings. Whenever any cybercrime happens the network forensic collect and examination all indication from the strategies throughout or afterward the cybercrime.

Figure 6. Cyber Security

Cyber security is a way to protect the system or networking devices that is connected to the internet from hackers or cyber attackers. There are many ways to track the original device that was used for cybercrime. They are - T race back methods, Messaging-based strategies, the overlay network-based techniques, the hybrid trace back approaches etc. However, in this trace back methods the main goal is identify the real source or devices or location of attack, collect all the evidence and forward them. But the main challenging part from all of these steps is collecting evidences. (*Digital Evidence and Computer Crime - 2nd Edition*, n.d.)A foundation of the gathering indication using network traffic has main boundaries due to the lack of specialize tools obtainable for analysis huge amounts of circulation from a scientific position (D et al., 2017). The most and hard to collect indication is to discovery the meticulous location of an attacker. Identifying the strategies which are castoff in cybercrime is the greatest applicable indication. So, at first the forensic team need to identify the devices which are lead them to the attackers and the device identify can be used as evidence. But finding the network which one is used for cybercrime is not the solution of everything. So, reach to the attacker is for some cases is still big challenges for forensic investigation team.

4.1 The Quantum Risk to Cybersecurity

Problems those are too complex to solve or to figure out for classical computers, QCCS will be talented to resolve those difficulties. This contains explaining the procedures overdue encoding solutions that defend our information and the Cyberspace's organization.

In present time encryption is based on mathematical rules and formulas that's why it would take today's computers impossibly long time to decode (Bennett & Brassard, 1989). For example, to describing this let's think that we have two large number and we need to multiple them together. It's not that much difficult to come up with the multiple products but it is much harder to begin with the large number and factor it into its two-prime number. However, a quantum computer can without difficulty factor those numbers and breakdown the code.

Figure 7. Cyber Security in Quantum Era

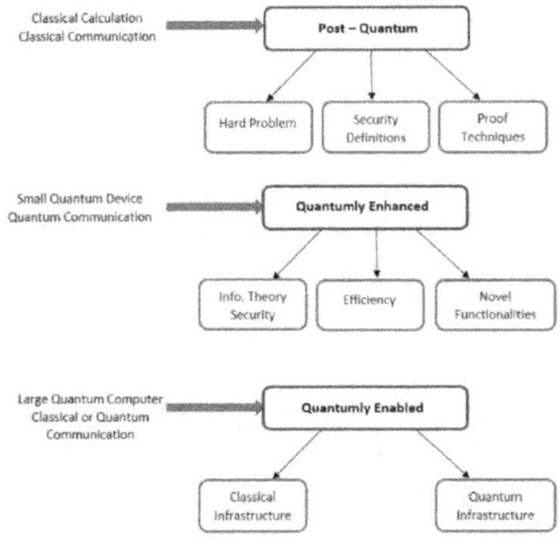

In figure 7 red color boxes are represent the three categories of research steps. Then blue lines are denoting the resources that involved from the users. Green color specifies issues which are we focus in this review.

4.3 The Positives and Negatives Side of QCCs

(*Quantum Computing and Cybersecurity: How to Capitalize on Opportunities and Sidestep Risks | IBM*, n.d.)Significant quantum processers will suggestively increase computation influence, generating original chances for enlightening cybersecurity. QCCS will exert the control to perceive and bounce quantum-era cyberattacks beforehand they source damage. But it may perhaps convert an ambiguous blade, as quantum computation might also generate original experiences, such as the capability to rapidly resolve the problematic mathematics complications that are the foundation of approximately procedures of encoding. While post-quantum coding ordinaries are still actuality completed, industries and additional administrations can start formulating these days.

4.4 Bibliographic Analysis of Research Trends in Quantum Computing in Cyber Security

Bibliographic analysis of Quantum computing in cyber security (QCCS) with its associated terms is shown in this section. Author wise and document wise analysis of the research in this domain is also definite and designated.

Figure 8. Country wise research analysis

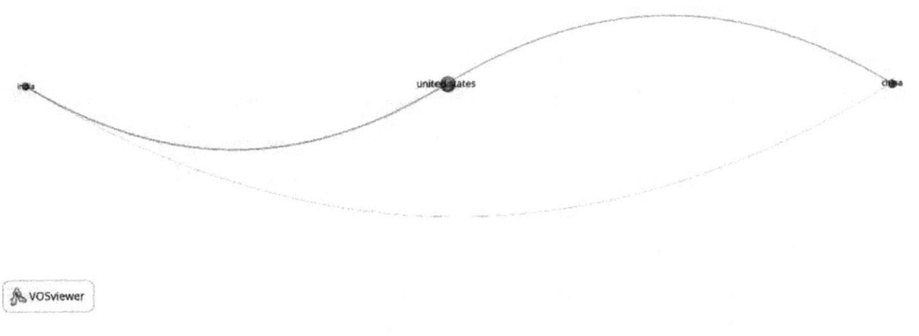

VOSviewer experimental data from figure 8 shown that United states has largely contributed in the worldly research in QCCS and India and China have strong connection with United State. But in this area, there are only three countries only participated in this research field. So, there is a lot of chance to do research work.

Figure 9. Document wise research analysis

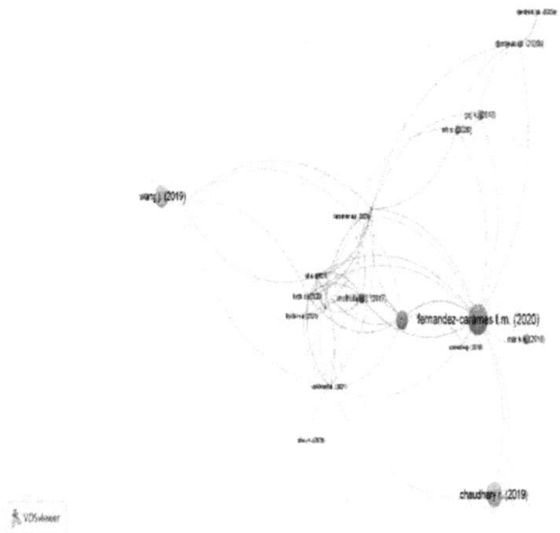

Based upon author, an experiment was also shown for research treads for each document based. From figure 9 it is cleared that various authors in the past not only research on Quantum computing but also on the Cyber security (Benioff, 1980). In the figure 9 it shows the document wise citation details for the same.

4.5 Conclusion

Quantum computing in Cyber security is now day by day more popular technique to build the security system for networking (Meunkaewjinda et al., 2008). The capability to interconnect securely and figure out capably is more significant than eternally to humanity. The Internet and progressively the IoTs, consumes had a radical influence on our earth. Future networks will positively contain of both traditional and quantum strategies and associations with functionalities of numerous complexities, fluctuating from unpretentious routers to servers performing universal quantum procedures. In the following article we will emphasis on how quantum computers will hypothetically interrupt the cyber security landscape. We will discuss about the threats and chances of quantum computers and study which fields are predicted to be typically affected.

5. INTRODUCTION OF ROUTING TECHNIQUES OF QUANTUM COMPUTING

The emerging development of Quantum communication is the latest technological advancement for secure data transmission in various application areas (T. Singh et al., 2021). It is observed that out of many practical sources of data, Wireless Sensor Network is an effective source of big data where the sensor node can merge a large volume of data by analyzing different routing protocol features. By using quantum communication techniques, various types of routing protocols are designed to transfer data from source to sink effectively and efficiently in a quantum computing network environment (Prasad et al., 2012). Considering different parameters like energy, distance, time routing techniques are designed for a large volume of data in quantum computing network. In this work, our main goal is to explain modern research approaches on routing techniques of quantum computing Network quantum communications. Around fifty-five (55) research articles of different regular journals are organized and reviewed in a chronological manner in order to complete our goal using Scopus and the Web of Science database (H.D., 2020). A variety of experiments are conducted for data analysis purposes taking different considerations like authorship, organizations, publication time, source, citation number, and documents independently with these Web of Science and Scopus databases in a global perspective (L. Shen & Bai, 2006).

Quantum computing and quantum communications are the two fundamental elements of the quantum networks (Fitzsimons, 2017). In order to transfer information among the different quantum processors, quantum bits that qubits are the form that is being used in this type of network (M. Pant et al., 2019). The quantum processor is defined as a small computer capable of performing different logic gates functions using qubits (Pirker & Dür, 2019). Optical fiber networks and free space networks are the two effective media for successful data transmission in a quantum computing environment where the traditional routing techniques can be implemented (Li et al., 2021). Similar to the classical computing network, routing protocols are designed and implemented considering the internal energy, performance analysis, reliability, etc. as parameters in quantum computing communication (Jiang et al., 2012). With the help of trusted repeater and quantum repeater, routing techniques can be implemented effectively in quantum computing networks (Caleffi, 2017). First, some researchers of QuTech Research center proposed the concept of the quantum network where various nodes called quantum nodes are connected with each other to pass the information in the form of qubits (Cacciapuoti et al., 2020). For big data transmission over a large geographic distance throughout the world at an enormous speed, quantum communication is an effective method (Mollazade et al., 2013). Nuclear Magnetic Resonance, Iron Trap, Quantum Dot are the different hardware and software technologies used in quantum computing communication. The main task of any quantum network is to provide large distance quantum entanglement between two remote communicating device (Mokhtar et al., 2016).

5.1 Background Study of Routing Techniques of Quantum Computing

Nowadays, quantum computing is such a growing research field where the various perspectives of this field are examined in order to enhance the security and privacy of the data and the medium of this network (Sethy et al., 2017). In some research works, applications of quantum computing are discussed for the designing of routing techniques in quantum computer networks (Madiwalar & Wyawahare, 2017). Research shows that logic gates can be used to construct protocols in the quantum networks for the betterment of service (Wiharto et al., 2021). In order to design a quantum router to transfer data in a quantum computing network, a quantum circuit is designed. There is a significant development on entanglement routing issues whose main goal is to design large distance entanglement for simultaneous source to sink pairs through multiple connecting devices (Kumar Sharma et al., 2021). Proper distribution of quantum keys is another important issue for quantum computing networks (L. Zhao et al., 2014). Considering this fact, some routing protocols are already proposed based on the reliable node vs. repeaters performance as various types of repeaters are the key connecting devices for this type of network (Wilhite, 2012). It is maybe mentioned here that according to the network model, a quantum transport protocol is a network layer protocol that is designed for the purposes of network layer functionalities. Another protocol is quantum walk protocol which is designed to perform distributed quantum computing operations of a quantum networks (Agrawal et al., 2012).

5.2 Bibliographical Research Analysis on Routing Techniques of Quantum Computing

In this section, we will observe the mode of raising the research approaches on different routing techniques in quantum computing networks tremendously day by day in various perspectives (Gu et al., 2019). This rising research trend in this field is also elaborated based on research subject-related parameters like specific keyword of the research field, year of research publication, research location, source of research, author, citation quantity, document, inter research collaboration, etc. In order to assess and express the entire research potency and topic-related study of key conditions in this field, a resourceful number of experiments are performed to justify data taking the last twenty years of research work on routing techniques of quantum computing networks. We have followed a bibliometric method in our research to examine all these data scientifically, theoretically, and systematically. The multi-focused research approaches are shown here in our research area using the Scopus database and Web of Science database separately.

5.3 Year-Based Research Trends in Routing Techniques of Quantum Computing

In order to determine the research growth in routing techniques of quantum computing networks, year (time) is one of the significant parameters for a particular period of time. In this year-based research approach of this area, all of the application fields and correlated research of our topic are analyzed. The experiment is performed taking all the relevant research article data of the Scopus database in the last twenty (2001-2021) years (Kaur & Bhatia, 2019). Using the VOSviewer software, figure 10 is shown. The highest number of research work was performed in 2019. Apart from this, in 2018, 2020, and 2021 a significant amount of research was conducted in this research field. Additionally, there were some research articles published in 2009, 2012, 2014, 2015 on routing techniques of quantum computing.

Figure 10. Year-based research trends in routing techniques of quantum computing

5.4 Global Research Analysis in Routing Techniques of Quantum Computing

From the perspective of global research tendency, country-based research is an important tool to determine the research growth in a particular research field. VOSviewer experimental data from figure 11 shows that China, Germany, India, and the USA have research contributions on this topic but they performed their research work independently without any research collaboration. Out of these four countries, the USA has the highest research contribution in this research field.

Figure 11. Country oriented research analysis in routing protocol of quantum computing

In the Table 2, the following experiment was conducted to check the international collaboration of research in routing protocol of quantum computing and it was noticed that only four (04) countries from the globe have participated in research related to our topic. Out of these four countries, the USA is the only country that has the highest number of publications and more than one hundred citations which is the maximum citation quantity in routing techniques of quantum computing networks (Ahmadi et al., 2017). On the other side, India, Germany, and China have less than ten publications and the citation quantity been between one to ten which is shown in Table 2.

Table 2. Global analysis of research tendency in routing protocol of quantum computing based on country (Source-Scopus Database)

Sl. No.	Country	Documents	Citation	Total Link Strength
1	China	3	1	0
2	Germany	3	8	0
3	India	5	10	0
4	USA	7	106	0

5.5 Web of Science Database-Based Research Analysis in Routing Techniques of Quantum Computing

Another experiment was conducted taking research data from the Web of Science database in order to show the variation of research approaches in routing protocol of quantum computing research field (Nithya et al., 2016). It is noticed that in the last twenty years (2001-2021) of the research articles in this area are based on the publication's quantity. The assessment explains a variety of research works published in different standard journals and conferences. Figure 12 certainly states that in today's century, there is incredible progress in routing techniques of quantum computing-related research works (Cacciapuoti et al., 2020). It is obviously observed that there is the maximum number of research publications done in 2019 out of the last twenty-one years (2000 to 2021) in this field (Mustafa et al., 2011).

Figure 12. Research analysis on data management in cloud computing (year vs. no of publication)

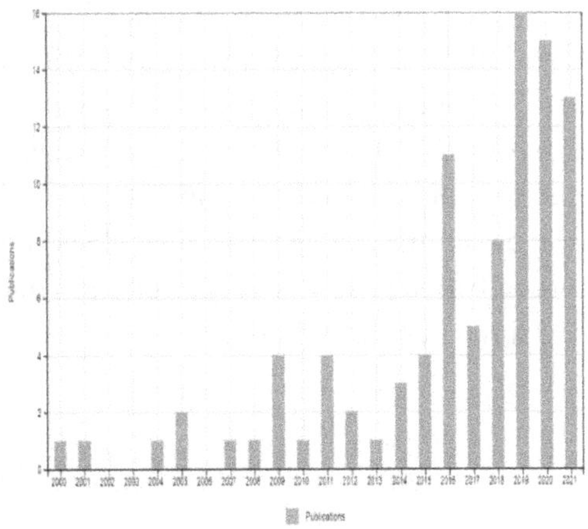

From the above figure 12, it is noticed that in the last three years (2019 to 2021) research in routing protocol of quantum computing field is tremendously increased, (Source: Web of Science Database). Research tendency in this field increased significantly from 2017. From 2000 to 2015, every year the number of publications in this field was maximum from zero (0) to four (04). From 2016 there is a dramatic development in research of this field (Y. Pant & Bhadauria, 2017).

Figure 13. Year and citation research trends in routing techniques of quantum computing (Source: Web of Science)

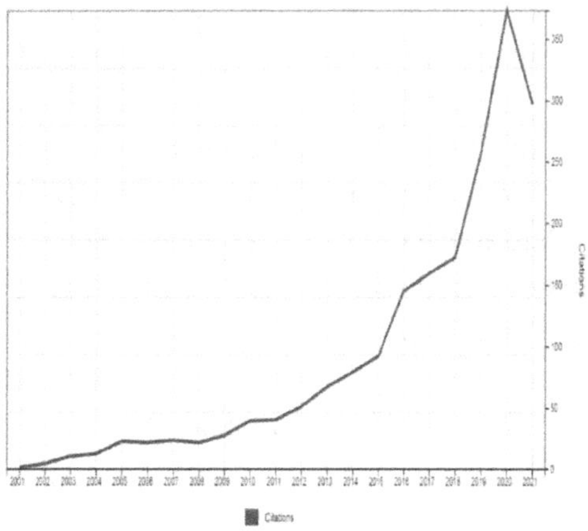

Figure 13 undoubtedly shows the research tendency in the routing techniques of the quantum computing field in the last twenty years (2001 to 2021) based on the year and citation quantity of the published research articles (Cosma et al., 2017). From this figure 13, it is visualized that there is a numerous growth of citations in this research area throughout this period (Alazzam et al., 2013). This indicates that routing techniques in the quantum computing field are becoming a popular research field in the standard research area progressively as citation quantity is increasing in the last twenty years (Madhavi et al., 2017).

6. CONCLUSION

Routing Techniques in Quantum Computing is the most recent research field of the modern quantum network in today's world. The rule of routing techniques in quantum Computing is to provide efficient and effective data transmission in a quantum computing network environment. In our research work, a scient metric study of the modern research trends in the routing protocol of quantum computing is analyzed. In order to conduct this research successfully, various types of assessments are performed with the help of the Scopus database and Web of Science databases independently. The VOSviewer software is also being used for data analysis to show research development in the last twenty years (2001-2021) taking different parameters like a year of publication, organization, source, citation quantity, document, inter research collaboration in our research domain 'Routing techniques of Quantum Computing'. The verified outcomes of all experiments are very audacious for the researchers involved in the area of routing protocol in quantum computing.

REFERENCES

A universal two-bit gate for quantum computation. (1995). *Proc. R. Soc., 449*(1937), 679–683. 10.1098/rspa.1995.0066

Adalı, E. (n.d.). *3. Uluslararası Bilgisayar Bilimleri ve Mühendisliği Konferansı = 3rd International Conference on Computer Science and Engineering : Sarajevo - Bosnia&Herzegovina, 20-23 Eylül (September) 2018.* IEEE.

Agarwal, M., Singh, A., Arjaria, S., Sinha, A., & Gupta, S. (2020). ToLeD: Tomato Leaf Disease Detection using Convolution Neural Network. *Procedia Computer Science*, 167, 293–301. 10.1016/j.procs.2020.03.225

Agrawal, D., Das, S., & El Abbadi, A. (2012). Data Management in the Cloud: Challenges and Opportunities. *Synthesis Lectures on Data Management*, 4(6), 1–138. 10.1007/978-3-031-01895-4

Ahmadi, P., Muharam, F. M., Ahmad, K., Mansor, S., & Seman, I. A. (2017). Early detection of ganoderma basal stem rot of oil palms using artificial neural network spectral analysis. *Plant Disease*, 101(6), 1009–1016. 10.1094/PDIS-12-16-1699-RE30682927

Alazzam, A., & AlOmar, B. (2018). Using Average Uniform Algorithm to model educational data. *ITT 2017 - Information Technology Trends: Exploring Current Trends in Information Technology, Conference Proceedings,* 30–34. 10.1109/CTIT.2017.8259563

Alazzam, A., Yuzgec, E., & Lewis, H. W. (2013). A new optimization algorithm for non-convex problems. *IIE Annual Conference and Expo 2013*, 2784–2791.

Amer, O., Krawec, W. O., & Wang, B. (2020). Efficient Routing for Quantum Key Distribution Networks. *Proceedings - IEEE International Conference on Quantum Computing and Engineering,* 137–147. 10.1109/QCE49297.2020.00027

Awad, M., & Khanna, R. (2015). Efficient learning machines: Theories, concepts, and applications for engineers and system designers. *Efficient Learning Machines: Theories, Concepts, and Applications for Engineers and System Designers*, 1–248. 10.1007/978-1-4302-5990-9

Bansal, A., & Singhal, A. (n.d.). *Proceedings of the 7th International Conference Confluence 2017 on Cloud Computing, Data Science and Engineering : 12th-13th January 2017, Amity University, Noida, Uttar Pradesh, India.* Academic Press.

Benioff, P. (1980). The computer as a physical system: A microscopic quantum mechanical Hamiltonian model of computers as represented by Turing machines. *Journal of Statistical Physics*, 22(5), 563–591. 10.1007/BF01011339

Bennett, C. H., & Brassard, G. (1989). Experimental quantum cryptography: the dawn of a new era for quantum cryptography: the experimental prototype is working]. *SIGACT News*, 20(4), 78–82. 10.1145/74074.74087

Bernardino, H. S., & Barbosa, H. J. (2009). Artificial immune systems for optimization. *Studies in Computational Intelligence*, 193, 389–411. 10.1007/978-3-642-00267-0_14

Blum, C., & López-Ibáñez, M. (2007). Ant colony optimization. *Scholarpedia Journal*, 2(3), 1461. 10.4249/scholarpedia.1461

Cacciapuoti, A. S., Caleffi, M., Tafuri, F., Cataliotti, F. S., Gherardini, S., & Bianchi, G. (2020). Quantum Internet: Networking Challenges in Distributed Quantum Computing. *IEEE Network*, 34(1), 137–143. 10.1109/MNET.001.1900092

Caleffi, M. (2017). Optimal Routing for Quantum Networks. *IEEE Access : Practical Innovations, Open Solutions*, 5, 22299–22312. 10.1109/ACCESS.2017.2763325

Cosma, G., Brown, D., Archer, M., Khan, M., & Graham Pockley, A. (2017). A survey on computational intelligence approaches for predictive modeling in prostate cancer. *Expert Systems with Applications*, 70, 1–19. 10.1016/j.eswa.2016.11.006

D, D., JG, L., S, D., W, M., A, C., & V, S. (2017). Cardiotocography versus intermittent auscultation of fetal heart on admission to labour ward for assessment of fetal wellbeing. *The Cochrane Database of Systematic Reviews, 1*(1). 10.1002/14651858.CD005122.pub5

de Andrade, M. G., Dai, W., Guha, S., & Towsley, D. (2021). *A quantum walk control plane for distributed quantum computing in quantum networks*. Academic Press.

Digital Evidence and Computer Crime - 2nd Edition. (n.d.). Retrieved September 7, 2021, from https://www.elsevier.com/books/digital-evidence-and-computer-crime/casey/978-0-08-047250-8

Du, K. L., & Swamy, M. N. S. (2016). Search and optimization by metaheuristics: Techniques and algorithms inspired by nature. *Search and Optimization by Metaheuristics: Techniques and Algorithms Inspired by Nature*, 1–434. 10.1007/978-3-319-41192-7

Fitzsimons, J. F. (2017). Private quantum computation: An introduction to blind quantum computing and related protocols. *NPJ Quantum Information*, 3(1), 1–10. 10.1038/s41534-017-0025-3

G, J. M., & R., S. D. (2020). Automatic Tomato Plant Leaf Disease Classification using Multi-Kernel Support Vector Machine. *International Journal of Engineering and Advanced Technology, 9*(5), 560–565. 10.35940/ijeat.E9689.069520

Gu, Q., Sheng, L., Zhang, T., Lu, Y., Zhang, Z., Zheng, K., Hu, H., & Zhou, H. (2019). Early detection of tomato spotted wilt virus infection in tobacco using the hyperspectral imaging technique and machine learning algorithms. *Computers and Electronics in Agriculture*, 167, 105066. Advance online publication. 10.1016/j.compag.2019.105066

H.D., G. (2020). Machine Learning Approach towards Tomato Leaf Disease Classification. *International Journal of Advanced Trends in Computer Science and Engineering*, 9(1), 490–495. 10.30534/ijatcse/2020/67912020

Jiang, H., Wang, M., Liu, M., Yan, J., & Engineering, I. (2012). *A Quantum-inspired Ant-Based Routing algorithm forWSNs*. Academic Press.

Kaur, M., & Bhatia, R. (2019). Development of an improved tomato leaf disease detection and classification method. *2019 IEEE Conference on Information and Communication Technology, CICT 2019*. 10.1109/CICT48419.2019.9066230

Khan, S., & Narvekar, M. (2020). Novel fusion of color balancing and superpixel based approach for detection of tomato plant diseases in natural complex environment. *Journal of King Saud University. Computer and Information Sciences*. Advance online publication. 10.1016/j.jksuci.2020.09.006

Kumar Sharma, D., Sreenivasa Chakravarthi, D., Ara Shaikh, A., Al Ayub Ahmed, A., Jaiswal, S., & Naved, M. (2021). The aspect of vast data management problem in healthcare sector and implementation of cloud computing technique. *Materials Today: Proceedings*, xxxx. Advance online publication. 10.1016/j.matpr.2021.07.388

Li, C., Li, T., Liu, Y. X., & Cappellaro, P. (2021). Effective routing design for remote entanglement generation on quantum networks. *NPJ Quantum Information*, 7(1), 10. Advance online publication. 10.1038/s41534-020-00344-4

Madhavi, R., Karri, R. R., Sankar, D. S., Nagesh, P., & Lakshminarayana, V. (2017). Nature inspired techniques to solve complex engineering problems. *Journal of Industrial Pollution Control*, 33(1), 1304–1311.

Madiwalar, S. C., & Wyawahare, M. V. (2017). Plant disease identification: A comparative study. *2017 International Conference on Data Management, Analytics and Innovation, ICDMAI 2017*, 13–18. 10.1109/ICDMAI.2017.8073478

Meunkaewjinda, A., Kumsawat, P., Attakitmongcol, K., & Srikaew, A. (2008). Grape leaf disease detection from color imagery using hybrid intelligent system. *5th International Conference on Electrical Engineering/Electronics, Computer, Telecommunications and Information Technology, ECTI-CON 2008, 1*, 513–516. 10.1109/ECTICON.2008.4600483

Mo, H., & Xu, L. (2011). Biogeography migration algorithm for traveling salesman problem. *International Journal of Intelligent Computing and Cybernetics*, 4(3), 311–330. 10.1108/17563781111160002

Mokhtar, U., Ali, M. A. S., Hassenian, A. E., & Hefny, H. (2016). Tomato leaves diseases detection approach based on Support Vector Machines. *2015 11th International Computer Engineering Conference: Today Information Society What's Next? ICENCO 2015*, 246–250. 10.1109/ICENCO.2015.7416356

Mollazade, K., Omid, M., Akhlaghian Tab, F., Kalaj, Y. R., Mohtasebi, S. S., & Zude, M. (2013). Analysis of texture-based features for predicting mechanical properties of horticultural products by laser light backscattering imaging. *Computers and Electronics in Agriculture*, 98, 34–45. 10.1016/j.compag.2013.07.011

Mustafa, H. M., Badran, S. M., Al-Hamadi, A., & Al-Somani, T. F. (2011). On mathematical modeling of cooperative e-learning performance during face to face tutoring sessions (Ant Colony System approach). *2011 IEEE Global Engineering Education Conference, EDUCON 2011*, 338–346. 10.1109/EDUCON.2011.5773158

Nithya, S., & Dhanuja, B., & AshaJerlin, M. (2016). Evolutionary algorithm – A review. *International Journal of Pharmacy and Technology*, 8(4), 25959–25966.

Pant, M., Krovi, H., Towsley, D., Tassiulas, L., Jiang, L., Basu, P., Englund, D., & Guha, S. (2019). Routing entanglement in the quantum internet. *NPJ Quantum Information*, 5(1), 1–9. 10.1038/s41534-019-0139-x

Pant, Y., & Bhadauria, H. S. (2017). Performance Study of Routing Protocols in Wireless Sensor Network. *Proceedings - 2016 8th International Conference on Computational Intelligence and Communication Networks, CICN 2016*, 134–138. 10.1109/CICN.2016.32

Patil, R. Y., & Devane, S. R. (2019). Network Forensic Investigation Protocol to Identify True Origin of Cyber Crime. *Journal of King Saud University. Computer and Information Sciences*. Advance online publication. 10.1016/j.jksuci.2019.11.016

Pemberton-Ross, P. J., & Kay, A. (2011). Perfect quantum routing in regular spin networks. *Physical Review Letters*, 106(2), 1–4. 10.1103/PhysRevLett.106.02050321405213

Pirker, A., & Dür, W. (2019). A quantum network stack and protocols for reliable entanglement-based networks. *New Journal of Physics*, 21(3), 033003. Advance online publication. 10.1088/1367-2630/ab05f7

Prasad, S., Kumar, P., Hazra, R., & Kumar, A. (2012). Plant leaf disease detection using Gabor wavelet transform. *Lecture Notes in Computer Science (Including Subseries Lecture Notes in Artificial Intelligence and Lecture Notes in Bioinformatics), 7677 LNCS*, 372–379. 10.1007/978-3-642-35380-2_44

Quantum Computing and Communications - Google Books. (n.d.). Retrieved September 7, 2021, from https://books.google.co.in/books?hl=en&lr=&id=RQXpBwAAQBAJ&oi=fnd&pg=PR4&dq=quantum+computing+definitions&ots=BHTZvRoueA&sig=pr7ABq6SeII0jd4siDiOZh2CxLE&redir_esc=y#v=onepage&q&f=false

Quantum computing and cybersecurity: How to capitalize on opportunities and sidestep risks | IBM. (n.d.). Retrieved September 8, 2021, from https://www.ibm.com/thought-leadership/institute-business-value/report/quantumsecurity#Ramentol10.1007/s10115-011-0465-6

Rekaya, R., Robbins, K., Spangler, M., Smith, S., Hay, E. H., & Bertrand, K. (2013). *Ant Colony Algorithm with Applications in the Field of Genomics*. Ant Colony Optimization - Techniques and Applications. 10.5772/52051

Sardogan, M., Tuncer, A., & Ozen, Y. (2018). Plant Leaf Disease Detection and Classification Based on CNN with LVQ Algorithm. *UBMK 2018 - 3rd International Conference on Computer Science and Engineering*, 382–385. 10.1109/UBMK.2018.8566635

Sethy, P. K., Negi, B., Behera, S. K., Barpanda, N. K., & Rath, A. K. (2017). An Image Processing Approach for Detection, Quantification, and Identification of Plant Leaf Diseases -A Review. *IACSIT International Journal of Engineering and Technology*, 9(2), 635–648. 10.21817/ijet/2017/v9i2/170902059

Shen, L., & Bai, L. (2006). A review on Gabor wavelets for face recognition. *Pattern Analysis & Applications*, 9(2–3), 273–292. 10.1007/s10044-006-0033-y

Shen, L. L., Bai, L., & Fairhurst, M. (2007). Gabor wavelets and General Discriminant Analysis for face identification and verification. *Image and Vision Computing*, 25(5), 553–563. 10.1016/j.imavis.2006.05.002

Shi, S., & Qian, C. (2019). *Modeling and Designing Routing Protocols in Quantum Networks*. Academic Press.

Singh, D., & Singh, B. (2020). Investigating the impact of data normalization on classification performance. *Applied Soft Computing*, 97, 105524. Advance online publication. 10.1016/j.asoc.2019.105524

Singh, T., Kumar, K., & Bedi, S. S. (2021). A review on artificial intelligence techniques for disease recognition in plants. *IOP Conference Series. Materials Science and Engineering*, 1022(1), 012032. Advance online publication. 10.1088/1757-899X/1022/1/012032

Srivastava, A., & Chaudhury, P. (2021). Application of Nature-Inspired Swarm Optimization Algorithms in Artificial Neural Networks. *Proceedings - International Conference on Artificial Intelligence and Smart Systems, ICAIS 2021*, 1–6. 10.1109/ICAIS50930.2021.9395806

Srivastava, S., & Sahana, S. K. (2020). A survey on traffic optimization problem using biologically inspired techniques. *Natural Computing*, 19(4), 647–661. 10.1007/s11047-019-09731-z

Steane, A. (1998). Quantum computing. *Reports on Progress in Physics*, 61(2), 117–173. 10.1088/0034-4885/61/2/002

Von Solms, R., & Van Niekerk, J. (2013). From information security to cyber security. *Computers & Security*, 38, 97–102. 10.1016/j.cose.2013.04.004

Wiharto, N., Nashrullah, F. H., Suryani, E., Salamah, U., Prakisy, N. P. T., & Setyawan, S. (2021). Texture-Based Feature Extraction Using Gabor Filters to Detect Diseases of Tomato Leaves. *Revue d'Intelligence Artificielle*, 35(4), 331–339. 10.18280/ria.350408

Wilhite, S. E. (2012). Cloud computing? *HDA Now*, 12. 10.4018/IJCAC.2015040104

Yusuf, I. I., Thomas, I. E., Spichkova, M., Androulakis, S., Meyer, G. R., Drumm, D. W., Opletal, G., Russo, S. P., Buckle, A. M., & Schmidt, H. W. (2015). Chiminey: Reliable Computing and Data Management Platform in the Cloud. *Proceedings - International Conference on Software Engineering. International Conference on Software Engineering*, 2, 677–680. 10.1109/ICSE.2015.221

Zhao, L., Sakr, S., Liu, A., & Bouguettaya, A. (2014). Cloud data management. In *Cloud Data Management* (Vol. 9783319047). 10.1007/978-3-319-04765-2

Zhao, Y., & Qiao, C. (2021). *Quantum Transport Protocols for Distributed Quantum Computing*. Academic Press.

Chapter 2
Fundamentals of Quantum Computation and Basic Quantum Gates

Swathi Mummadi
Department of Information Technology, National Institute of Technology, Karnataka, India & Department of Computer Science and Engineering, B V Raju Institute of Technology, India

Bhawana Rudra
Department of Information Technology, National Institute of Technology, Karnataka, India

ABSTRACT

Quantum computing plays a major role in modern computation. It can perform operations exponentially faster when compared to classical computation. It has applications in various areas like secure communication, drug design, artificial intelligence, cyber security, etc. Thus, the researchers and students are showing interest to perform experiments in quantum computing to design novel architectures. But to learn/understand quantum computing, one should have strong knowledge of its basics. Because quantum computing performs operations at the atomic level, so the learners need to understand basic concepts like Qubits, superposition, quantum gates, etc. This chapter gives a clear idea about the basic concepts of quantum computing like Qubits, superposition, entanglement, and quantum gates.

DOI: 10.4018/978-1-7998-9522-0.ch002

INTRODUCTION

Quantum Computing plays a major role in future computation and also changes the entire paradigm of computation, communication, and encryption. It has applications in various areas like secure communication, Finance, Drug design, simulations, Quantum Chemistry, Smart city transportation, Agriculture, etc. (Gill et al., 2020). A revolution in computing technology i.e., 'Quantum computing' is the next big thing that is going to change the whole computing process as we know it today. Quantum computers are highly potential compared to the classical computers (Feynman, 1986). Classical computer deals with the binary information i.e., 0 and 1 but Quantum computer performs the operations on atomic and subatomic particles. An atom performs operations based on principles of Physics. Hence in Quantum computing, the information is represented in the form of qubits and it can store an infinite range of values between 0 and 1 in multiple states so the Quantum computers can perform multiple operations at a time (Benioff, 1982). Due to this, one can say that quantum computers are highly efficient than the world's best Supercomputers. Efficient quantum algorithms like Shor's (1995) and Grover's (Long, 2001) solves the classical security and searching problems in very little time. The advantages of quantum computing are attracting the researchers to work in this area. But the information regarding this technology is very less so it is difficult for the people to understand the concepts especially to work in quantum technology one should have a strong knowledge in quantum basics. Hence this chapter gives a clear idea about the basic concepts of Quantum computing.

QUANTUM BITS (QUBITS)

Qubits are generated from the light beam which is a collection of electrons or photons based on the principles of quantum physics. In classical computation, information is quantified in the form of binary bits i.e., 0's and 1's but Quantum information is quantified in the form of Quantum bits or Qubits. A classical bit can be in either state 0 or 1 at a time but a qubit can be superposition of both i.e., it can be in both the states at a time but when we measure it, it will be collapsed into either state 0 or state 1. Quantum bits are represented with bra and ket notation as $|0\rangle$ and $|1\rangle$ and its basis vector representations are $|0\rangle = \begin{pmatrix} 1 \\ 0 \end{pmatrix}$, $|1\rangle = \begin{pmatrix} 0 \\ 1 \end{pmatrix}$. Single qubit system or multi qubit system are implemented by performing tensor operations on quantum states (Jaeger, 2007; Swathi & Rudra, 2022). These operations are performed on the basis vectors of $|0\rangle$, $|1\rangle$. For example to implement two qubit state $|00\rangle$, the tensor operation should be performed on states $|0\rangle$, $|0\rangle$ which is expressed as

$$|0\rangle \otimes |0\rangle = \begin{pmatrix}1\\0\end{pmatrix} \otimes \begin{pmatrix}1\\0\end{pmatrix} = \begin{pmatrix}1\\0\\0\\0\end{pmatrix}$$

The tensor operation on two qubits represents the four possible outputs as shown in the above result.

SUPERPOSITION

In a Quantum Computer, information is calculated in the form of Qubits. Qubit is a superposition of both the states $|0\rangle$, $|1\rangle$ at a time due to this it can perform multiple operations at a time (de Ronde, 2018; Swathi & Rudra, 2022). A quantum computer with n qubits will be represented as superposition of 2^n states. Hence compared to the classical computer, a quantum computer is exponentially faster and can perform more operations at a time. A pure Quantum state $|\psi\rangle$ can be represented with the equation

$$|\psi\rangle = \alpha|0 + \beta|1\rangle \tag{1}$$

Here α, β are the probabilities of $|0\rangle$ or $|1\rangle$ and $|\alpha|^2+|\beta|^2=1$.

Initially, a pure quantum state will be represented with equal probabilities, i.e. α and β values with probability 50% due to superposition but when we measure it, it will be in any one of the state. A pure quantum state is geometrically represented using the Bloch sphere as shown in Figure1.

Figure 1. The arbitrary representation of a pure quantum state $|\Psi$

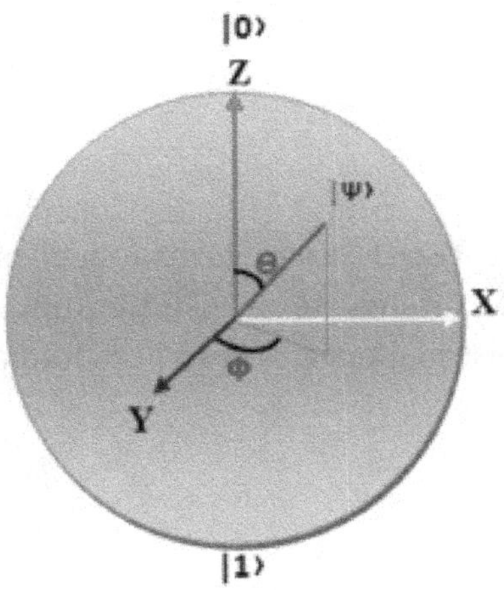

From this we can write an equation for an arbitrary single qubit as

$$|\psi\rangle = \cos\frac{\theta}{2}|0 + e^{i\Phi}\sin\frac{\theta}{2}|1\rangle \quad (2)$$

Here θ, Φ are the real numbers and $0 \leq \theta \leq \pi$, $0 \leq \Phi \leq 2\pi$ and equation (1) is a simplest form of equation (2). Bloch Sphere represents the basis vectors of states $|0\rangle$ and $|1\rangle$, which are also known as spin-up and spin-down of a photon. The new pure state of a photon can be measured based on its position in the Bloch sphere as in Figure 1. Here a new Qubit state $|\psi\rangle$ which is making angle θ with Z-axis and Φ with X-axis can be measured using the equation(2), where the value of $\cos\frac{\theta}{2}$ gives the possibility of state $|\psi\rangle$ becoming $|0\rangle$ and the value of $e^{i\Phi}\sin\frac{\theta}{2}$ represents the possibility of becoming $|1\rangle$. We will measure the final state of $|\psi\rangle$ based on the higher probability.

ENTANGLEMENT

It is a special property of a Quantum particle. According to the Quantum mechanical principles, a pair or group of particles are generated with similar properties i.e. the Quantum states of these particles cannot be described independently. If the state of one particle is modified then it affects the other particle. With this property, Quantum information can be transmitted from one place to another without any physical medium (Horodecki et al., 2009; Swathi & Rudra, 2022). If we know the state of one entangled particle then we can measure the other even without knowing it. Hence entanglement plays a major role in secure data communication. Entangled pairs are represented in the form of Bell states. The possible bell states for four-dimensional Hilbert space with two entangled Qubits are

$$|\Psi^{\pm}\rangle = \frac{1}{\sqrt{2}}(|0\rangle_{A11} \oplus |0\rangle_{B11} \pm |1\rangle_{A12} \oplus |1\rangle_{B12}) \quad (3)$$

$$|\Phi^{\pm}\rangle = \frac{1}{\sqrt{2}}(|0\rangle_{A11} \oplus |1\rangle_{B11} \pm |1\rangle_{A12} \oplus |0\rangle_{B12}) \quad (4)$$

Quantum Entanglement can be useful in many applications like, Quantum Error Correction, Quantum teleportation, secure data transmission, Quantum Information process, Quantum key distribution, etc. (Swathi & Rudra, 2022).

IMPORTANCE OF REVERSIBLE OPERATIONS

Classical computers performs the irreversible operations i.e., after performing the operations due to system failure if we lost the input information then we cannot retrieve it form the produced output. Due to this it is necessary to store the input information which increases the power consumption. Hence it is clear that the computers use a lot of energy because of irreversible operations. The loss of every bit of information in irreversible computing will generate the ln2kT Jouls of heat energy (Takeuchi & Yoshikawa, 2018). Later it has proved that if the reversible computation is used instead of irreversible computation then the energy consumption will be less. In irreversible computation, each output logic will be represented with a unique input combination. The classical set of gates like AND, OR, and EXOR are irreversible as they're all multiple-input and single-output logic gates with this system can run in forward direction only but with reversible operations system can run in forward as well as in backward direction with this the input can also be retrieved from the output (Swathi & Rudra, 2021). With reversible logical gates we can implement various sequential and combinational circuits which will have several advantages over the conventional gates. Various reversible gates are discussed in following section.

QUANTUM GATES

In Quantum computers to perform the reversible operation, reversible Quantum circuits need to be implemented. To implement quantum circuits the single qubit and multi qubit quantum gates are required usually quantum computation can be described in terms of Quantum logic gates (Sleator & Weinfurter, 1995). Quantum gates are used to implement the Quantum circuits. Quantum gates are reversible in nature. The basic Single and multi qubit Quantum gates and its matrix representation are given below

Single-Qubit Gates

Single-Qubit gate plays a vital role in quantum circuit implementation. These gates operates on a single qubit i.e., it accepts single input and produces single output (Docs and Resources, n.d.). The single qubit gates which are used to produce superposition and to perform bit and phase shifts, etc are discussed as follows.

- **Hadamard Gate:** It is the fundamental method for creating the superposition of a qubit. It performs the operation on a single Qubit i.e., it takes single input and produces single output. It represents the input state which is either $|0\rangle$ or $|1\rangle$ to a superposition of $|0\rangle$ and $|1\rangle$ with the equal probabilities to become state 0 and 1. It plays a major role in many quantum circuits to generate superposition. It is graphically represented with the symbol H. Hadamard operation on a single qubit system with quantum states $|0\rangle$ and $|1\rangle$ can be represented as $H = \frac{|0\rangle \pm |1\rangle}{\sqrt{2}}$ for both positive and negative phases. It's matrix depiction is $H = \frac{1}{\sqrt{2}}\begin{bmatrix} 1 & 1 \\ 1 & -1 \end{bmatrix}$
- **Pauli Gates:** Pauli gates are 3 types, X, Y and Z gates. These are represented based on the π angle rotation around the x, y and z axis in Bloch sphere.

    - **Pauli X Gate:** It also performs the operations on a single Qubit. It is same as NOT gate in a classical computation hence it is also called as Quantum NOT gate. It is visually represented with the symbol X. If the input is given as $|0\rangle$ then X gate converts it into $|1\rangle$ and $|1\rangle$ will be converted to $|0\rangle$ hence it can also be called as bit-flip gate. Its matrix representation is $X = \begin{bmatrix} 1 & 0 \\ 0 & 1 \end{bmatrix}$
    - **Pauli Y Gate:** It also accepts single input and produces single output. It represents state $|0\rangle$ to state $i|1\rangle$ and state $|1\rangle$ to state $-i|0\rangle$. It is pictorially represented with symbol Y. It's matrix depiction is $Y= \begin{bmatrix} 0 & -i \\ i & 0 \end{bmatrix}$
    - **Pauli Z Gate:** It also performs operations on a single Qubit. After applying this gate on a qubit will not changes the state if the input is $|0\rangle$ but if it is $|1\rangle$ then it will be modified to $-|1\rangle$ due to this, it can also be called as phase shift/phase flip gate. It is pictorially represented with symbol Z and its matrix representation is $Z = \begin{bmatrix} 1 & 0 \\ 0 & -1 \end{bmatrix}$

- **Phase Gates:** Phase gates are used to represent the phase shifts while performing the operations. These are also single qubit gates. A phase gate is represented with symbol P. If the phase gate is applied on a qubit with $|0\rangle$ then it will not modify anything the state will be as it is but if the input is $|1\rangle$ then it will be modified to $e^{i\Phi}|1\rangle$ i.e., it only modifies the phase of a qubit but the probabilities of $|0\rangle$ and $|1\rangle$ will not be modified. It changes the phase of a qubit along the Z-axis. Its matrix representation is $P(\Phi) = \begin{bmatrix} 1 & 0 \\ 0 & e^{i\Phi} \end{bmatrix}$. Based on the phase ($\Phi$) value the phase gates are divided into Z, S, and T gates. If a P gate is applied on a qubit then it rotates along Z-axis with phase $\Phi = 2\pi$. If $\Phi = \pi$ phase gate acts as a Z gate which is also called as Pauli Z gate as discussed above.

- o **S Gate:** Phase gate acts as a S gate for $\Phi = \frac{\pi}{2}$ Its matrix representation is S = $\begin{bmatrix} 1 & 0 \\ 0 & e^{\frac{i\pi}{2}} \end{bmatrix}$ If the phase is negative i.e., for $\Phi = -\frac{\pi}{2}$, then it is represented as $S^{\dagger}$ and termed as S-dagger and its matrix representation is S = $\begin{bmatrix} 1 & 0 \\ 0 & e^{-\frac{i\pi}{2}} \end{bmatrix}$
- o **T Gate:** Phase gate acts as a T gate for $\Phi = \frac{\pi}{4}$ Its matrix representation is S = $\begin{bmatrix} 1 & 0 \\ 0 & e^{\frac{i\pi}{4}} \end{bmatrix}$ If the phase is negative i.e., for $\Phi = -\frac{\pi}{4}$, then it is represented as $T^{\dagger}$ and termed as T-dagger and its matrix representation is S = $\begin{bmatrix} 1 & 0 \\ 0 & e^{-\frac{i\pi}{4}} \end{bmatrix}$

- **U Gate:** U gate also performs the operation on a single qubit and it is also called as parameterized gate. Till now we have discussed about single qubit gates with two parameters θ, Φ, but U gate represented with three parameters θ, Φ, δ. Hence initially it is termed as U3 gate. Its matrix representation is

$$U(\theta, \Phi, \delta) = \begin{bmatrix} \cos\frac{\theta}{2} & -e^{i\delta}\sin\frac{\theta}{2} \\ e^{i\Phi}\sin\frac{\theta}{2} & e^{i(\Phi+\delta)}\cos\frac{\theta}{2} \end{bmatrix}.$$

U gate with parameters (0, 0, δ) acts as a P gate and with parameters ($\frac{\pi}{2}$, 0, π) acts as a Hadamard Gate.

Multi-Qubit Gates

As of now we discussed about the gates which operates on a single qubit. To perform operations on multi qubits we need multi qubit gates. Hence this section explains the various quantum gates which operate on two or more qubits.

- **Swap Gate:** It is used to swap the qubits. It operates on two qubits i.e., it accepts two inputs and produces the two outputs as shown in Figure 3. If the inputs (I1, I2) of Swap gate are $|0\rangle$ and $|1\rangle$ then the outputs of this gate are $|1\rangle$, $|0\rangle$.

Figure 2. Graphical Representation of SWAP Gate

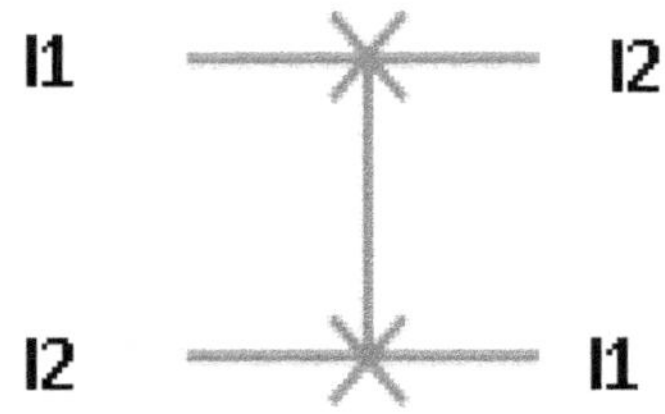

- **Controlled-NOT Gate:** This gate also referred as CNOT gate and Feynman gate. It operates on two qubits i.e., it accepts two inputs called as control and target inputs and produces two outputs called as control and target outputs. For an example, consider the inputs of this gate as (I_{n1}, I_{n2}) and outputs as (O_{t1}, O_{t2}) as shown in Figure 4. The outputs can be retrieved from the equations $O_{t1}=I_{n1}$, $O_{t2}=I_{n1}\oplus I_{n2}$. The output of CNOT gate depends on the control input I_{n1}, if the I_{n1} is $|0\rangle$ then the outputs are same as inputs but if I_{n1} is $|1\rangle$ then the output O_{t1} is same as I_{n1} and O_{t2} is complement of I_{n2}. All possible input and output combinations of CNOT gate are represented in Table1.

Figure 3. Schematic Representation of CNOT Gate

Table 1. Truth Values of CNOT Gate

Inputs		Outputs	
I_{n1}	I_{n2}	O_{t1}	O_{t2}
0	0	0	0
0	1	0	1

Inputs		Outputs	
I_{n1}	I_{n2}	O_{t1}	O_{t2}
1	0	1	1
1	1	1	0

- **Double Controlled-NOT gate:** This gate also referred as Double Feynman Gate. It operates on three qubits i.e., it accepts three inputs and produces three outputs. For an example, consider the inputs of this gate as (I_{n1}, I_{n2}, and I_{n3}) and outputs as (O_{t1}, O_{t2}, and O_{t3}) as shown in Figure5. The output equations for Double CNOT gate are $O_{t1}= I_{n1}$, $O_{t2}= I_{n1} \oplus I_{n2}$, $O_{t3}= I_{n1} \oplus I_{n3}$. Here to retrieve the outputs two CNOT gates are used hence it is called as Double Controlled-NOT gate. All possible input and output combinations of Double CNOT gate are represented in Table2.

Figure 4. Schematic Representation of Double CNOT Gate

Table 2. Truth Values of Double CNOT Gate

Inputs			Outputs		
I_{n1}	I_{n2}	I_{n3}	O_{t1}	O_{t2}	O_{t3}
0	0	0	0	0	0
0	0	1	0	0	1
0	1	0	0	1	0
0	1	1	0	1	1
1	0	0	1	1	1
1	0	1	1	1	0
1	1	0	1	0	1

Inputs			Outputs		
I_{n1}	I_{n2}	I_{n3}	O_{t1}	O_{t2}	O_{t3}
1	1	1	1	0	0

- **Fredkin Gate:** It also operates on three qubits i.e., it accepts three inputs and produces three outputs. The input and output operations for input qubits (I_{n1}, I_{n2}, and I_{n3}) and output qubits (O_{t1}, O_{t2}, and O_{t3}) are represented in Figure 6. Two CNOT gates and one NOT gate, one Toffoli gate are used to implement the Fredkin gate. All possible input and output combinations of Fredkin gate are represented in Table3.

Figure 5. Schematic Representation of Fredkin Gate

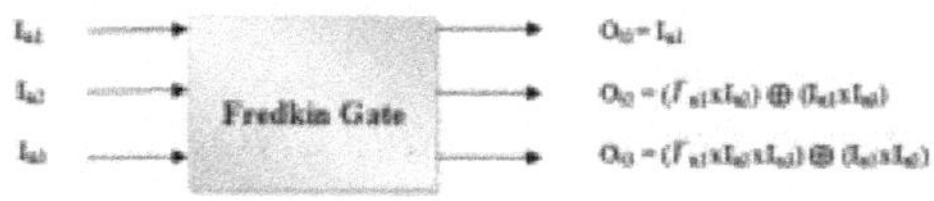

Table 3. Truth Values of Fredkin Gate

Inputs			Outputs		
I_{n1}	I_{n2}	I_{n3}	O_{t1}	O_{t2}	O_{t3}
0	0	0	0	0	0
0	0	1	0	0	1
0	1	0	0	1	0
0	1	1	0	1	1
1	0	0	1	0	0
1	0	1	1	1	0
1	1	0	1	0	1
1	1	1	1	1	1

- **R Gate:** It also operates on three qubits i.e., it accepts three inputs and produces three outputs. When a signal is transmitted through a channel, if we want to make a duplicate of it then R gate can be useful and also it is used to make an invert of a signal. To implement this gate, NOT gate, Toffoli gate, CNOT gate and Swap gates are required. The input and output operations for input qubits (I_{n1}, I_{n2}, and I_{n3}) and output qubits (O_{t1}, O_{t2}, and O_{t3}) are represented in Figure 7. All possible input and output combinations of R gate are represented in Table4.

Figure 6. Schematic Representation of R Gate

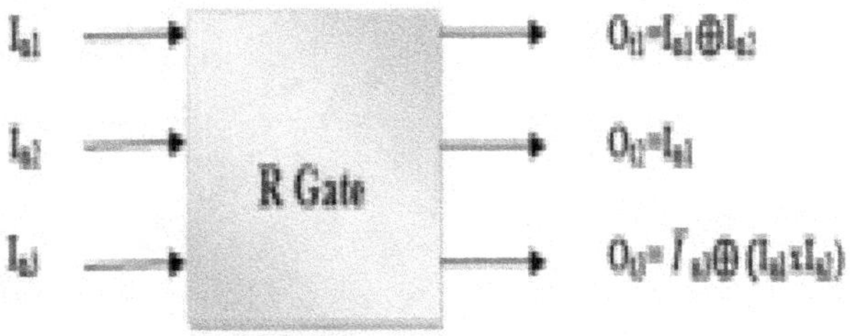

Table 4. Truth Values of R Gate

Inputs			Outputs		
I_{n1}	I_{n2}	I_{n3}	O_{t1}	O_{t2}	O_{t3}
0	0	0	0	0	1
0	0	1	0	0	0
0	1	0	1	0	1
0	1	1	1	0	0
1	0	0	1	1	1
1	0	1	1	1	0
1	1	0	0	1	0
1	1	1	0	1	1

- **Toffoli Gate:** It operates on three qubits i.e., it accepts three inputs and produces three outputs. Let us consider the inputs as (I_{n1}, I_{n2}, and I_{n3}) and outputs as (O_{t1}, O_{t2}, and O_{t3}). The Outputs can be retrieved from the equations $O_{t1}= I_{n1}$, $O_{t2}= I_{n2}$, $O_{t3}= (I_{n1} x\ I_{n2}) \oplus I_{n3}$ as shown in Figure8. All possible input and output combinations of Toffoli gate are represented in Table5.

Figure 7. Schematic Representation of Toffoli Gate

Table 5. Truth Table of Toffoli Gate

Inputs			Outputs		
I_{n1}	I_{n2}	I_{n3}	O_{t1}	O_{t2}	O_{t3}
0	0	0	0	0	0
0	0	1	0	0	1
0	1	0	0	1	0
0	1	1	0	1	1
1	0	0	1	0	0
1	0	1	1	0	1
1	1	0	1	1	1
1	1	1	1	1	0

- **Peres Gate:** It also operates on three qubits i.e., it accepts three inputs and produces three outputs. The input and output operations for input qubits (I_{n1}, I_{n2}, and I_{n3}) and output qubits (O_{t1}, O_{t2}, and O_{t3}) are represented in Figure 9. To implement the Peres gate, CNOT gate and Toffoli gates are required. All possible input and output combinations of Peres gate are represented in Table6.

Figure 8. Graphical Representation of Peres Gate

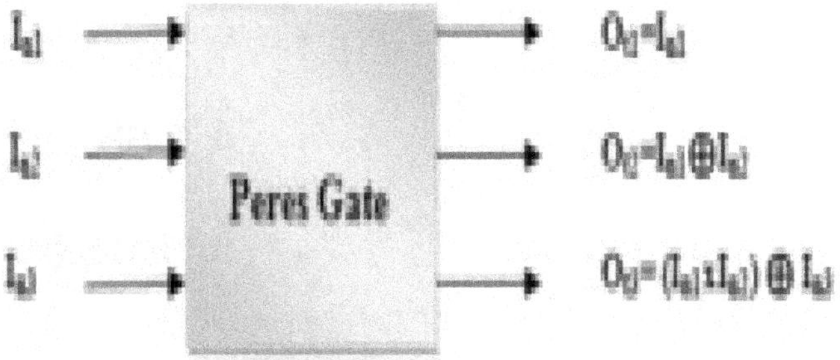

Table 6. Truth Table of Peres Gate

Inputs			**Outputs**		
I_{n1}	I_{n2}	I_{n3}	O_{t1}	O_{t2}	O_{t3}
0	0	0	0	0	0
0	0	1	0	0	1
0	1	0	0	1	0
0	1	1	0	1	1
1	0	0	1	1	0
1	0	1	1	1	1
1	1	0	1	0	1
1	1	1	1	0	0

- **TR Gate:** It also operates on three qubits i.e., it accepts three inputs and produces three outputs. The input and output operations for input qubits (I_{n1}, I_{n2}, and I_{n3}) and output qubits (O_{t1}, O_{t2}, and O_{t3}) are represented in Figure 10. To implement the TR gate, NOT gate, CNOT gate and Toffoli gates are required. All possible input and output combinations of Peres gate are represented in Table7.

Figure 9. Graphical Representation of TR Gate

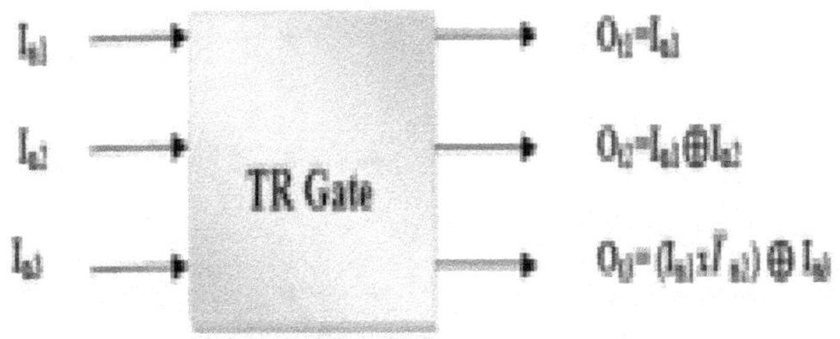

Table 7. Truth Table of TR Gate

Inputs			Outputs		
I_{n1}	I_{n2}	I_{n3}	O_{t1}	O_{t2}	O_{t3}
0	0	0	0	0	0
0	0	1	0	0	1
0	1	0	0	1	0
0	1	1	0	1	1
1	0	0	1	1	1
1	0	1	1	1	0
1	1	0	1	0	0
1	1	1	1	0	1

APPLICATIONS

Superposition and Entanglement are the two major properties of a quantum particle and these play a major role in many applications like drug design, weather forecasting, secure communication and etc. Now a day's, it is difficult to protect the sensitive information while transmitting from one place to other. In such cases quantum technology is a best solution for sensitive data transmission to the longer distances with the help of entanglement. The use of entangled qubits and how these will be teleported between the source and destination are discussed in following section.

QUANTUM TELEPORTATION

Quantum teleportation is a process of transmitting Quantum information from one place to another with the help of entangled pairs and classical communication as shown in Figure 2.

Figure 10. Quantum Teleportation Between Sender and Receiver

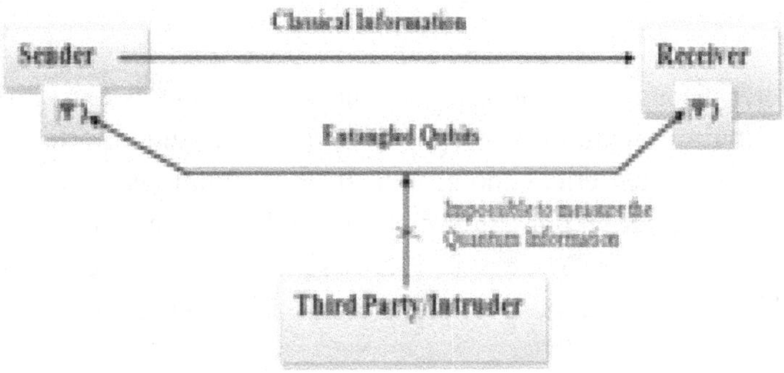

As shown in above Figure, if Sender wants to transmit the information to Receiver and they use only classical communication then there is a possibility of a Third party/Intruder to either measure or modify the information but if they use the Classical information along with Quantum bits then it is impossible for an intruder to measure or modify the information due to the quantum properties i.e., Decoherence (Schlosshauer, 2019) and No cloning theorem (Wootters & Zurek, 2009). In above figure sender sent the classical information to the receiver and also both shared an entangled pair as expressed in equation (4). If a third party tries to observe the information transmitted between sender and receiver then he may find the classical information but it is impossible for him to find the state of an entangled bit. Even though if he tries to measure the state of a Qubit then its actual state will be lost due to the disturbance made by the third party. Hence it is impossible to copy the Qubits. Hence we can say that the data which is transmitted in the form of Quantum states will be more secured. Quantum teleportation can be used in many applications where the RSA Encryption lost to protect the information.

FUTURE RESEARCH DIRECTIONS

The combination of various quantum gates can be used for efficient and cost effective quantum architectures. Hence there is a lot of scope for research in implementing hybrid architectures with mixed quantum gates. This chapter helps the researcher's who are willing to implement efficient architectures with the high performance.

CONCLUSION

This chapter discussed about the basic quantum principles like quantum states, qubit representation, quantum superposition, quantum teleportation, and various quantum gates with theoretical and graphical representations which helps the students who are interested in quantum research. In this we have discussed about the hybrid quantum gates like Fredkin, R, TR, and Peres which are the combination of NOT and CNOT gates and these all are useful in implementing efficient quantum architecture.

ACKNOWLEDGMENT

This research received no specific grant from any funding agency in the public, commercial, or not-for-profit sectors.

REFERENCES

Benioff, P. (1982). Quantum mechanical hamiltonian models of turing machines. *Journal of Statistical Physics*, 29(3), 515–546. 10.1007/BF01342185

de Ronde, C. (2018). Quantum Superpositions and the Representation of Physical Reality Beyond Measurement Outcomes and Mathematical Structures. *Foundations of Science*, 23(4), 621–648. 10.1007/s10699-017-9541-z

Docs and Resources. (n.d.). *IBM Quantum Experience IBM*. Available: https://quantum-computing.ibm.com/docs/

Feynman, R. P. (1986). Quantum mechanical computers. *Foundations of Physics*, 16(6), 507–531. 10.1007/BF01886518

Gill, S. S., Kumar, A., Singh, H., Singh, M., Kaur, K., Usman, M., & Buyya, R. (2020). Quantum Computing: A Taxonomy, Systematic Review and Future Directions. *Emerging Technologies*. DOI=arXiv:2010.15559 [cs.ET]

Horodecki, Horodecki, Horodecki, & Horodecki. (2009). Quantum entanglements. *Rev. Mod. Phys., 81*(2), 865-942. .10.1103/RevModPhys.81.865

Jaeger, G. (2007). *Quantum information*. Springer.

Long. (2001). Grover algorithm with zero theoretical failure rate. *Physical Review A, 64*(2).

Schlosshauer, M. (2019). Quantum decoherence. *Physics Reports*, 831, 1–57. 10.1016/j.physrep.2019.10.001

Shor, P. W. (1995, October). Scheme for reducing decoherence in Quantum computer memory. *Physical Review A: General Physics*, 52(4), R2493–R2496. 10.1103/PhysRevA.52.R24939912632

Sleator & Weinfurter. (1995). Realizable universal quantum logic gates. *Physical Review Letters, 74*(20).

Swathi, M., & Rudra, B. (2021). Implementation of Reversible Logic Gates with Quantum Gates. *2021 IEEE 11th Annual Computing and Communication Workshop and Conference (CCWC)*. IEEE. 10.1109/CCWC51732.2021.9376060

Swathi, M., & Rudra, B. (2022). A Novel Architecture for Binary Code to Gray Code Converter Using Quantum Cellular Automata. In *Edge Analytics* (pp. 43–61). Springer. 10.1007/978-981-19-0019-8_4

Swathi, M., & Rudra, B. (2022). A Novel Approach for Asymmetric Quantum Error Correction With Syndrome Measurement. *IEEE Access : Practical Innovations, Open Solutions*, 10, 44669–44676. 10.1109/ACCESS.2022.3170039

Swathi & Rudra. (2022). An efficient approach for quantum entanglement purification. *International Journal of Quantum Information, 20*(4).

Swathi, M., & Rudra, B. (2022). Novel Encoding method for Quantum Error Correction. *2022 IEEE 12th Annual Computing and Communication Workshop and Conference (CCWC)*. IEEE. 10.1109/CCWC54503.2022.9720880

Takeuchi & Yoshikawa. (2018). Minimum energy dissipation required for a logically irreversible operation. *Physical Review E, 97*(1).

Wootters, W. K., & Zurek, W. H. (2009). The no-cloning theorem. *Physics Today*, 62(2), 76–77. 10.1063/1.3086114

Chapter 3
Quantum Key Distribution Protocols

Shyam Sihare
https://orcid.org/0000-0003-2096-8273
Dr. A.P.J. Abdul Kalam Government College, India

ABSTRACT

The author will discuss different quantum protocols in this chapter for guided media and open space communication. The author examines already developed quantum protocols for photons and electrons. Earlier developed quantum protocols are the basis of recently developed quantum protocols. The author makes a study on quantum protocols with the help of quantum mechanics features such as entanglement, superposition, uncertainty principle, and no cloning.

1. INTRODUCTION

Quantum protocols act as a guide to transfer a message from one party to more than one parties. Quantum protocols are completely different than classical protocols as functionally and operationally. There are different types of quantum protocols (Colbeck R.; 2009). Among different types of protocols, the BB84 protocol plays a significant role and guide the development of the subsequent protocol e.g. decoy, SARG04, E91, KMB09 (Malathy et al.; 2022).

Photons or electrons can practice for message encryption and apply on them different orientation. Furthermore, the encrypted message has been transferred from Alice to Bob in the presence of Eve (intruder) (Williams C P & Williams C P; 2011). The quantum protocols give the guarantee of absolute security due to quantum mechanics features (Van De Graaf J; 1997). Particle orientation has been performed by the PBS, Filter, Polarizer, Merger, and Amplifier (Tsekeri et al.; 2021).

DOI: 10.4018/978-1-7998-9522-0.ch003

Oriented particles send through guided media, or open space, or both medias. Before sending oriented particles, Bob and Eve share her key either private or public way (Chun W H K; 2008).

A classical network protocol contains a set of rules enabling the exchange messages from one computer to another (Bonaventure; 2011). A well-known classical protocol is the TCP-IP protocol. Other classical protocols include FTP, SMTP, Telnet, POP, IMAP, Bitcoin, and VoIP. Currently, few protocols are used (Yildirim; 2010).

Different quantum protocols exist for quantum communication and quantum information. The operation of quantum protocols is entirely different from classical network protocols (Cacciapuoti et al.; 2019). The quantum protocol operation depends on quantum mechanics features, whereas the classical network protocol operation depends on classical physics features (Perseguers et al.; 2010).

Quantum protocols are used for exchanging quantum keys between Alice and Bob in the presence of Eve by using a public channel (Parakh A; 2013). The protocols are used to establish a link between Alice and Bob (Mödersheim S; 2009, March). This method is known as private key cryptosystem because the communication is conducted privately with high security. After a connection is established, message exchange is performed through a public channel without worrying about the presence of an unauthorised person. For communication, a control channel is required throughout the communication process (Dzung et al.; 2005).

The possibility of errors during quantum communication is more than the possibility during classical communication (Buhrman H, Cleve R, & Wigderson A; 1998, May). During classical communication, errors are detected and corrected by classical algorithms such as CRC, Hash function $H(x)$, Hamming code, and parity bits (Babar et al.; 2018). These algorithms are applied for error detection and correction while using IPv4, IPv6, user datagram protocol (UDP), and TCP-IP and during deep-space communications, satellite broadcasting, and data storage. The sender and receiver are not involved in the error correction and detection (De Cola et al.; 2011). A sender sends messages over a classical channel without the knowledge of a receiver's operational pattern. Approximately, 10% reverse communication occurs between a sender and receiver for error correction. During the communication between Alice and Bob, error correction and detection algorithms are used to detect bit-, stream-, message-, and block-level errors (Chiueh T D & Tsai P Y; 2008).

Quantum errors are detected and corrected with the assistance of Alice, Bob, or both by sharing of private key. Without mutual understanding, quantum errors cannot be controlled (El Ashmawy M S; 2021). Alice sends qubits and Bob receives the qubits without the knowledge of Eve presence. Every qubit measurement is conducted at the receiver end. After the qubits are transferred, the polarised qubits are checked by Bob with the help of Alice (Valivarthi V R R; 2017). If the error in the checked qubits is below a threshold value, then the qubits are accepted otherwise the receiver rejects the qubits. If the error rate crosses a threshold value, then the possibility that Eve has obtained the qubits and has tried to modify the messages is high (Djordjevic I; 2012). If the quantum error is below a threshold value, then Bob verifies the basis states with the basis states originally transferred by Alice (Lo H K; 2001). If Eve gains access to the messages and modifies them, then the communication is rejected and the messages are again transferred between Alice and Bob (Chin E et al.; 2011, June) (Figure 1).

Figure 1. Critical Components and Relationships in Quantum Communication Protocols

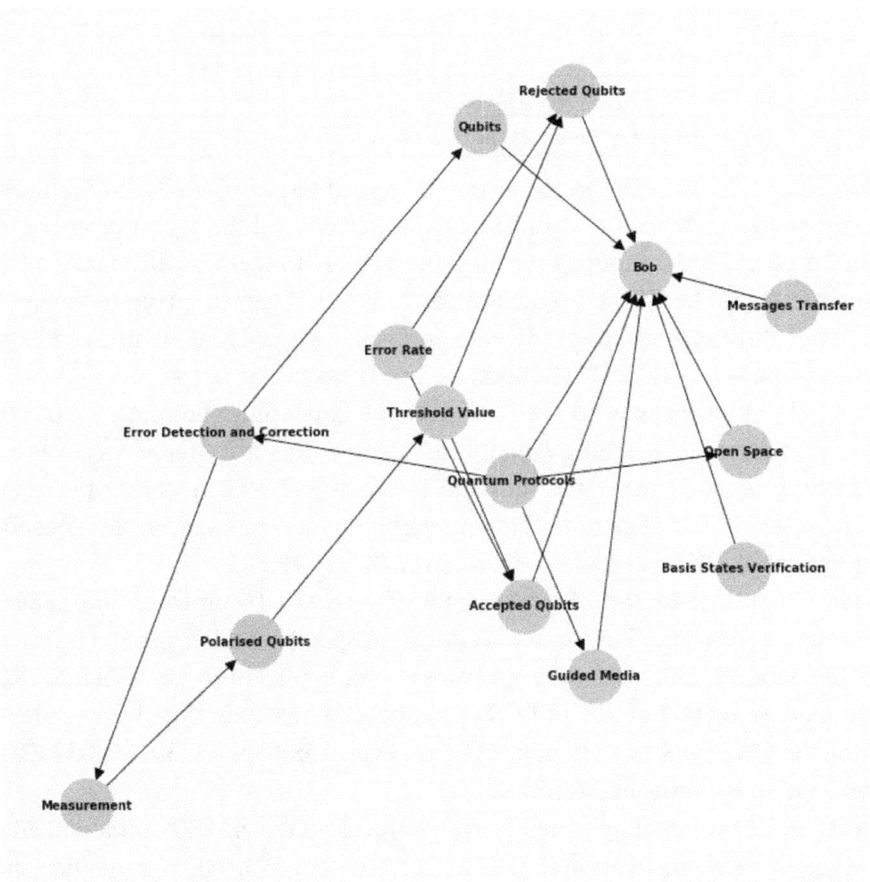

2. EVOLUTION AND HISTORICAL BACKGROUND

The domain of QKD protocols has witnessed a profound evolution, marked by significant milestones that have shaped the landscape of secure quantum communication. A historical perspective unveils the foundational pillars and the chronological development timeline of these protocols, providing essential context for a comprehensive understanding of their current state.

Table 1. Historical Development of QKD Protocols

Year	Milestone
1984	BB84 Protocol by Bennett and Brassard
1991	E91 Protocol by Ekert
2004	SARG04 Protocol by Scarani et al.
2009	KMB09 Protocol by Koashi, Mabuchi, and Barrett

The inception of QKD protocols can be traced back to 1984 with the groundbreaking work of Bennett and Brassard, who introduced the BB84 protocol. This seminal protocol laid the foundation for subsequent advancements in quantum secure communication. The BB84 protocol utilized the principles of quantum mechanics, such as quantum superposition and entanglement, to enable the secure exchange of cryptographic keys (Table 1) (Bennett C H & Brassard G; 2014).

In 1991, Ekert introduced the E91 protocol, expanding the horizons of QKD. The E91 protocol leveraged quantum entanglement as a resource for secure key distribution, demonstrating the intricate interplay between quantum phenomena and cryptographic applications. This marked a pivotal moment in the conceptual development of QKD protocols (Ekert, Artur K.; 1991).

The year 2004 witnessed the emergence of the SARG04 protocol, introduced by Scarani and colleagues. This protocol addressed specific challenges and limitations of earlier models, enhancing the robustness and applicability of QKD. SARG04 contributed to the maturation of QKD protocols by incorporating decoy states and refining the QKD process (Branciard, Cyril, et al.; 2005; p. 032301-1- 032301-15; Scarani, Aćin, Ribordy, and Gisin; 2004).

In 2009, Koashi, Mabuchi, and Barrett introduced the KMB09 protocol, adding another layer of sophistication to QKD. This protocol focused on dynamic aspects of QKD, adapting to changing conditions in the quantum channel. KMB09 marked a shift towards more adaptive and resilient QKD protocols, addressing real-world challenges (Koashi, Mabuchi, and Ben-Or, 2009).

The landscape of QKD continues to evolve, with ongoing research exploring novel protocols, quantum technologies, and integration with practical communication systems. Current endeavors aim to enhance the scalability, efficiency, and security of QKD protocols, paving the way for their integration into mainstream communication infrastructure.

In conclusion, the historical development of QKD protocols exemplifies a journey from foundational concepts to sophisticated, adaptive frameworks. The evolution of these protocols mirrors the dynamic nature of quantum communication research, where each milestone builds upon the knowledge and achievements of its predecessors. Understanding this historical trajectory is indispensable for researchers, practitioners, and enthusiasts seeking to navigate the complexities of quantum secure communication (Figure 2).

Figure 2. Historical development of quantum protocols

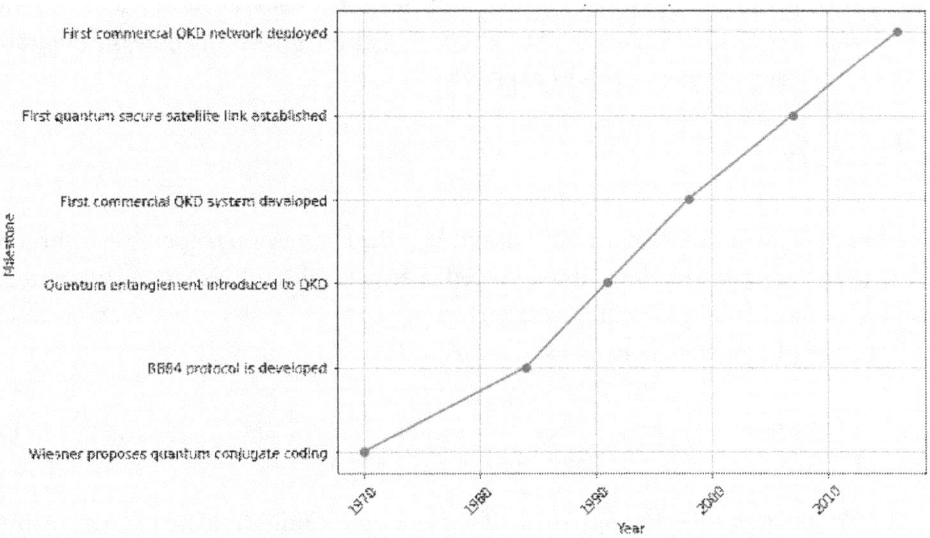

3. BASIC PRINCIPLES OF QUANTUM MECHANICS

Quantum mechanics, a fundamental branch of modern physics, provides a precise and mathematical explanation of the behavior of microscopic particles. The wave-particle duality principle posits that particles, such as electrons or photons, exhibit both wave-like and particle-like attributes. This duality is encapsulated in the de Broglie wavelength equation:

$$\lambda = \frac{h}{p} \tag{1}$$

Here, λ signifies the wavelength, h is Planck's constant, and p denotes the momentum of the particle. Eq. (1) underscores the inherent wave nature intrinsic to particles. Heisenberg's uncertainty principle is a fundamental principle, asserting the intrinsic limits to the precision with which certain pairs of properties, like position (Δx) and momentum (Δp), can be simultaneously known:

$$\Delta x \cdot \Delta p \geq \frac{\hbar}{2} \tag{2}$$

Here, $\hbar$ represents the reduced Planck's constant. This principle challenges classical determinism, introducing inherent indeterminacy at the quantum scale. Quantization is manifested in the Schrödinger equation, governing how the quantum state of a physical system evolves over time:

$$i\hbar\frac{\partial}{\partial t}\Psi = \hat{H}\Psi \tag{3}$$

Here, Ψ signifies the wave function, and $\hat{H}$ is the Hamiltonian operator. Solutions to Eq. (3) offer understandings into allowed energy levels and associated wave functions. Quantum entanglement, characterized by correlated states between particles irrespective of distance, is expressed as:

$$|\psi = \frac{1}{\sqrt{2}}(|0 \otimes |1 - |1 \otimes |0) \tag{4}$$

Eq. (4) illustrates the entanglement of two qubits, highlighting their superposition of states. Quantum mechanics is a framework based on principles like wave-particle duality, uncertainty, energy quantization, and entanglement, used to understand the intricate behavior of matter and energy (Figure 3).

Figure 3. Example of wave function and wave-particle de Broglie wavelength

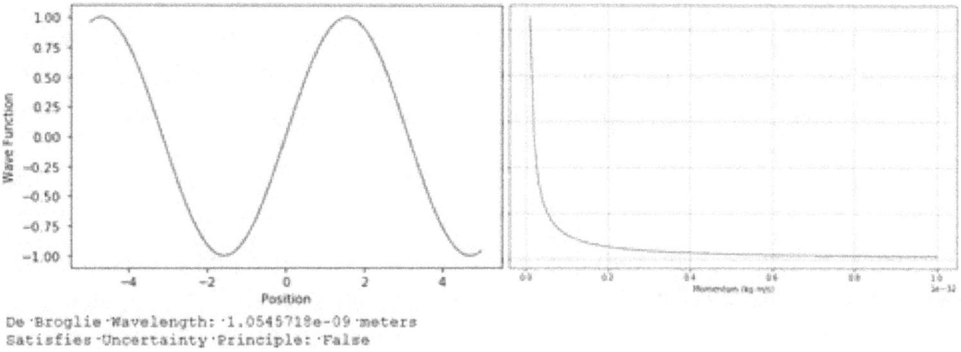

4. QUANTUM KEY DISTRIBUTION PROTOCOLS

Within this section, we expound upon the foundational principles of QKD. Subsequently, an in-depth analysis is undertaken, investigating into diverse fundamental quantum protocols. These protocols encompass the influential BB84, the sophisticated SARG04, the innovative KMB09, and the entangled E91, protocols employing decoy states, the intricate domain of quantum teleportation, and the dynamic framework of MQKA.

4.1 Introduction to QKD

QKD utilizes the principles of quantum mechanics to enable secure communication through the establishment of an unbreakable cryptographic key between two parties, usually referred to as Alice and Bob. One of the prominent methods in QKD is entanglement-based protocols. Entangled particles exhibit correlations that are exploited for secure key distribution. The entangled state is typically represented as:

$$|\psi = \frac{1}{\sqrt{2}}(|0 \otimes |1 - |1 \otimes |0) \quad (5)$$

Here, $|0$ and $|1$ represent the basis states. Alice and Bob share entangled particles, and by measuring the entangled particles in the appropriate basis, they can establish a secret key. Furthermore, BBM92 protocol, a renowned entanglement-based QKD protocol, involves the use of Bell states. The shared entangled state is expressed as:

$$|\psi\rangle = \frac{1}{\sqrt{2}}(|00\rangle + |11\rangle) \tag{6}$$

Bell measurements are conducted, and outcomes are used to derive a secret key between Alice and Bob. The process involves the application of Pauli operators, denoted as X and Z, and the creation of a key from the correlated measurement results.

$$X|0\rangle = |1\rangle, Z|1\rangle = -|1\rangle \tag{7}$$

QKD ensures security through the detection of eavesdropping attempts. The security is quantified using the QKD security parameter, often represented as ε. The relationship between ε and the error rate (Q) is given by:

$$\varepsilon \geq Q - \eta \tag{8}$$

Here, η represents the fraction of errors attributed to the quantum channel. The key generation rate (R) in QKD protocols is crucial for practical applications.

$$R = f \cdot (1 - H_2(\varepsilon)) \cdot Q \tag{9}$$

Here, f is the efficiency of the system, and $h_2(\varepsilon)$ is the binary entropy function. QKD protocols, especially those based on entanglement, leverage the principles of quantum mechanics to establish secure communication channels. The choice of entangled states, measurements, and the quantification of security parameters through error rates contribute to the robustness of QKD systems, making them pivotal in the quest for secure communication in the quantum domain (Figure 4).

Figure 4. QKD by key rate (R), Error rate (ε) and efficiency (f)

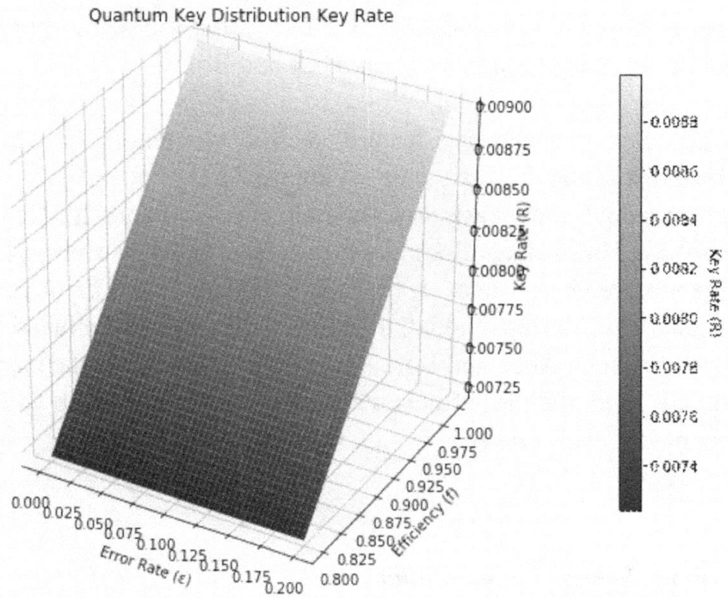

4.2 BB84 Protocol and Its Limitations

Three users are present during quantum communication-sender (Alice), receiver (Bob), and an intruder (Eve). These users have different motives during communication. The detection of an intruder is crucial for quantum security. Alice sends messages to the receiving end; Bob receives those messages at the receiving end in the form of a sequence sent by Alice. While Alice and Bob are communicating, Eve attempts to obtain the messages. On a classical network, Eve makes the following efforts during message transmission between Alice and Bob (Bennett, Charles H., et al.; 2014; p. 7-11):

1. Eve attempts to copy the encrypted message and store it for analysing and manipulating the message later. By conducting investigation and crosschecking of the encoded messages, Eve acquires the encrypted message pattern and develops algorithms or prepares drafts for decrypting the encrypted message.
2. Eve copies all encrypted messages without letting Alice and Bob know. Subsequently, Eve attempts to alter the message according to her requirements. The detection of Eve's presence is complex at the initial stage. However, her

presence is detected after some time. At this point, she might have already obtained many messages.

During quantum communication, Eve makes the following efforts to acquire messages during transmission between Alice and Bob:

1. Due to quantum mechanics feature, such as no-cloning, copying a message without disturbing the basis states is impossible.
2. During readout, most basis states are altered automatically. This implies that Bob received basis states dissimilar to those sent by Alice. Bob performs readout on the received basis states. If the QBER (Quantum Bit Error Rate) crosses a threshold value during readout, then Eve's presence is concluded during transmission between Alice and Bob. Eve attempts to readout the basis states, and then, she alters the basis states when received by Bob. Bob and Alice could easily detect Eve's presence.

Table 2. Bloch Sphere Representation

Send State	Corresponding Superposition State	Output
$\vert\nwarrow$	+	$\frac{1}{\sqrt{2}}(\vert\uparrow - \vert\rightarrow)$
$\vert\nearrow$	+	$\frac{1}{\sqrt{2}}(\vert\uparrow + \vert\rightarrow)$
$\vert\uparrow$	×	$\frac{1}{\sqrt{2}}(\vert\nearrow + \vert\nwarrow)$
$\vert\rightarrow$	×	$\frac{1}{\sqrt{2}}(\vert\nearrow - \vert\nwarrow)$

Bloch sphere representation of Table 2 is presented in Figure 5. When basis states are transferred from Alice to Bob, Eve attempts to obtain and copy the encrypted message. Table 3 indicates the basis states at Alice's and Bob's end and Eve is not included. Clearly, Bob has not received the basis states as sent by Alice. The probability of half of the basis states at Bob's end would be correct. The remaining half, less than half, or more than half basis states are not corrected. These states have to be corrected by mutual understanding between Alice and Bob. The received uncorrected basis states are corrected by agreement of private key between Alice and Bob in the presence of Eve. Bob obtains permission for correcting the states by conducting quantum key exchange with Alice. Eve never receives 101010111100 original messages even after many iterations, whereas Bob achieves 101010111100 original messages after finite iterations owing to the quantum key. Each iteration requires retransmission of the uncorrected basis states from Alice to Bob.

Figure 5. Bloch's Representation of Table 2

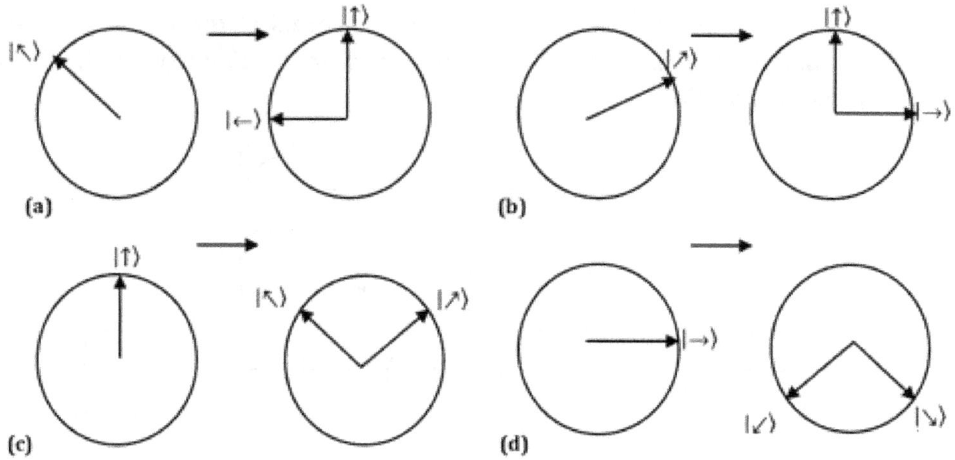

Let BB84 protocol representation for a message 101010111100 as follows:

Table 3. BB84 protocol representation for message 101010111100 at Alice's and Bob's ends

Bit Number	1	2	3	4	5	6	7	8	9	10	11	12
Random bits at Alice's end	1	0	1	0	1	0	1	1	1	1	0	0
Random bases at Alice's end	+	+	×	×	+	×	+	+	×	×	+	×
Alice sends basis states	→	→	↖	↗	↑	↖	↑	→	↖	↖	↑	↖
Random bases at Bob's end	+	×	+	×	×	×	+	+	+	×	+	+
Bob observes basis states	→	↖	↗	↗	→	↖	↑	→	↗	↖	↑	↗
Received bits at Bob's end	1	0	0	1	0	1	1	1	0	1	1	0

Figure 6 represents the barriers in quantum communication during basis state transmission between Alice and Bob. In Figure 6 (a–f), Figure 6 (f) is usually completely irrelevant for quantum communication, but Figure 6 (a), (b), and (d) is appropriate. Eve presence is easily detected at Alice's and Bob's in Figure 6 (d) and (f).

Figure 6. Quantum communication barriers: (a) Basis states are transferred from Alice to Bob in the absence of Eve. (b) Basis states are interrupted due to the surrounding effect without the involvement of Eve. (c) Basis states are obtained by Eve during quantum communication. (d) Basis states are modified by Eve. (f) Basis states are revised by Eve.

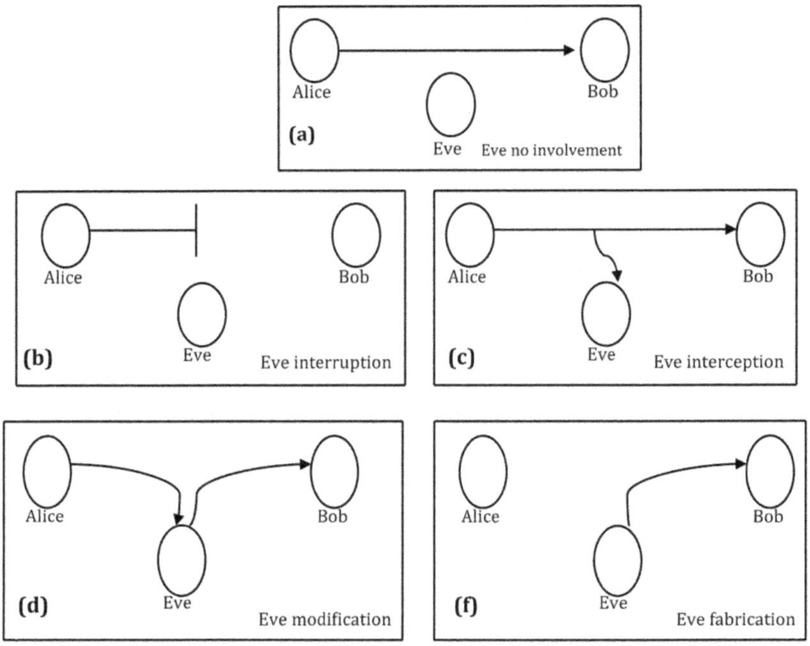

4.3 SARG04 Quantum Protocol

The SARG04 protocol (Branciard, Cyril, et al.; 2005; p. 032301-1- 032301-15) is also a superset of BB84 and was proposed by Scarani, Aćin, Ribordy, and Gisin in 2004. SARG04 enhances quantum security by imposing additional constraints pertaining to quantum mechanics and prevents Eve from controlling basis states and the quantum channel. BB84 is uses a single photon as the basis state and send the photon to Bob in the presence of Eve. However, while using SARG04, multiple same basis states are sent from Alice to Bob. Multiple same basis states imply that Alice sends more than two same basis states at a time to Bob (Branciard, Cyril, et al.; 2005; p. 032301-7-032301-8). Sending more than two same basis states is a complicated process. Two basis states are interprets as 2^2= 4 distinct basis states. This protocol is used for all four same basis states. These basis states are sent by

Alice to Bob in the presence of Eve. Every basis state is assigned a different intensity. Due to the different intensities, it is difficult for Eve to differentiate between the individual qubits, thus increasing the complexity at Eve's end. Qubit differentiation can be performed using quantum computation and Hadamard computation. At the initial phase, it is uncertain whether the received qubits are Hadamard or quantum computational basis states. Due to this uncertainty, it is more complex for Eve to accurately predict a polarised photon without disturbing them. In this case, Bob possesses an additional key that is unknown to Eve. During quantum error correction, this key is crucial for correcting the uncorrected quantum basis states. Eve attempts to obtain the basis states, but she cannot do so due to the multiple same basis states sent by Alice. Qubits are interchangeably used as basis states. Suppose that Alice sends the first qubit as follows:

$$|\psi_a = |\psi^a_{t(i)}. \tag{10}$$

Here, $t(i)$ implies the time i, $a$indicates Alice's end and ψ presents the wave function. Alice sends qubits to Bob as follows:

$$|\psi_b = |\psi^b_{t(i+1)}. \tag{11}$$

For the SARG04 protocol, Eqs. (10) and (11) is rewritten as below:

$$|\psi^a_{t(i)} \equiv |\psi^b_{t(i+1)}. \tag{12}$$

Another qubit is send by Alice to Bob as follows:

$$|\phi_a = |\phi^a_{t(i+2)}. \tag{13}$$

Bob receives the qubit as follows:

$$|\phi_b = |\phi^b_{t(i+3)}. \tag{14}$$

Thus,

$$|\phi_a \not\equiv |\phi_b.$$

The qubit presented in Eq. (14) depends on Eq. (13). Hence, Eq. (14) and Eq. (13) is rewritten as follows:

$$|\phi_a \otimes |\phi_b \; = \; |\psi^a_{t(i)} \otimes |\psi^b_{t(i+1)}. \tag{15}$$

Suppose that $|\psi_a = |\phi_a$ and $t(i) = t(i+1)$, then Eq. (15) is rewritten as follows:

$$|\psi_{aa'} = |\psi_a|\psi_{a'}. \tag{16}$$

Here, $|\psi_{a'}$ either acts as quantum computational or Hadamard computational term. If $a = a'$, then Eq. (16) is

$$|\psi_{aa'} = |\psi_{aa'} \text{ h } a=\{0,1\}. \tag{17}$$

If a photon is transferred using Eq. (17) through OSQC for quantum key exchange, then Eq. (16) is written as

$$|\psi_{aa'} = \xi(|\psi_a|\psi_{a'}). \tag{18}$$

Here, ξ is a noise factor when qubits interact in open space. Assume ξ is absent during quantum computation and quantum communication. The mean quantum error would be $\xi < 1\%$ when qubits are transferred by OSQC. This method is also applicable to an n-bit long string.

Alice publically announces the $|\psi_a$ qubit, and Bob receives the $|\psi_b$ qubit; $|\psi_a \cong| \psi_b$. Eve finds it difficult to readout the second basis state. Hence, this protocol is more secure than the aforementioned proposed protocols (i.e. BB84 and B92). When this method is used, Eve fails to readout the basis states. Even if Eve is successful to readout a few of basis states, the remaining basis states cannot be read correctly by her because of the special key possessed by Bob.

4.4 KMB09 Quantum Protocol

The KMB09 protocol (Khan, Muhammad Mubashir, et al.; 2009; p. 063043(1-17)) was proposed by Khan, Murphy, and Beige in 2009. This is a QKD protocol and is slightly similar to BB84. Compared with when aforementioned QKD protocols are used, the transfer efficiency is higher and the error rate is considerably lower when this protocol is used. In KMB09, Alice announces the basis index key and sends qubits $|e_i$ and $|f_i$ to Bob in the presence of Eve (Khan, Muhammad Mubashir, et al.; 2009; p. 063043(4-5)).

At Bob's end, parameters $|\eta_{ij}$ and $|\mu_{ij}$ are received; the values of these parameters could be '1', '0', or '×'. Here, '×' indicates *donotcare* condition which values are any of '0', '1', 'nothing'. Hence,

$|e_i = |f_i$ i i is equal in both states; otherwise,

$$|e \neq |f \tag{19}$$

i.e.

$$|e_j^a \equiv |\eta_{ij}^b \text{ w en lice}(j) = \text{Bob}(j). \qquad (20)$$

Moreover,

$$|f_j^a \equiv |\mu_{ij}^b \text{w en lice}(j) = \text{Bob}(j). \qquad (21)$$

At Bob's end, the photon measurement states are

$$\eta_{ij}^b \equiv \{0,1,\times\}. \qquad (22)$$

Similarly,

$$\mu_{ij}^b \equiv \{0,1,\times\}. \qquad (23)$$

Consider that,

Alice sends

$$|e_i^a \equiv 0 \qquad (24)$$

then bob receives

$$\eta_{ij}^b \equiv 0 \text{ r} \times. \qquad (25)$$

When $|f_i = 1$, the correct key is sent to Bob. If $|\mu_{ij}^b = 1$ or $\times$ (iff $i = 1$) and $|f_{j\neq 1}^a$, then $|\mu_{ij}^b \equiv |e_i^a$. Similarly, $|f_i^a \equiv |f_{j\neq 1}^b$ if Alice transmitted $i = 1$ (Table 4 and Table 5).

Table 4. Index key transmitted by Alice and the outcome at Bob's end on the basis of the index key i as $|\eta_{ij}$ and $|\mu_{ij}$ (Khan, Muhammad Mubashir, et al.; 2009; p. 063043 (5))

Index Transmitted by Alice	State Measured by Bob									
	$|e_1$	$|e_2$	…	…	$|e_N$	$|f_1$	$|f_2$	…	…	$|f_N$
1	$|\eta_{11}$	$|\eta_{12}$	…	…	$|\eta_{1N}$	$|\mu_{11}$	$|\mu_{12}$	…	…	$|\mu_{1N}$

Index Transmitted by Alice	State Measured by Bob									
	$\|e_1$	$\|e_2$	...	...	$\|e_N$	$\|f_1$	$\|f_2$	...	...	$\|f_N$
2	$\|\eta_{21}$	$\|\eta_{22}$	...	...	$\|\eta_{2N}$	$\|\mu_{21}$	$\|\mu_{22}$	...	...	$\|\mu_{2N}$
...	...	...	...	...	...	...	...	...	...	...
...	...	...	...	...	...	...	...	...	...	...
N	$\|\eta_{N1}$	$\|\eta_{N2}$	...	...	$\|\eta_{NN}$	$\|\mu_{N1}$	$\|\mu_{N2}$	...	...	$\|\mu_{NN}$

Table 5. Index key transmitted by Alice and the outcome at Bob's end on the basis of the index key i as $|\eta_{ij}$ and $|\mu_{ij}$ in OSQC, where $\alpha, \beta, \gamma,$ **and** *δ are environmental components when the photon interacts with the environment. Here, γ* **and** *δ are components that are present in $|\eta_{ij}$ and $|\mu_{ij}$ when bit measurement is conducted on the encoded photons at Bob's end*

Indexed Transmitted by Alice	State Measured by Bob									
	$\alpha\|e_1$	$\alpha\|e_2$	...	...	$\alpha\|e_N$	$\beta\|f_1$	$\beta\|f_2$	...	...	$\beta\|f_N$
1	$\gamma\|\eta_{11}$	$\gamma\|\eta_{12}$	...	...	$\gamma\|\eta_{1N}$	$\delta\|\mu_{11}$	$\delta\|\mu_{12}$	...	...	$\delta\|\mu_{1N}$
2	$\gamma\|\eta_{21}$	$\gamma\|\eta_{22}$	...	...	$\gamma\|\eta_{2N}$	$\delta\|\mu_{21}$	$\delta\|\mu_{22}$	...	...	$\delta\|\mu_{2N}$
...	...	...	...	...	...	...	...	...	...	...
...	...	...	...	...	...	...	...	...	...	...
N	$\gamma\|\eta_{N1}$	$\gamma\|\eta_{N2}$	...	...	$\gamma\|\eta_{NN}$	$\delta\|\mu_{N1}$	$\delta\|\mu_{N2}$	...	...	$\delta\|\mu_{NN}$

Eve's presence can easily be detected while exchanging the quantum-encoded bits. The ITER and QBER are very low at the final phase in quantum communication.

This protocol can be experimentally used for OSQC. Moreover, the protocol can be used for long-distance quantum communication because of the index key. New quantum protocols for key exchange in OSQC can be developed by using this protocol. The primary difficulty is the management of the index key at Alice's end during OSQC. At Bob's end, knowing the exact index key sent by Alice is difficult. During OSQC, numerous atmospheric components interact with the publically transmitted basis state index. Moreover, the error rate is higher when OSQC is used than that when a fiber optic channel is used (Khan, Muhammad Mubashir, et al.; 2009; p. 063043(5)). Hence, the following is stated:

$$|\alpha|^2 + |\beta|^2 < 1\% \text{ and } |\gamma|^2 + |\delta|^2 < 1\% \tag{26}$$

4.5 E91 Quantum Protocol

The E91 quantum protocol is used for exchanging a quantum key between Alice and Bob (Eve present in between). E91 is an extension of BB84 and B92 protocol features. However, the photon polarisation method of the above protocols differs. E91 was proposed by Ekert in 1991 (in E91, 'E' represents Ekert or entanglement and '91' is used for the proposed year) (Ekert, Artur K.; 1991; p. 661). In this protocol, photons are transformed in the entanglement form. Entangled photons are sent by Alice to Bob in the rectilinear or orthogonal form. The orthogonal form is preferred for obtaining high security. The combination of rectilinear and orthogonal basis states provides high security during quantum communication. Entangled photon states used while using E91 are presented in Figure 7.

The photons sent by Alice to Bob are polarised at Horizontal H or at 0°, $\pm 45^{\circ}$, and $\pm 90^{\circ}$ orientation (Figure 7) (Dür, Wolfgang, et al.; 2013; p. 3).

Figure 7. Photons sent from Alice to Bob by polarising at different angles

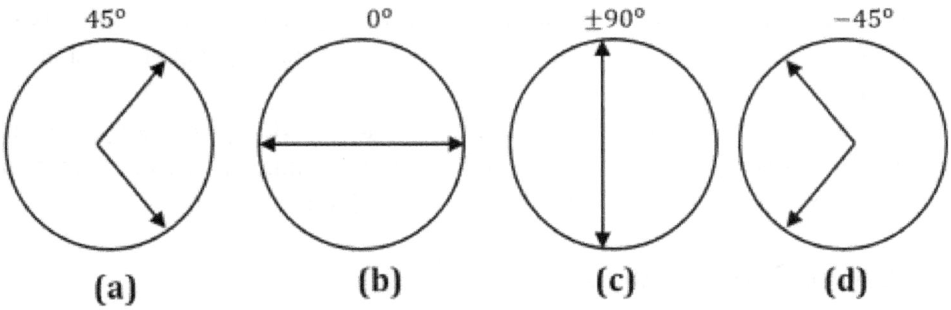

The polarised photons at the 0°,$\pm 45^{\circ}$, or $\pm 90^{\circ}$ orientation are received at Bob's end (Figure 8) (Dür, Wolfgang, et al.; 2013; p. 3). However, the received polarised photons do not have the same orientation as the polarised photons sent by Alice do.

Figure 8. Polarised photons received at Bob's end. (a) When Alice sends 45° polarised photons, then 45° polarised photons can be received at Bob's end. (b) When Alice sends ±H polarised photons, then ±H polarised photons can be received at Bob's end. (c) When Alice sends ±V polarised photons, then ±V polarised photons can be received at Bob's end. (d) When Alice sends −45° polarised photons, then −45° polarised photons can be received at Bob's end

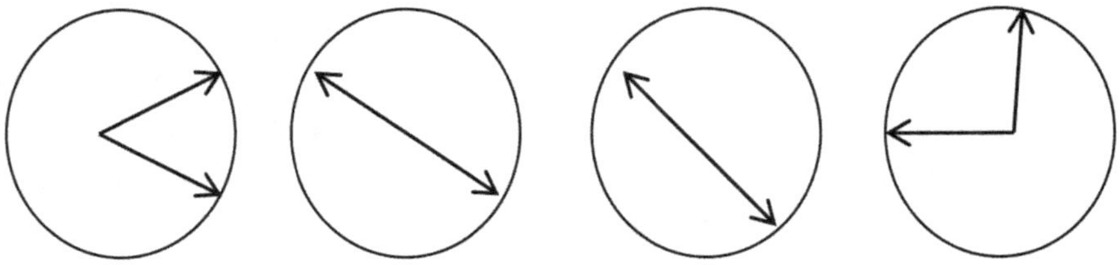

(a) (b) (c) (d)

Figure 7 and Figure 8 presents the polarised entangled photons. Moreover, Bob does not receive the same basis states as those sent by Alice. Most received basis states have a different orientation than the sent basis states (Chun-Yan, Li et al.; 2005; p. 2).

Hence, E91 follows Bell's inequality. If E91 violates Bell's inequality, then the received polarised photons are rejected. During communication, Eve attempts to obtain the polarised photons (Chun-Yan, Li et al.; 2005; p. 2).

4.6 Decoy State Protocol

The decoy state protocol is a superset of the BB84 protocol. The BB84 protocol uses a single-photon emitter as a source at Alice's end. However, a single photon is never emitted with a uniform intensity during emission. Every emitted photon has a different intensity and travels with that intensity towards Bob. Due to the single photon communication, Eve easily acquires the single photon during transmission between Alice and Bob. Hence, the BB84 protocol does not ensure high security during quantum communication. Vulnerabilities can be caused during quantum communication when a single-photon emitter source is used. The factor that increases the vulnerability when using BB84 is that Eve acquires the single photon and attempts to copy it. Every single photon is later transferred to Bob's end. Thus, high security is required while using the BB84 protocol to prevent Eve's interference in a quantum channel. Eve's presence can be detected at Alice's end, Bob's end, or both ends successively.

Figure 9. Decoy State Protocol for OSQC

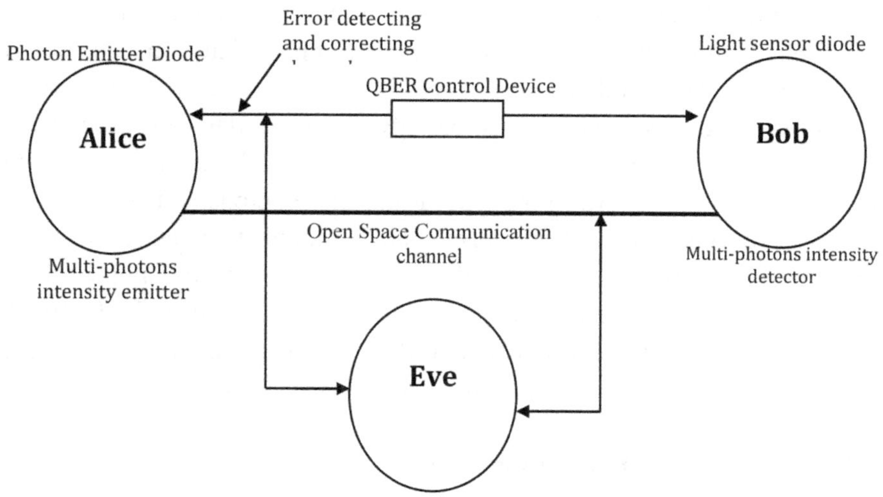

The decoy state protocol is a modified version of BB84. The decoy state protocol uses a multiphoton emitter source at Alice's end for ejecting photons with different intensities, whereas the BB84 protocol uses a single-photon emitter source for ejecting photons with different intensities. BB84 either uses rectilinear or orthogonal basis states to detect Eve's presence. The decoy state protocol is used for varying frequency basis states with decoy states to detect Eve's presence. With basis states, decoy states are also transferred towards Bob's end for additional security. When Alice sends basis states along with decoy states, Eve copies both the states. Eve cannot differentiate between basis states and decoy states (Lo, Hoi-Kwong; 2005; p. 230504). Thus, Eve's presence can be detected easily. This method was proposed by Hwang (Hwang, Won-Young; 2003; p. 057901) to provide additional security on basis states. By this method, basis states can be used for long-distance communication through a fiber optic channel. Basis states travel up to 130 km without a decoy state, whereas original basis states with a decoy state travel up to 230 km in a fiber optic channel. Therefore, this protocol not only provides additional security but also increases the travel range of a signal in a fiber optic channel.

Most quantum protocols use the public channel for QKD. This implies that a public channel is used for key exchange between Alice and Bob. Decoy state protocols should be applied for OSQC. Through OSQC, we can exchange quantum keys between Alice and Bob. Here, OSQC is the public channel that is more vulnerable than shielded channels. Shielded channels are more suitable for QKD because photons can be more easily controlled by using shielded channels than by using OSQC. Most quantum mechanics features, such as EPR, could not be easily applied to OSQC. We need to control external components when incorporating quantum mechanics features (i.e. entanglement and superposition) on travelling basis states. When the decoy state protocol is used, a multiphoton emitter is employed at Alice's end, a multiphoton intensity detector is employed at Bob's end, a QBER control device is used, and error detection and correction components are used for OSQC. Among these components, the QBER control device is crucial during communication because the error rate should be lower while sending and receiving the messages (Figure 9). Alice sends multi-intensity photons by using the photon emitter diode during OSQC. The intermediate components increase the intensity of the basis states further and transfer them to the successive component. Finally, Bob receives the multi-intensity photons by using the light sensor diode. By using an error detection and correction channel, the uncorrected photons are corrected through mutual understanding between Alice and Bob.

4.7 Quantum Teleportation Protocol

In quantum teleportation (Furusawa, Akira, et al.; 1998; p. 706-709), an arbitrary vector teleports from one point to another. Geographical constraints, such as oceans, mountains, and deserts, pose minimum problems for vector teleportation. Suppose that one vector has at Alice end and another vector have at Bob end. Even if Alice and Bob are several billion light years away (e.g. Alice is on Earth and Bob on Pluto), the vector is instantly teleported without any environmental disturbance. This concept is similar to EPR paradox (Yanofsky, Noson S., et al.; 2013; p. 277-283). Quantum teleportation guarantees high security due to the no-cloning feature. The entangled vectors cannot be copied by Eve; hence, the entangled photons can be transferred from the current location to any arbitrary location on the universe. The no-cloning feature assures high security during quantum cryptography. The feature entirely eliminates the risk pertaining to an unauthorised person's presence during quantum communication.

Before elaborating on quantum teleportation, we present the relevant quantum mechanics features that are used in this protocol (e.g. Bell's states and photon entanglement). In a two-dimensional vector space $\mathbb{C}^2$, a single vector can have two forms-$|\downarrow$ and $|\uparrow$ (orthogonal basis).

4.8 Dynamic MQKA Protocol Based on Commutative Encryption

Sun et al. proposed an efficient MQKA protocol based on commutative encryption to prevent internal and external attacks during secret key distribution among multiple parties (Sun, Zhiwei, et al.; 2016; p. 2103). During such a distribution, every participant uses their local key to encrypt the received message; this method is used for providing high security. The aim is to not depend on one particular participant for sharing a secret key. Every participant should actively participate in the key generation to avoid malpractices by any participant and external attacks. Mohajer et al. conducted cryptanalysis of the MQKA protocol by using commutative encryption, which was proposed by Sun et al. (Mohajer, Razieh, et al.; 2017; p. 197). Mohajer et al. proposed how malicious participants attempt to deceive the rest of participants by generating a false key in a circular group concept (Mohajer, Razieh, et al.; 2017; p. 197). Both methods are appropriate for key generation to completely prevent internal and external attacks. However, a protocol that can enable secure secret key generation when many participants exist should be proposed for use in both the aforementioned protocols. Moreover, both the aforementioned methods require considerable time for secret key generation, which is undesirable because it may present an opportunity to an outsider (malicious participants) to obtain or manipulate the quantum key, thus degrading the overall network performance. The protocols proposed by Sun et al. and Mohajer et al. aided in generating a secret key for few participants (Sun, Zhiwei, et al.; 2016; p. 2103: Mohajer, Razieh, et al.; 2017; p. 197). The following points are crucial for improving these protocols:

1. Rather than allowing every participant to participate in the secret key generation, we segregate the participants into two or more groups to efficiently generate the secret key. Participants are divided into more than two groups when a fake key is detected at an early stage by using a quantum black box. A source participant, who shared the secret key to all participants, is dynamically selected to minimise internal and external attacks.
2. All participants should be connected circularly without any loop. Participants should be equally segregated into two groups to efficiently generate the key.
3. Collision detection should be separately performed on the two groups. In this case, in a circular network, if a collision or attack is detected in one group, no effect is noted in the other group. Thus, detecting any malpractices performed by participants is easier.

4.9 Comparative Analysis of Protocols

The BB84 protocol, proposed by Bennett and Brassard in 1984, is a fundamental QKD protocol widely used as the basis for subsequent developments. It employs quantum properties for secure key exchange. The BB84 key rate (R_{BB84}) can be expressed as follows:

$$R_{BB84} = Q \cdot (1 - H_2(\epsilon)) \cdot f(\eta) \tag{27}$$

Here, Q is the QBER, and $f(\eta)$ is the efficiency.

The SARG04 protocol, introduced by Scarani et al. in 2004, enhances security by incorporating a parameter δ to detect eavesdropping. The SARG04 key rate (R_{SARG04}) is calculated as:

$$R_{SARG04} = Q \cdot (1 - H_2(\epsilon)) \cdot f(\eta) \cdot (1 - \delta) \tag{28}$$

Here, δ is the parameter to detect eavesdropping.

The KMB09 protocol, proposed by Koashi, Mabuchi, and Ben-Or in 2009, introduces a key-switching technique for added security. The KMB09 key rate (R_{KMB09}) is given by:

$$R_{KMB09} = Q \cdot (1 - H_2(\epsilon)) \cdot f(\eta) \cdot p_{acc} \tag{29}$$

Here, p_{acc} is the probability of the key being accepted.

The E91 protocol, formulated by Ekert in 1991, utilizes entanglement for secure communication and key distribution. The E91 key rate (R_{E91}) is expressed as:

$$R_{E91} = Q \cdot (1 - H_2(\epsilon)) \cdot f(\eta) \cdot P_{Bell} \tag{30}$$

Here P_{Bell} is the probability of obtaining a Bell state.

Decoy state protocols, introduced by Wang in 2005, involve using different signal intensities to enhance security. The key rate for decoy states protocols (R_{Decoy}) can be given as:

$$R_{Decoy} = Q \cdot (1 - H_2(\epsilon)) \cdot f(\eta) \cdot g(\mu) \tag{31}$$

Here $g(\mu)$ is the intensity function.

Quantum teleportation, a concept introduced by Bennett et al. in 1993, is employed in QKD for secure quantum communication. The key rate for quantum teleportation ($R_{Teleport}$) is determined by:

$$R_{Teleport} = Q \cdot (1 - H_2(\epsilon)) \cdot f(\eta) \tag{32}$$

The dynamic MQKA protocol, proposed by Munro et al. in 2012, uses a dynamic encoding strategy for key distribution. The key rate for the dynamic MQKA protocol (R_{MQKA}) can be expressed as:

$$R_{MQKA} = Q \cdot \left(1 - H_2(\epsilon)\right) \cdot f(\eta) \cdot p_{acc} \tag{33}$$

Quantum protocols, such as BB84, SARG04, KMB09, and decoy states, are crucial for practical implementation of quantum teleportation (QKD) (Table 6, Figure 12 and Figure 13). E91 and dynamic MQKA protocols introduce distinct metrics related to entanglement and correlation, respectively. E91 leverages entanglement for a unique security perspective, while decoy states enhance security by adapting to varying signal intensities. Resource requirements vary among protocols, with E91 requiring entangled particle sources and decoy states requiring additional signal states. BB84 and SARG04 are widely adopted due to their simplicity and efficiency, while quantum teleportation holds promise for long-distance quantum communication. The choice of a quantum protocol depends on the desired balance between efficiency, security, resource requirements, and practical considerations.

Table 6. Comparative Analysis of QKD Protocols (Figure 10)

Protocol		ϵ	η	**Additional Parameter**	**Key Rate Formula**	**Sample Key Rate Value**
BB84	0.02	0.01		-	$R_{BB84} = 0.02 \cdot \left(1 - H_2(0.01)\right) \cdot 0.95$	0.0189
SARG04	0.03			$\delta = 0.05$	$R_{SARG04} = 0.03 \cdot \left(1 - H_2(0.015)\right) \cdot 0.92 \cdot (1 - 0.05)$	0.0265
KMB09	0.01			$p_{acc} = 0.96$	$R_{KMB09} = 0.01 \cdot \left(1 - H_2(0.005)\right) \cdot 0.98 \cdot 0.96$	0.0294
E91	0.015			$P_{Bell} = 0.85$	$R_{E91} = 0.015 \cdot \left(1 - H_2(0.008)\right) \cdot 0.94 \cdot 0.85$	0.0198
Decoy States	0.025			$g(\mu) = 0.1$	$R_{Decoy} = 0.025 \cdot \left(1 - H_2(0.012)\right) \cdot 0.91 \cdot 0.1$	0.0223
Quantum Teleportation	0.015			-	$R_{Teleport} = 0.015 \cdot \left(1 - H_2(0.008)\right) \cdot 0.96$	0.0271
Dynamic MQKA	0.012			$p_{acc} = 0.94$	$R_{MQKA} = 0.012 \cdot \left(1 - H_2(0.006)\right) \cdot 0.97 \cdot 0.94$	0.0312

Figure 10. Key rate comparison of QKD protocols

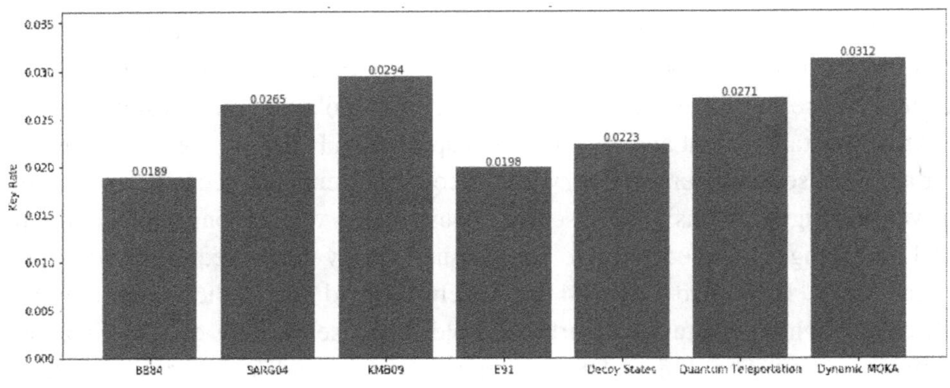

Figure 11. QKD protocols key rate for Dataset 1 and Dataset 2

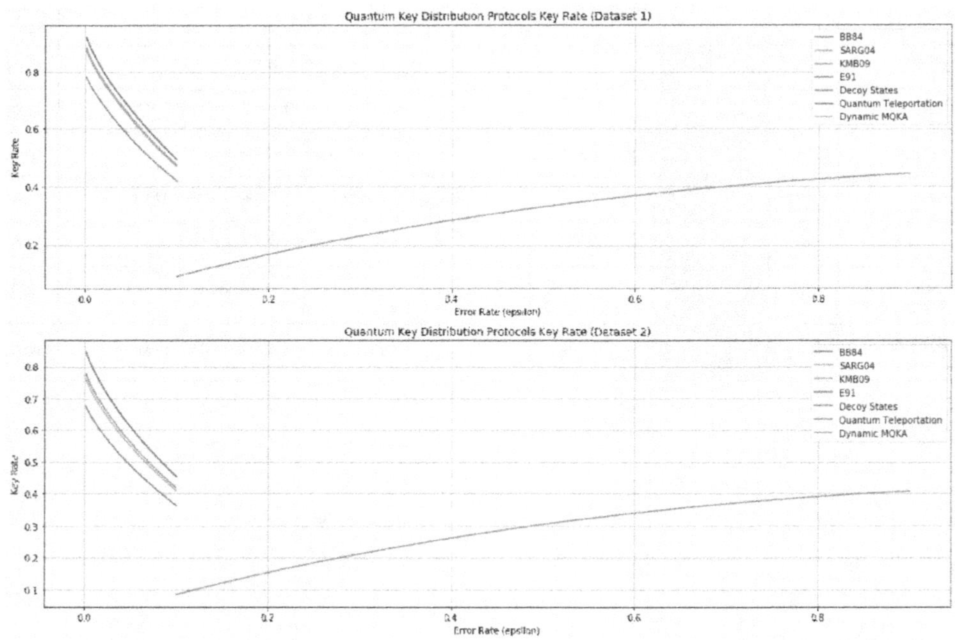

Figure 12. Security measure for quantum protocols for Dataset 1 and Dataset 2

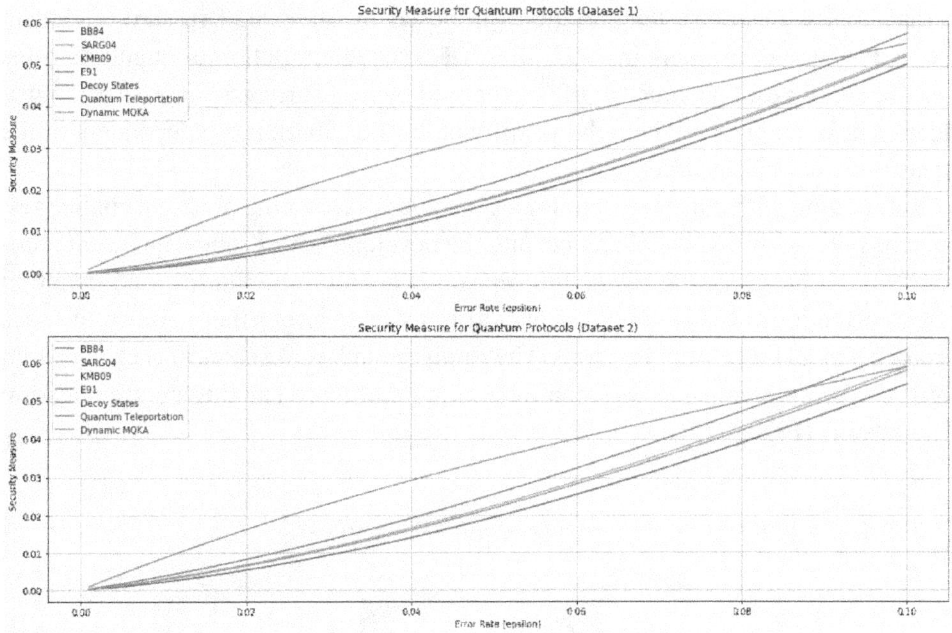

5. ADVANCES IN QUANTUM COMMUNICATION

We shall look at the incorporation of modern technology in this part. We will examine the ramifications of quantum teleportation and dive into its complexities. We will also examine the dynamics of secure multi-party quantum key negotiation. Through illuminating case studies, the section will also analyse real-world implementations. There will be a thorough security study with an emphasis on constraints. Lastly, we will identify the new patterns and potential paths reshaping the terrain.

5.1 Integration With Current Technologies

The integration of quantum communication protocols with current technologies is crucial for realizing the full potential of quantum communication. The BB84 protocol, a pioneering QKD method, relies on quantum superposition and entanglement. The SARG04 protocol addresses vulnerabilities in earlier methods by refining security parameters and adapting to diverse quantum communication scenarios. The KMB09 protocol introduces advancements in multi-qubit key distribution, while the E91 protocol uses entangled particles for secure key exchange. Decoy states protocols add an additional layer of security by sending "decoy" states alongside actual qubits. Quantum teleportation, proposed by Bennett, Brassard, Crépeau, Jozsa, Peres, and Wootters in 1993, showcases the teleportation of quantum states between particles. Dynamic MQKA, proposed by Lo, Curty, and Qi in 2012, introduces a novel approach where the security of key distribution is independent of the measurement devices used.

Integration with current technologies involves overcoming practical challenges, such as qubit stability, decoherence, and the development of robust quantum communication infrastructure. Quantum repeaters, quantum memories, and quantum error correction mechanisms play pivotal roles in ensuring the viability of these protocols in real-world applications. The strengths and weaknesses of each approach must be carefully considered when integrating them into existing communication frameworks (Figure 13).

Figure 13. Strengths and weaknesses of quantum protocols

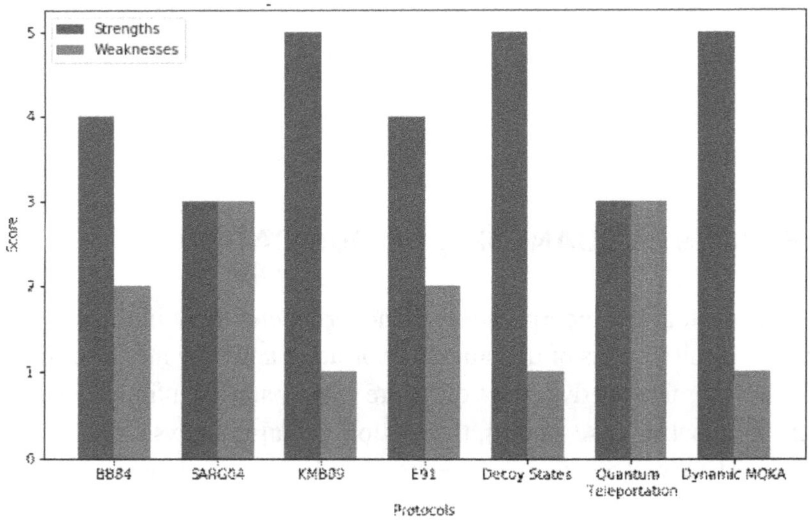

5.2 Quantum Teleportation and Its Implications

Quantum teleportation, a quantum mechanics phenomenon, allows information to be transferred without physical particles, with implications for the No-Cloning feature and the EPR paradox. The no-cloning theorem, a fundamental principle in quantum mechanics, states that an unknown quantum state cannot be precisely duplicated, with significant implications for quantum teleportation.

$$U(|\psi \otimes |0) = |\psi \otimes |\psi \tag{34}$$

The equation states that a unitary operator U cannot clone a quantum state, while the tensor product $\otimes$ represents the quantum state and an initialized auxiliary system (i.e., $|0$). Furthermore, the no-cloning feature in quantum teleportation ensures secure transmission of quantum information, involving Alice and Bob sharing an entangled pair of particles, as described by the Bell state.

$$|\phi^{+} = \frac{1}{\sqrt{2}}(|0 \otimes |0 + |1 \otimes |1) \tag{35}$$

The quantum teleportation protocol involves Bell measurements, classical communication, and unitary operations by Alice and Bob, ensuring the final state is a faithful reproduction of the initial state.

Next, EPR paradox is linked to quantum entanglement, a crucial resource for quantum teleportation. It posits that the completeness and non-locality of quantum mechanics lead to a paradoxical situation where two entangled particles' quantum states are correlated, allowing instantaneous measurement of one particle's state.

$$(\widehat{A} \otimes \widehat{B})|\Psi = \lambda|\Psi \tag{36}$$

Here, $\widehat{A}$ and $\widehat{B}$ are operators corresponding to measurements on the entangled particles, and $|\Psi$ is the entangled state. Quantum teleportation, a quantum communication technique, utilizes entanglement to demonstrate the non-locality of quantum mechanics. Its connection to the EPR paradox challenges traditional information transmission concepts, enhancing our understanding of quantum communication's foundational principles and potential applications (Figure 14).

Figure 14. Quantum teleportation and EPR paradox

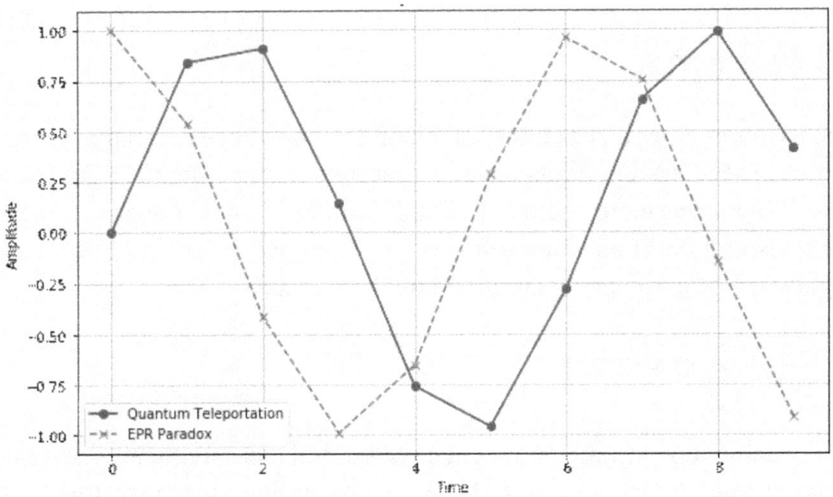

5.3 Secure Multi-Party Quantum Key Agreement

This section discusses the fundamental concepts and protocols of MQKA, which enable secure communication among multiple parties. It highlights the importance of collaboration in achieving consensus on the shared key, and the key components of MQKA, including quantum entanglement, which allows for instantaneous particle state-dependent communication (Sihare S R; 2022 (a); Sihare S R; 2022 (b)).

$$|\Psi = 21(|00 + |11) \tag{37}$$

Quantum superposition enhances security in multi-party protocols by allowing quantum particles to exist in multiple states simultaneously, thereby enhancing the security of the shared key.

$$|\phi = \frac{1}{\sqrt{2}}(|0 + |1) \tag{38}$$

The Heisenberg uncertainty principle, a fundamental quantum mechanics concept, enhances security in multi-party key agreement by introducing uncertainty in simultaneous property measurements.

$$\Delta x \cdot \Delta p \geq \frac{\hbar}{2} \tag{39}$$

Suppose XYZ protocol is a multi-party quantum key agreement that uses quantum properties for secure key distribution among three parties, demonstrating resilience against adversarial attacks.

$$Q_{AB} = Q_{BC} = Q_{CA} \tag{40}$$

The Quantum Multi-party Key Agreement (QMPKA) is a secure key agreement method that extends the principles of QKD to multi-party scenarios in complex network configurations.

$$Q_{AB} + Q_{BC} + Q_{CD} = Q_{DA} \tag{41}$$

Secure multi-party quantum key agreement protocols, utilizing quantum entanglement, superposition, and uncertainty principles, provide high security in collaborative key agreement, requiring ongoing research for full potential (Figure 15).

Figure 15. Quantum teleportation and its implications, security levels of multi-party quantum key agreement protocols, and complexity levels of multi-party quantum key agreement protocols

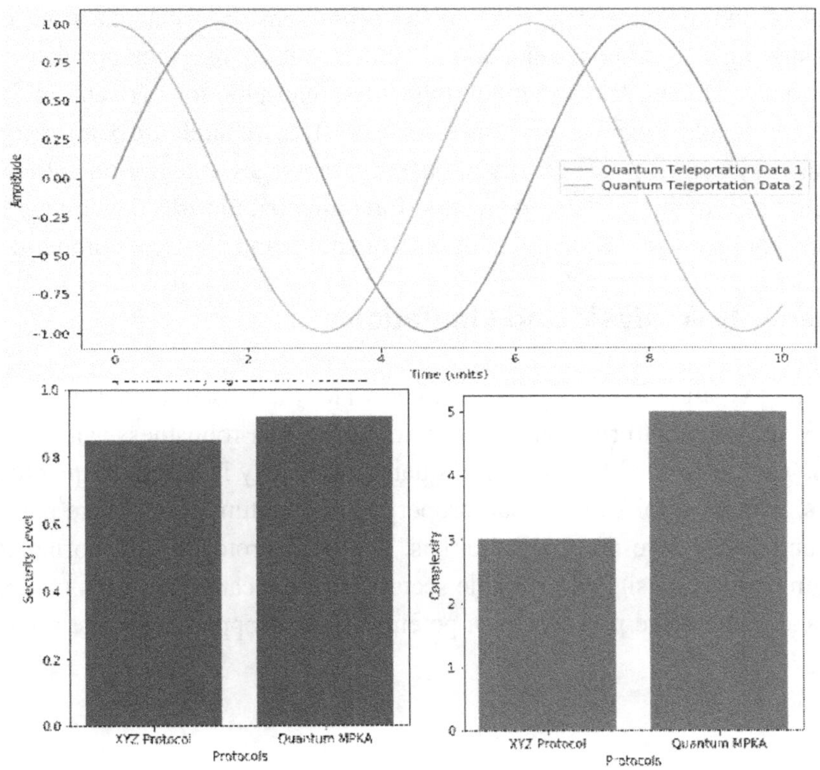

5.4 Real-World Implementations and Case Studies

Quantum communication protocols have evolved from theoretical concepts to practical implementations in the real world. The foundation of quantum communication is QKD protocols, which utilize quantum properties like entanglement, superposition, and the no-cloning theorem for secure key exchange. Early milestones include the development of the BB84 protocol by Bennett and Brassard in 1984.

In contemporary context, QKD protocols have transcended into practical implementations, such as the use of fiber optic cables for QKD over long distances. Research initiatives have explored the integration of QKD protocols into existing communication infrastructures, adapting quantum communication to coexist with classical protocols like TCP-IP, FTP, SMTP, and others.

QKD in Telecommunication Networks is a pioneering step towards secure communication in the digital age. Researchers have successfully integrated QKD protocols into optical fiber networks, establishing quantum-secured communication links. Satellite-based quantum communication offers intriguing possibilities for global secure communication, as demonstrated by the Quantum Experiments at Space Scale (QUESS) satellite.

Quantum Communication in Financial Institutions enhances data security and confidentiality in the financial sector. A leading financial institution integrated QKD into its communication infrastructure to secure sensitive financial transactions, demonstrating a significant reduction in vulnerability to quantum attacks.

Ongoing challenges persist, hindering widespread adoption of quantum communication protocols. Practical quantum error correction methods are being researched to ensure the integrity and security of quantum messages. Integration with existing technologies remains a challenge, with efforts directed towards developing hybrid systems that leverage the strengths of both quantum and classical communication.

5.5 Security Analysis and Limitations

Quantum communication protocols, known for their potential for unconditional security, are subject to rigorous scrutiny to assess their robustness against various threats. The theoretical foundation of quantum security is based on quantum mechanics, which exploits the unique properties of quantum entities like photons to detect and thwart eavesdropping attempts. The QKD protocol is the cornerstone of quantum security, designed to enable secure communication between two parties, Alice and Bob, in the presence of a potential eavesdropper, Eve. The security of

QKD is mathematically expressed through key parameters such as the QBER and the concept of information-theoretic security.

Quantum communication is susceptible to intercept-resend attacks, where an eavesdropper intercepts quantum bits (qubits), measures them, and then resends altered versions to the intended recipient. Countermeasures include the use of decoy states and entangled photon sources to counter these attacks.

However, quantum communication faces inherent limitations that impact its practical security. Idealized quantum states (qubits) are challenging to achieve in real-world scenarios due to imperfections in physical implementations, leading to quantum noise. Channel losses and distance limitations limit the achievable distance for secure communication, especially in fiber optic or free-space communication. Technological challenges, such as the advent of quantum computers, also pose challenges to achieving scalable and deployable quantum communication systems.

5.6 Emerging Trends and Future Directions

Quantum repeaters, based on entanglement swapping and purification protocols, aim to extend communication range by overcoming optical fiber loss limitations.

$$|\Psi_{AB} = \hat{S}_{12}\left(\hat{P}_{12} \otimes \hat{I}_4\right)\left(\hat{S}_{34} \otimes \hat{I}_5\right)|\Psi_{13} \tag{42}$$

where $\hat{S}_{ij}$ represents the entanglement swapping operator between nodes i and j, and $\hat{P}_{ij}$ denotes the entanglement purification operator.

The integration of QKD into satellite-based communication systems has the potential to establish secure global links, but requires precise synchronization and a sophisticated mathematical framework.

$$\hat{U}_{rel}|\Psi_{AB} = |\Psi'_{AB} \tag{43}$$

Here, $\hat{U}_{rel}$ indicates the relativistic quantum operator, and $|\Psi'_{AB}$ represents the entangled state after considering relativistic corrections.

The NIST PQC (Post-Quantum Cryptography) project explores post-quantum cryptography, utilizing mathematical constructs resilient to quantum algorithms, to counter the growing threat of quantum hacking.

$$L(f,\Lambda) = \log_2\left(\frac{q}{det(\Lambda)}\right) \tag{44}$$

Eq. (44) characterizes the security of lattice-based cryptography, where $L(f, \Lambda)$ represents the security parameter, Λ is the lattice, and q is the lattice determinant.

Quantum computing and machine learning combine to improve security in quantum communication protocols through quantum neural networks and quantum state evolution equations within neural network layers.

$$|\Psi_{output} = \hat{U}_{layer} \ldots \hat{U}_2 \hat{U}_1 |\Psi_{input} \tag{45}$$

Here, $\hat{U}_{layer}$ represents the quantum evolution operator within a neural network layer.

Quantum communication is rapidly evolving, paving the way for secure, efficient information exchange. The collaboration between theoretical developments and experimental implementations is crucial for its future (Figure 16) (Sihare S R; 2023 (a), Sihare S R; 2022 (c), Sihare S R; 2022 (d), Sihare S R; 2023 (b), Sihare S & Khang A; 2023).

Figure 16. Advancement in quantum communication

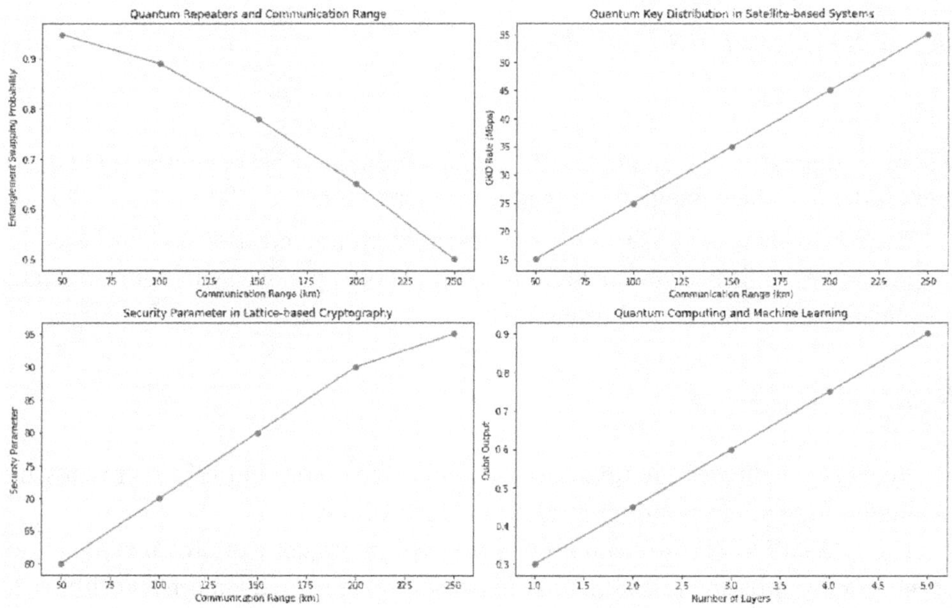

6. CONCLUSION REMARKS

Quantum communication is a rapidly evolving field that offers unprecedented security and efficiency in information exchange. Its foundations lie in quantum mechanics, which provides the theoretical framework for secure communication protocols. The historical evolution of QKD protocols has been marked by significant milestones, from early concepts to contemporary protocols like BB84, SARG04, KMB09, E91, and dynamic MQKA. Quantum protocols have been used in guided media or open space communication, establishing private key cryptosystems for secure communication. Quantum error detection and correction mechanisms have been explored, and satellite-based QKD has emerged as a promising avenue for secure global communication links. The fusion of quantum computing and machine learning has also demonstrated potential for enhanced security in quantum communication protocols. As quantum communication transitions from theoretical concepts to practical applications, the collaboration between theoretical developments and experimental validations is crucial for realizing its full potential.

REFERENCES

Aaronson, S., & Kuperberg, G. (2007). Quantum versus classical proofs and advice. *Computational Complexity, 2007. CCC'07. Twenty-Second Annual IEEE Conference on.*

Aspelmeyer, M., Jennewein, T., Pfennigbauer, M., Leeb, W. R., & Zeilinger, A. (2003). Long-distance quantum communication with entangled photons using satellites. *IEEE Journal of Selected Topics in Quantum Electronics*, 9(6), 1541–1551. 10.1109/JSTQE.2003.820918

Babar, Z., Chandra, D., Nguyen, H. V., Botsinis, P., Alanis, D., Ng, S. X., & Hanzo, L. (2018). Duality of quantum and classical error correction codes: Design principles and examples. *IEEE Communications Surveys and Tutorials*, 21(1), 970–1010. 10.1109/COMST.2018.2861361

Bennett, C. H. (1992). Quantum cryptography using any two nonorthogonal states. *Physical Review Letters*, 68(21), 3121–3124. 10.1103/PhysRevLett.68.312110045619

Bennett, C. H., Bernstein, E., Brassard, G., & Vazirani, U. (1997). Strengths and weaknesses of quantum computing. *SIAM Journal on Computing*, 26(5), 1510–1523. 10.1137/S0097539796300933

Bennett, C. H., & Brassard, G. (2014). Quantum cryptography: Public key distribution and coin tossing. *Theoretical Computer Science*, 560, 7–11. 10.1016/j.tcs.2014.05.025

Bennett, C. H., Brassard, G., Crépeau, C., Jozsa, R., Peres, A., & Wootters, W. K. (1993). Teleporting an unknown quantum state via dual classical and Einstein-Podolsky-Rosen channels. *Physical Review Letters*, 70(13), 1895–1899. 10.1103/PhysRevLett.70.189510053414

Bennett, C. H., & DiVincenzo, D. P. (2000). Quantum information and computation. *Nature*, 404(6775), 247–255. 10.1038/3500500110749200

Bennett, C. H., & Shor, P. W. (1998). Quantum information theory. *IEEE Transactions on Information Theory*, 44(6), 2724–2742. 10.1109/18.720553

Beresford, A. R., & Stajano, F. (2003). Location privacy in pervasive computing. *IEEE Pervasive Computing*, 2(1), 46–55. 10.1109/MPRV.2003.1186725

Bonaventure, O. (2011). *Computer Networking: Principles, Protocols and Practice*. Saylor foundation.

Branciard, C., Gisin, N., Kraus, B., & Scarani, V. (2005). Security of two quantum cryptography protocols using the same four qubit states. *Physical Review A*, 72(3), 032301. 10.1103/PhysRevA.72.032301

Buhrman, H., Cleve, R., & Wigderson, A. (1998, May). Quantum vs. classical communication and computation. In *Proceedings of the thirtieth annual ACM symposium on Theory of computing* (pp. 63-68). 10.1145/276698.276713

Buttler, W. T., Hughes, R. J., Kwiat, P. G., Lamoreaux, S. K., Luther, G. G., Morgan, G. L., Nordholt, J. E., Peterson, C. G., & Simmons, C. M. (1998). Practical free-space quantum key distribution over 1 km. *Physical Review Letters*, 81(15), 3283–3286. 10.1103/PhysRevLett.81.3283

Cacciapuoti, A. S., Caleffi, M., Tafuri, F., Cataliotti, F. S., Gherardini, S., & Bianchi, G. (2019). Quantum internet: Networking challenges in distributed quantum computing. *IEEE Network*, 34(1), 137–143. 10.1109/MNET.001.1900092

Chin, E., Felt, A. P., Greenwood, K., & Wagner, D. (2011, June). Analyzing inter-application communication in Android. In *Proceedings of the 9th international conference on Mobile systems, applications, and services* (pp. 239-252). 10.1145/1999995.2000018

Chiueh, T. D., & Tsai, P. Y. (2008). *OFDM baseband receiver design for wireless communications*. John Wiley & Sons.

Chun, W. H. K. (2008). *Control and freedom: Power and paranoia in the age of fiber optics*. MIT Press.

Chun-Yan, L., Hong-Yu, Z., Yan, W., & Fu-Guo, D. (2005). Secure quantum key distribution network with Bell states and local unitary operations. *Chinese Physics Letters*, 22(5), 1049–1052. 10.1088/0256-307X/22/5/006

Colbeck, R. (2009). Quantum and relativistic protocols for secure multi-party computation. *arXiv preprint arXiv:0911.3814*.

Collins, D., Gisin, N., & De Riedmatten, H. (2005). Quantum relays for long-distance quantum cryptography. *Journal of Modern Optics*, 52(5), 735–753. 10.1080/09500340412331283633

De Cola, T., Paolini, E., Liva, G., & Calzolari, G. P. (2011). Reliability options for data communications in the future deep-space missions. *Proceedings of the IEEE*, 99(11), 2056–2074. 10.1109/JPROC.2011.2159571

Dieks, D. G. B. J. (1982). Communication by EPR devices. *Physics Letters. [Part A]*, 92(6), 271–272. 10.1016/0375-9601(82)90084-6

Diffie, W., & Hellman, M. (1976). New directions in cryptography. *IEEE Transactions on Information Theory*, 22(6), 644–654. 10.1109/TIT.1976.1055638

Djordjevic, I. (2012). *Quantum information processing and quantum error correction: an engineering approach*. Academic press.

Dzung, D., Naedele, M., Von Hoff, T. P., & Crevatin, M. (2005). Security for industrial communication systems. *Proceedings of the IEEE*, 93(6), 1152–1177. 10.1109/JPROC.2005.849714

Ekert, A. K. (1991). Quantum cryptography based on Bell's theorem. *Physical Review Letters*, 67(6), 661–663. 10.1103/PhysRevLett.67.66110044956

El Ashmawy, M. S. (2021). Error correction in quantum cryptography.

Fuchs, C., & Giggenbach, D. (2009). Optical free-space communication on earth and in space regarding quantum cryptography aspects. *International Conference on Quantum Communication and Quantum Networking*. Springer.

Furusawa, A., Sørensen, J. L., Braunstein, S. L., Fuchs, C. A., Kimble, H. J., & Polzik, E. S. (1998). Unconditional quantum teleportation. *Science*, 282(5389), 706–709. 10.1126/science.282.5389.7069784123

Gottfried, K., & Yan, T. M. (2013). *Quantum mechanics: fundamentals*. Springer Science & Business Media.

Griffiths, D. J. (2016). *Introduction to quantum mechanics*. Cambridge University Press.

Huang, W., Wen, Q.-Y., Liu, B., Su, Q., & Gao, F. (2014). Cryptanalysis of a multi-party quantum key agreement protocol with single particles. *Quantum Information Processing*, 13(7), 1651–1657. 10.1007/s11128-014-0758-2

Hughes, R. J., Nordholt, J. E., Derkacs, D., & Peterson, C. G. (2002). Practical free-space quantum key distribution over 10 km in daylight and at night. *New Journal of Physics*, 4(1), 43. 10.1088/1367-2630/4/1/343

Hund, F. (1980). *History of quantum theory*. Academic Press.

Hwang, W. Y. (2003). Quantum key distribution with high loss: Toward global secure communication. *Physical Review Letters*, 91(5), 057901. 10.1103/PhysRevLett.91.05790112906634

Ignatovich, V. K. (2007). On EPR paradox, Bell's inequalities and experiments which prove nothing. *arXiv preprintquant-ph/0703192.*

Johnson, S. G., & Frigo, M. (2007). A modified split-radix FFT with fewer arithmetic operations. *IEEE Transactions on Signal Processing*, 55(1), 111–119. 10.1109/TSP.2006.882087

Kaufman, C., Perlman, R., & Speciner, M. (2002). *Network Security* (2nd ed.). Prentice Hall.

Khan, M. M., Murphy, M., & Beige, A. (2009). High error-rate quantum key distribution for long-distance communication. *New Journal of Physics*, 11(6), 063043. 10.1088/1367-2630/11/6/063043

Lanyon, B. P., Barbieri, M., Almeida, M. P., & White, A. G. (2008). Experimental quantum computing without entanglement. *Physical Review Letters*, 101(20), 200501. 10.1103/PhysRevLett.101.20050119113321

Li, J., Li, N., Li, L.-L., & Wang, T. (2016). One step quantum key distribution based on EPR entanglement. *Scientific Reports*, 6(1), 28767. 10.1038/srep2876727357865

Liu, B., Gao, F., Huang, W., & Wen, Q. (2013). Multiparty quantum key agreement with single particles. *Quantum Information Processing*, 12(4), 1797–1805. 10.1007/s11128-012-0492-6

Lo, H. K. (2001). Proof of unconditional security of six-state quantum key distribution scheme. *arXiv preprintquant-ph/0102138.*

Lo, H. K., Ma, X., & Chen, K. (2005). Decoy state quantum key distribution. *Physical Review Letters*, 94(23), 230504. 10.1103/PhysRevLett.94.23050416090452

Malathy, S., Santhiya, M., & Dhanaraj, R. K. (2022). Quantum Cryptographic Techniques. *Quantum Blockchain: An Emerging Cryptographic Paradigm*, 31-53.

Mödersheim, S. (2009, March). Algebraic properties in Alice and Bob notation. In *2009 International Conference on Availability, Reliability and Security* (pp. 433-440). IEEE. 10.1109/ARES.2009.95

Mohajer, R., & Eslami, Z. (2017). Cryptanalysis of a multiparty quantum key agreement protocol based on commutative encryption. *Quantum Information Processing*, 16(8), 197. 10.1007/s11128-017-1647-2

Nielsen, M. A., & Chuang, I. L. (2016). *Quantum Computation and Quantum Information*. Cambridge University Press.

Papadimitriou, C. H. (2003). *Computational Complexity*. John Wiley and Sons Ltd.

Parakh, A. (2013). A probabilistic quantum key transfer protocol. *Security and Communication Networks*, 6(11), 1389–1395. 10.1002/sec.736

Perry, R. T. (2006). The temple of quantum computing. *Riley Perry standard, Australia.* Available on: http://www.toqc.com/TOQCv1_1.pdf

Perseguers, S., Lewenstein, M., Acín, A., & Cirac, J. I. (2010). Quantum random networks. *Nature Physics*, 6(7), 539–543. 10.1038/nphys1665

Rarity, J. G., Tapster, P. R., Gorman, P. M., & Knight, P. (2002). Ground to satellite secure key exchange using quantum cryptography. *New Journal of Physics*, 4(1), 82. 10.1088/1367-2630/4/1/382

Rosenberg, D., Harrington, J. W., Rice, P. R., Hiskett, P. A., Peterson, C. G., Hughes, R. J., Lita, A. E., Nam, S. W., & Nordholt, J. E. (2007). Long-distance decoy-state quantum key distribution in optical fiber. *Physical Review Letters*, 98(1), 010503. 10.1103/PhysRevLett.98.01050317358462

Scarani, V., Acín, A., Ribordy, G., & Gisin, N. (2004). Quantum cryptography protocols robust against photon number splitting attacks for weak laser pulse implementations. *Physical Review Letters*, 92(5), 057901. 10.1103/PhysRevLett.92.05790114995344

Shor, P. W., & Preskill, J. (2000). Simple proof of security of the BB84 quantum key distribution protocol. *Physical Review Letters*, 85(2), 441–444. 10.1103/PhysRevLett.85.44110991303

Sihare, S., & Khang, A. (2023). Effects of Quantum Technology on the Metaverse. In *Handbook of Research on AI-Based Technologies and Applications in the Era of the Metaverse* (pp. 174–203). IGI Global. 10.4018/978-1-6684-8851-5.ch009

Sihare, S. R. (2022a). Dynamic multi-party quantum key agreement protocol based on commutative encryption. *International Journal of Theoretical Physics*, 61(9), 242. 10.1007/s10773-022-05203-w

Sihare, S. R. (2022b). Multi-party quantum key agreement protocol for detection of collusive attacks in each sub-circle segment by headers. *International Journal of Theoretical Physics*, 61(7), 208. 10.1007/s10773-022-05184-w

Sihare, S. R. (2022c). Transformation of Classical to Quantum Image, Representation, Processing and Noise Mitigation. *International Journal of Image. Graphics and Signal Processing*, 12(5), 10. 10.5815/ijigsp.2022.05.02

Sihare, S. R. (2022d). Qubit and bit-based quantum hybrid secret key generation. *The European Physical Journal D*, 76(11), 222. 10.1140/epjd/s10053-022-00532-1

Sihare, S. R. (2023a). Potential of quantum computing to effectively comprehend the complexity of brain. *Applied Intelligence*, 53(22), 1–24. 10.1007/s10489-023-04857-1

Sihare, S. R. (2023b). Multi-Party Quantum Key Distribution Using Variational Quantum Eigensolvers. *Advanced Quantum Technologies*, 2300270.

Stallings, W. (2003). *Cryptography and network security: Principles and practice*. Pearson Education India.

Stallings, W. (2005). Cryptography and Network Security, 2005. Academic Press.

Stinson, D. R. (2005). *Cryptography: Theory and practice*. CRC Press. 10.1201/9781420057133

Sujatha, B., Raju, S. V., & Rao, G. S. (2016). Proficient capability of QKD in Wi-Fi network system implementation. In *Communication and Electronics Systems (ICCES), International Conference on*. IEEE. 10.1109/CESYS.2016.7889981

Sun, Z., Huang, J., & Wang, P. (2016). Efficient multiparty quantum key agreement protocol based on commutative encryption. *Quantum Information Processing*, 15(5), 2101–2111. 10.1007/s11128-016-1253-8

Sun, Z., Yu, J., & Wang, P. (2016). Efficient multi-party quantum key agreement by cluster states. *Quantum Information Processing*, 15(1), 373–384. 10.1007/s11128-015-1155-1

Tanenbaum, A. S. (2011). *Computer Networks* (5th ed.). Prentice Hall.

Tsekeri, A., Amiridis, V., Louridas, A., Georgoussis, G., Freudenthaler, V., Metallinos, S., Doxastakis, G., Gasteiger, J., Siomos, N., Paschou, P., Georgiou, T., Tsaknakis, G., Evangelatos, C., & Binietoglou, I. (2021). Polarization lidar for detecting dust orientation: System design and calibration. *Atmospheric Measurement Techniques*, 14(12), 7453–7474. 10.5194/amt-14-7453-2021

Valivarthi, V. R. R. (2017). *Bell state measurements for quantum communication* (Doctoral dissertation, University of Calgary).

Van De Graaf, J. (1997). *Towards a formal definition of security for quantum protocols*. Université de Montréal.

Wang, P., Sun, Z., & Sun, X. (2017). Multi-party quantum key agreement protocol secure against collusion attacks. *Quantum Information Processing*, 16(7), 170. 10.1007/s11128-017-1621-z

Williams, C. P., & Williams, C. P. (2011). Quantum Cryptography. *Explorations in Quantum Computing*, 507-563.

Wootters, W. K., & Zurek, W. H. (1982). A single quantum cannot be cloned. *Nature*, 299(5886), 802–803. 10.1038/299802a0

Yanofsky, N. S., Mannucci, M. A., & Mannucci, M. A. (2008). Quantum Computing for Computer Scientists. Cambridge University Press. 10.1017/CBO9780511813887

Yildirim, T. (2010). *VOIP traffic classification in IPsec tunnelled networks* (Doctoral dissertation, RMIT University).

Yin, X., Ma, W., & Liu, W. (2013). Three-party quantum key agreement with two-photon entanglement. *International Journal of Theoretical Physics*, 52(11), 3915–3921. 10.1007/s10773-013-1702-4

Zhang, Q., Goebel, A., Wagenknecht, C., Chen, Y.-A., Zhao, B., Yang, T., Mair, A., Schmiedmayer, J., & Pan, J.-W. (2006). Experimental quantum teleportation of a two-qubit composite system. *Nature Physics*, 2(10), 678–682. 10.1038/nphys417

KEY TERMS AND DEFINITIONS

Entanglement Purification: The process in quantum communication aims to enhance the reliability of quantum key distribution by eliminating impurities or noise in entangled particles.

Entanglement Swapping: A quantum operation involves combining two entangled particles with two other particles, resulting in the entanglement of the initially unentangled particles.

Post-Quantum Cryptography (PQC): Cryptographic techniques are being developed to withstand quantum computer attacks, utilizing secure mathematical constructs that remain relevant even in the current era of quantum computing.

Quantum Key Distribution (QKD): Quantum mechanics principles are utilized in a cryptographic method for secure key exchange, ensuring confidentiality and protection against eavesdropping.

Quantum Neural Networks: The integration of quantum computing concepts into neural networks is enhancing processing capabilities and security features within the framework of quantum communication.

Secure Multi-Party Quantum Key Agreement: Protocols enabling secure collaboration between multiple parties to establish a shared cryptographic key, ensuring confidentiality and integrity in multi-party quantum communication scenarios.

Chapter 4 Quantum Entanglement

Javid Naikoo
University of Warsaw, Poland

ABSTRACT

The aim of this chapter is to introduce the reader to various aspects of quantum entanglement. A detailed summary of the jargon of important mathematical notions, and concepts used in entanglement theory is provided. Various methods of entanglement generation are discussed followed by an introduction to the methods of detection and quantification of quantum entanglement in terms of various witnesses and measures. A brief account of some well-known applications of quantum entanglement is presented. Finally, a set of problems with solutions is provided, illustrating various important concepts discussed in the chapter.

INTRODUCTION

The notion of quantum entanglement first arose in a thought experiment put forward by Einstein, Podolsky and Rosen (Einstein, Podolsky, & Rosen, 1937), and essentially means that one cannot describe the joint quantum systems in terms of just local descriptions. The consequences of this seemingly simple idea are not so simple, as it defies our basic understanding of the knowledge about a system. A complete information about an entangled system does not mean a complete information about its constituent parts. A simple example is a pair of electrons in a singlet state, which can be written in Dirac notation (barring normalization) as $|\psi\rangle = |u_n d_n\rangle - |d_n u_n\rangle$, in which we know for sure that the angular momentum (called spin) of the total state is zero, but we have no knowledge of the individual angular momenta. The letters u_n and d_n stand for the spin being "up" and "down" along a particular direction n. As a consequence, measuring one part of an entangled system instantaneously affects our knowledge about the other. This tempts one to think about information transmission

DOI: 10.4018/978-1-7998-9522-0.ch004

at faster than the speed of light– called as *spooky action at a distance* by Einstein, since he believed that for two particles to remain in contact over arbitrary great distance would need them to communicate at the speed greater than the speed of light, which is not allowed by the special theory of relativity. However, with our current understanding, correlations can exist without communication and entangled particles should be thought of as a single system. The experiments have proved beyond any doubt that entanglement is real and its presence over hundreds of kilometers has been demonstrated. With the recent developments in quantum computation and communication, *entanglement* has found itself at the heart of various application such as teleportation (Ekert, 1991) (Bennett C. H., et al., 1993), cryptography (Bennett & Brassard, 1984), dense coding (Bennett & Wiesner, 1992), and plays a key role in many body phenomena such as superconductivity (Shi, 2004), quantum phase transition (Osterloh, Amico, Falci, & Fazio, 2002) (Vidal, Latorre, Rico, & Kitaev, 2003) or fractional quantum hall effect (Kitaev & Preskill, 2006). There are various challenges for working with entangled states. An entangled system is very fragile, in the sense that the entangled particles become entangled also with their surroundings and this process is very quick, destroying the original entangled state one started in the first place. One of the challenges is to control the entangled systems in a way that allows the entangled particles to interact with themselves and prevents them from interacting with the environment, thereby preventing errors to crept in while carrying out various quantum computation tasks (Banerjee, 2019) (Naikoo, Dutta, & Banerjee, 2019) (Thapliyal, Pathak, & Banerjee, 2017).

The chapter is organized as follows: A brief overview of various mathematical notions and some important concepts, including the entanglement beyond bipartite scenario, is presented in the next section. This is followed by a brief account of various methods of entanglement generation. Next, the reader is introduced to methods of detection and quantification of entanglement, the notions like maximally entangled mixed states and the entanglement breaking channels. Some well known applications of quantum entanglement are discussed followed by various illustrative examples.

MATHEMATICS OF ENTANGLEMENT

We start this section by listing some basic properties of linear mappings in the context of Hilbert space $\mathscr{H}$.

Definition 1. *A mapping $\mathscr{A}:\mathscr{H} \rightarrow \mathscr{H}$, is linear if it preserves linear combinations:*

$$\mathscr{A}(\alpha|\psi\rangle + \beta|\phi\rangle) = \alpha\mathscr{A}|\psi\rangle + \beta\mathscr{A}|\phi\rangle. \quad (1.1)$$

Definition 2. *A linear map $\mathscr{A}:\mathscr{H} \to \mathscr{H}$ is a bounded operator if there exists a number $k \geq 0$, such that:*

$$\frac{\|\mathscr{A}\psi\|}{\|\psi\|} \leq k. \quad (1.2)$$

Here $\|.\|$ denotes the norm of the vector enclosed. Roughly speaking, a bounded operator maps a vector in $\mathscr{H}$ into the interior of a finite shell of "radius" k (Heathcote, 1990).

Definition 3. *A bounded operator $\mathscr{A}$* is a trace class operator if $Tr[|\mathscr{A}|] < \infty$. *The set of trace class operators, acting on Hilbert space $\mathscr{H}$*, is denoted by $T(\mathscr{H})$.

In quantum mechanics, the state of a system is described by a density matrix ρ, which by definition is *Hermitian*, i.e., $\rho^\dagger = \rho$, and *positive semi-definite*, i.e., $\rho \geq 0$ and has *unit trace*, $Tr[\rho] = 1$, acting on a Hilbert space $\mathscr{H}$. The positive semidefinite condition ensures that ρ has no negative eigenvalues. The set of density matrices of dimension N will be represented by $S(\mathscr{H})$. We write the state space of a quantum system as:

$$S(\mathscr{H}) = \{\rho \in T(\mathscr{H})|\rho \geq 0, Tr[\rho] = 1\}. \quad (1.3)$$

A density matrix has a *canonical decomposition* of the form projection operators Π_j satisfying $\Pi_j\Pi_k = \Pi_j\delta_{jk}$, such that

$$\rho = \sum_j \lambda_j \Pi_j,$$

with $\lambda_j \geq 0$, and $\sum_j \lambda_j = 1$. In Dirac notation $\Pi_j = |\psi_j\rangle\langle\psi_j|$, for some unit vector $|\psi_j\rangle \in \mathscr{H}$. As a special case, when , $\lambda_j = 1$, we have

$$\rho = |\psi_j\rangle\langle\psi_j|,$$

and is called as a *pure state*. Thus pure state represents the situation when a collection of physical systems are all in the same state $|\psi_j\rangle$, in contrast to a *mixed state* where a fraction of the members with relative population λ_j are described by $|\psi_j\rangle$. One can define the notion of *purity* $\mathbb{P}$ of state ρ as

$$\mathbb{P}(\rho) = Tr[\rho^2] = \sum_j \lambda_j,$$

where λ_j are the eigenvalues of ρ. Since projections satisfy $\rho^2 = \rho$, they have unit purity.

Consider the situation when a system comprises of two subsystems A and B such that we can manipulate these sub-systems separately by performing measurements on them individually. This means there must exist a mappings

$$\mathcal{S}: (S(\mathcal{H}_A), S(\mathcal{H}_B)) \rightarrow S(\mathcal{H}_{AB}),$$

and

$$\mathcal{M}: (M_A, M_B) \rightarrow M_{AB}$$

with measurements M_A, M_B, such that

$$Tr[\mathcal{S}(\rho_A, \rho_B)\mathcal{M}(M_A, M_B)] = Tr[\rho_A M_A] Tr[\rho_B M_B].$$

This can be achieved by *tensor product* by writing $\mathcal{H}_{AB} = \mathcal{H}_A \otimes \mathcal{H}_B$, $\mathcal{S}(\rho_A, \rho_B) = \rho_A \otimes \rho_B$, and $\mathcal{M}(M_A, M_B) = M_A \otimes M_B$. This can be understood as the motivation for choosing tensor product structure as a mathematical description of the composite system $A+B$. The state ρ_{AB} of a combined system $A+B$, determines the state of its subsystems (*reduced state*) via the *partial trace.*

Definition 4. *The reduced state of subsystem A is defined as*

$$\rho_A = Tr_B[\rho_{AB}] = \sum_k \left(\mathcal{I}_A \otimes \langle \psi_B{}^k| \right) \rho_{AB} \left(\mathcal{I}_A \otimes |\psi_B{}^k\rangle\right),$$

where $\{|\psi_B{}^k\rangle\}$ *is an orthonormal basis on subsystem B .*

Converse to the reduction in one of the subsystems, one may add an extra system called *ancillary system* such that any mixed state can be viewed as a reduced state of some pure state of the composite system. *For a state* $\rho \in \mathcal{H}$, a pure state σ on a composite system $\mathcal{H} \otimes \mathcal{H}_{ancilla}$ is a purification of ρ *if* $Tr_{ancilla}[\sigma] = \rho$. We will discuss this in little more detail ahead.

We are now in a position to define *entangled states.*

Definition 5. *A state vector* $|\xi\rangle \in \mathcal{H}_A \otimes \mathcal{H}_B$ is entangled if it cannot be written as a product of two vectors as $|\xi\rangle = |\psi\rangle \otimes |\phi\rangle$.

As an example, consider two orthogonal unit vectors $|\psi_1\rangle, |\psi_2\rangle \in \mathscr{H}$. The following unit vector represents an entangled state:

$$|\xi\rangle = \alpha|\psi_1\rangle \otimes |\psi_1\rangle + \beta|\psi_2\rangle \otimes |\psi_2\rangle \in \mathscr{H} \otimes \mathscr{H}. \tag{1.4}$$

Here, α and β are non-zero complex numbers satisfying $|\alpha|^2 + |\beta|^2 = 1$. This can be shown by noting that the a general factorized vector in $\mathscr{H} \otimes \mathscr{H}$ has the form

$$|\psi\rangle \otimes |\phi\rangle = \sum_j c_j|\psi_j\rangle \otimes \sum_k d_k|\psi_k\rangle,$$

which on comparing with Eq. (8), yields $c_1 d_1 = \alpha, c_2 d_2 = \beta, c_1 d_2 = c_2 d_1 = 0$. These lead to $\alpha\beta = 0$, which is in contradiction to our assumption that α and β are non-zero. Hence, $|\xi\rangle$ is entangled.

Let us now consider a more general scenario with the bases $\{|\psi_j\rangle\}^{d_A}{}_{j=1}$ for $\mathscr{H}_A$ and $\{|\phi_k\rangle\}^{d_B}{}_{k=1}$ for $\mathscr{H}_B$. One can write the non-entangled (product) state as

$$|\chi\rangle = |\psi\rangle \otimes |\phi\rangle = \sum_j c_j|\psi_j\rangle \otimes \sum_k d_k|\phi_k\rangle$$

$$= \sum_{j,k} c_j d_k|\psi_j\rangle \otimes |\phi_k\rangle$$

where c_j, d_k are complex number. It is not necessary for $|\psi\rangle$ to be one of $|\psi_j\rangle$ and for $|\phi\rangle$ to be one of $|\phi_k\rangle$. They may be any superposition of their respective basis vectors. In that case, there must exist at least two c_j and similarly at least two d_k with $j \neq k$ such that $c_j d_k \neq 0$. This implies that all vectors of the form $\sum_l c_l d_l|\psi_l\rangle \otimes |\phi_l\rangle$ are entangled. This leads us to the following theorem:

Schmidt Decomposition: For each vector $|\psi\rangle \in \mathscr{H}_A \otimes \mathscr{H}_B$, there exist orthonormal bases $\{|e_j\rangle\}^{d_A}{}_{j=1}$ and $\{|f_j\rangle\}^{d_B}{}_{j=1}$ such that

$$|\psi\rangle = \sum_j \sqrt{\lambda_j}|e_j\rangle \otimes |f_j\rangle,$$

where $d = \min[d_A, d_B]$, and $\lambda_1, \lambda_2, \cdots, \lambda_d$ are decreasingly ordered non negative numbers forming the so called Schmidt vector $\vec{\lambda}_\psi$. The number of non-vanishing elements in $\vec{\lambda}_\psi$ is called Schmidt rank $SR(\psi)$ of vector $|\psi\rangle$, such that $|\psi\rangle$ is entangled if and only if $SR(\psi) \geq 2$.

An immediate consequence of Schmidt decomposition is that the reduced states are given by

$$\rho_A = \lambda_j|e_j\rangle\langle e_j|,.$$

$$\rho_B = \lambda_j|f_j\rangle\langle f_j|$$

This means if the composite system is in pure state, the eigenvalues of the reduced states are same, and therefore they have same purity and entropy. With $|0\rangle$ and $|1\rangle$ denoting the eigenstates of Pauli σ_3 operator, one can write the famous Bell states as

$$|\Phi_\pm\rangle = \frac{1}{\sqrt{2}}(|00\rangle \pm |11\rangle),$$

$$|\Psi_\pm\rangle = \frac{1}{\sqrt{2}}(|01\rangle \pm |10\rangle) \tag{1.5}$$

with the reduced states $\rho_A = \mathcal{I}_A/2$, $\rho_B = \mathcal{I}_B/2$, being maximally mixed, i.e., a complete lack of knowledge of the state of individual sub-systems. It is in this sense, the Bell states are known to be maximally entangled.

So far, we have discussed the notion of entanglement at the level of state vectors, the rays in Hilbert space. We now turn our attention to the entanglement of positive operators, in particular the entanglement of density matrices.

Definition 6. A density matrix $\rho \in \mathcal{H}_A \otimes \mathcal{H}_B$ is factorized if $\rho = \rho_A \otimes \rho_B$, and the set of all factorized states is denoted by $\mathcal{S}_{fac}(\mathcal{H}_A \otimes \mathcal{H}_B)$. Further, ρ is separable if it is a convex combination of factorized states

$$\rho = \sum_j p_j \rho_1{}^j \otimes \rho_2{}^j, \tag{1.6}$$

with $\mathcal{S}_{sep}(\mathcal{H}_A \otimes \mathcal{H}_B)$ denoting the set of all separable states. If ρ is not separable, it is entangled.

Interestingly, the $|\xi_\pm\rangle = \alpha|\psi_1\rangle \otimes |\psi_1\rangle \pm \beta|\psi_2\rangle \otimes |\psi_2\rangle$ (see Eq. (1.4)) is entangled, but $\rho = \frac{1}{2}|\xi_+\rangle\langle\xi_+| + \frac{1}{2}|\xi_-\rangle\langle\xi_-|$ is separable.

In fact, $\rho = p|\xi_+\rangle\langle\xi_+| + (1-p)|\xi_-\rangle\langle\xi_-|$ is entangled for all p except $p = \frac{1}{2}$. This can be checked by using the positive partial transpose criterion discussed ahead.

The maximally entangled states are central to various quantum information tasks. However, at the same time they are also very fragile in the presence of any type of noise. Therefore, it is important to investigate to what extent the maximally entangled states can be "distilled" out of general mixed state by means of *local operations and classical communication* (LOCC). But before we discuss how exactly the entanglement distillation works, we briefly revisit the notion of quantum operations and LOCC.

Quantum Operations. The most commonly known quantum operations are *unitary evolutions* of closed systems governed by unitary operator Usuch that an initial state $\rho(0)$is evolved as to a final state $\rho(t)$as

$$\rho(t) = U\rho(0)U^{\dagger},$$
$$UU^{\dagger} = U^{\dagger}U = \mathcal{I}^{\cdot}$$

However, in realistic situations the system is almost invariably interacting with its environment. Let $\rho_s(0)$denote the initial state of the system and $\rho_E$be the state of the environment and the combined initial state is separable i.e., $\rho_S(0) \otimes \rho_E$. Further, let $\{|e_\mu\rangle\}$be the basis for the environment. Then the joint evolution of the system and environment is described by some unitary U_{SE}as:

$$\rho_{SE} = U_{SE}(\rho_S(0) \otimes \rho_E)\, U^{\dagger}{}_{SE}.$$

Assuming $\rho_E = |e_0\rangle\langle e_0|$, we have:

$$\rho_{SE} = U_{SE}(\rho_S(0) \otimes |e_0\rangle\langle e_0|)\, U^{\dagger}{}_{SE}.$$

Since we are often interested only in the state of the system, we trace over the environment degrees of freedom to obtain:

$$\rho_S = Tr_E[\rho_{SE}] = \sum_\mu \mathcal{K}_\mu \rho_S(0) \mathcal{K}_\mu{}^{\dagger}, \tag{1.7}$$

where $\mathcal{K}_\mu = \langle e_\mu | U_{SE} | e_0 \rangle$are the *Kraus* operator, satisfying the completeness condition $\sum_\mu \mathcal{K}^{\dagger}{}_\mu \mathcal{K}_\mu = \mathcal{I}$

The condition assures the trace preservation:

$$Tr\left[\sum_\mu \mathcal{K}_\mu \rho \mathcal{K}_\mu{}^{\dagger}\right] = Tr\left[\sum_\mu \mathcal{K}_\mu{}^{\dagger} \mathcal{K}_\mu \rho\right] = Tr[\rho].$$

The Eq. (1.7) can be viewed as a map

$$\rho_S(t) = \mathscr{E}\left[\rho_S(0)\right], \tag{1.8}$$

taking input state $\rho_S(0)$and giving output $\rho_S(t)$. In fact, Eq. (1.7) represents the general form of completely positive and trace preserving quantum operation.

In order to see how quantum operations affect the entanglement, consider the following example of a unitary operation described by:

$$\mathscr{U} = \frac{1}{\sqrt{2}}(|00\rangle + |11\rangle)\langle 00| + \frac{1}{\sqrt{2}}(|00\rangle - |11\rangle)$$

$$\langle 11| + |01\rangle\langle 01| + |10\rangle\langle 10| \tag{1.9}$$

This operation cannot be factorized as $\mathscr{U} = \mathscr{U}_1 \otimes \mathscr{U}_2$, and for this reason it represents an example of a *global* operation, which can be implemented only when the two system interact with each other. It is immediately clear that

$$\mathscr{U}|00\rangle = \frac{1}{\sqrt{2}}(|00\rangle + |11\rangle),$$

i.e., it takes a separable state to an entangled one. Thus, such a global operation is indeed capable of creating entanglement. However, this is not the case if one restricts to *local operations and classical communication* discussed next.

Local Operation and Classical Communication (LOCC)

An important notion in the theory of quantum entanglement is that of *local operations and classical communication* (LOCC), in which a multipartite system is distributed to various parties and they are restricted to perform operation locally on their respective subsystems (Dagmar & Gerd, 2019). However, the parties are allowed to communicate via classical channel to enhance their measurement strategies.

In bipartite scenario, the most general local operation that acts only on the first subsystem is described by:

$$\rho_{AB} \mapsto \sum_{\mu} \left(A_{\mu} \otimes \mathscr{I}\right)\rho\left(A^{\dagger}{}_{\mu} \otimes \mathscr{I}\right),$$

$$\sum_{\mu} A^{\dagger}{}_{\mu} A_{\mu} = \mathscr{I}$$

Similarly, one can define the action on the second subsystem alone. Note that such operations do not create any correlations, and map product states to product states:

$$\rho = \rho^{(1)} \otimes \rho^{(2)} \mapsto \left(\sum_j A_j \rho^{(1)} A_j^{\dagger}\right) \otimes \rho^{(2)},$$

and also map separable states to separable states:

$$\rho = \sum_j p_j \rho_j^{(1)} \otimes \rho_j^{(2)} \mapsto \sum_j p_j \left(\sum_k A_k \rho_j^{(1)} A_k^{\dagger}\right) \otimes \rho_j^{(2)}.$$

Now consider the case when such operations are applied in a correlated manner such that the application of an operation depends on the outcome of the previous operation as:

$$\rho \mapsto \sum_j \left(A_j \otimes \mathcal{I}\right) \rho \left(A_j^{\dagger} \otimes \mathcal{I}\right)$$

$$\mapsto \sum_{jk} \left(\mathcal{I} \otimes B_{jk}\right) \left(A_j \otimes \mathcal{I}\right) \rho \left(A_j^{\dagger} \otimes \mathcal{I}\right) \left(\mathcal{I} \otimes B_{jk}^{\dagger}\right)$$

$$\mapsto \sum_{jkl} \left(C_{jkl} \otimes \mathcal{I}\right) \left(\mathcal{I} \otimes B_{jk}\right) \left(A_j \otimes \mathcal{I}\right) \rho \left(A_j^{\dagger} \otimes \mathcal{I}\right)$$

$$\left(\mathcal{I} \otimes B_{jk}^{\dagger}\right) \left(C_{jkl}^{\dagger} \otimes \mathcal{I}\right)$$

$$\vdots$$

$$\mapsto \sum_{jkl\ldots p} \left(\mathcal{I} \otimes G_{jk\ldots p}\right) \cdots \left(A_j \otimes \mathcal{I}\right) \rho \left(A_j^{\dagger} \otimes \mathcal{I}\right) \cdots \left(\mathcal{I} \otimes G_{jk\ldots p}^{\dagger}\right)$$

The local operation A_j applied to the first party in first step can be viewed as an interaction of the system with an ancilla/environment followed by a measurement. The local operation characterized by B_{jk} is then applied on the second subsystem based on the outcome of the first operation. Conditioned on the outcome of this measurement, C_{jkl} operation is applied on the first subsystem, and so on. Such operations are called *local operations and classical communication*, since it involves the two individual subsystems being operated locally, performed by say Alice and Bob. However, the above protocol demands that they communicate their measurement results each other. This communication can be performed via classical channel and does not require to have quantum nature, hence the name classical. The LOCC operations can induce correlations between two subsystems but since these correlations are based on classical information, they remain classical in nature. As an example, Alice flips a coin communicates the outcome to Bob. Depending on the outcome being heads or tails, they prepare $|0\rangle\langle 0| \otimes |0\rangle\langle 0|$ and $|1\rangle\langle 1| \otimes |1\rangle\langle 1|$, respectively. If the coin is fair they generate the overall state:

$$\frac{1}{2}|0\rangle\langle 0| \otimes |0\rangle\langle 0| + \frac{1}{2}|1\rangle\langle 1| \otimes |1\rangle\langle 1|,$$

which is not a product sate.

Entanglement Distillation. We are now in a position to discuss entanglement distillation. Consider two parties sharing n copies of a bipartite mixed state ρ, which contains noisy entanglement. By performing LOCC they can obtain some (less) number of k copies of systems in state close to a singlet state which contains pure entanglement. A sequence of LOCC operations achieving this task is called entanglement purification or entanglement distillation protocol. An optimal distillation protocol is one that results in maximum k/n, for large n number of input states. The resulting nearly singlet states can be used to perform various quantum information tasks.

The following example is to illustrate the entanglement distillation scheme based on controlled NOT operation, and hence also known as CNOT-protocol (Bennett C., et al., 1996). Alice and Bob share a finite ensemble of n entangled qubit pairs, represented by the following state which is diagonal in Bell basis

$$\rho = a\rho_{\Phi_+} + b\rho_{\Psi_-} + c\rho_{\Psi_+} + d\rho_{\Phi_-}, \tag{1.10}$$

where $\rho_{\Phi_\pm} = |\Phi_\pm\rangle\langle\Phi_\pm|, \rho_{\Phi_\pm} = |\Psi_\pm\rangle\langle\Psi_\pm|$ are Bell states in density matrix form. The only assumption that is made here is that we start with the component ρ_{Φ_+} being dominant such that $a > 0.5$, since this is the regime in which state (1.10) is non-separable. This protocol involves processing two qubits pairs in one step. Alice holds qubit 1A and 2A, while as Bob has 1B and 2B in his possession. Following operations are made locally on these qubits:

1. Alice and Bob perform local operations $U_A = \frac{1}{\sqrt{2}}(\mathcal{I} - i\sigma_x)$, and $U_B = \frac{1}{\sqrt{2}}(\mathcal{I} + i\sigma_x)$, leading to state:

$$(U_A \otimes U_B)\rho(U_A{}^{\dagger} \otimes U_B{}^{\dagger}) = a\rho_{\Phi_+} + d\rho_{\Psi_-}$$

$$+c\rho_{\Psi_+} + b\rho_{\Phi_-} \tag{1.11}$$

exchanging the contribution of ρ_{Ψ_-} and ρ_{Φ_-}.

2. With qubits of pair 1 i.e., 1A and 1B as control and the other pair (2A and 2B) as target, Alice and Bob perform CNOT operations resulting in state:

$$\rho_{CNOT} = \left(a^2\rho_{\Phi_+} + cd\rho_{\Psi_-} + c^2\rho_{\Psi_+} + ab\rho_{\Phi_-}\right) \otimes \rho_{\Phi_+}$$

$$+\left(b^2\rho_{\Phi_+} + cd\rho_{\Psi_-} + d^2\rho_{\Psi_+} + ab\rho_{\Phi_-}\right) \otimes \rho_{\Phi_-}$$

$$+\left(bd\rho_{\Phi_+} + bc\rho_{\Psi_-} + bd\rho_{\Psi_+} + ad\rho_{\Phi_-}\right) \otimes \rho_{\Psi_-}$$

$$+\left(ac\rho_{\Phi_+} + ad\rho_{\Psi_-} + ac\rho_{\Psi_+} + bc\rho_{\Phi_-}\right) \otimes \rho_{\Psi_+} \tag{1.12}$$

3. Next, both Alice and Bob measure their target qubits (2A and 2B) in computational basis $\{|0\rangle, |1\rangle\}$, obtaining outcomes "0" or "1".
4. The results are exchanged using a classical channel. The distillation is successful if the combined result is "00" or "11", whence they keep pair 1. From state (1.12), we obtain the probability of success as:

$$\rho^{(s)} = \frac{1}{p^2/2}\ {}_2\langle 00|\rho_{CNOT}|00\rangle_2 = \frac{1}{p^2/2}\ {}_2\langle 11|\rho_{CNOT}|11\rangle_2$$

$$= \frac{1}{p^s}\left[(a^2+b^2)\rho_{\Phi_+} + 2cd\,\rho_{\Psi_-} + (c^2+d^2)\rho_{\Psi_+} + 2ab\,\rho_{\Phi_-}\right] \tag{1.13}$$

which is normalized by the probability of success $p^s = (a+b)^2 + (c+d)^2$, to find the results "00" or "11". However, if Alice and Bob obtain "10" or "01", the resulting state turns out to have Fidelity less than ½, and therefore Alice and Bob discard qubit 1 in this case.

Thus the successful step maps the Bell diagonal state (1.12) to Bell diagonal state (1.13) with Fidelity:

$$F(\rho^{(s)}) = Tr\left[\rho^{(s)}\,\rho_{\Phi_+}\right] = \frac{a^2+b^2}{(a+b)^2+(1-a-b)^2}.$$

ENTANGLEMENT BEYOND BIPARTITE SYSTEMS

Let $\mathscr{H} = \mathscr{H}_A \otimes \mathscr{H}_B \otimes \cdots \otimes \mathscr{H}_N$ denote the space of N-particle system, with $\mathscr{H}_k$ the d_k-dimensional space associated with k-th system. A pure state $|\psi\rangle \in \mathscr{H}$ is separable if it can be written as:

$$|\psi\rangle = |\psi_A\rangle \otimes |\psi_B\rangle \otimes \cdots \otimes |\psi_N\rangle,$$

otherwise it is entangled. However, this generalization is not the end of the story. It turns out that the bipartite scenario is very special in the following respect. Any pure bipartite state:

$$|\psi\rangle = \sum_{j,k=1}^{N} c_j d_k |e_j\rangle \otimes |f_k\rangle,$$

can be written as

$$|\psi\rangle = (U_A \otimes U_B)\sum_{j=1}^{N}\sqrt{\lambda_j}|e_j\rangle \otimes |f_j\rangle, \tag{1.14}$$

characterized by Schmidt vector $\vec{\lambda}_\psi$. However, for $N > 2$, *no expansion of the form given inEq. (1.6)is possible (Bengtsson & Życzkowski, 2006)*. This brings many novel features in multipartite entanglement and makes it richer than the bipartite scenario. A pictorial representation might help to appreciate this richness. In *Figure 1*(left), we represent the corners by bitstrings, say 10 which represents the product state $|10\rangle = |1\rangle \otimes |0\rangle$. An edge of the square represents the superposition of its end point and is always a separable state. A diagonal represents (an equal weight) superposition of the corners and pertains to the maximally entangled Bell states introduced in Eq. (1.5). For a three qubit system, the eight separable states form the corners of a unit cube, *Figure 1* (right). The superposition of 000 and 111 at diametrically opposite corners is known as Greenberger, Horne, and Zeilinger (GHZ) state

$$|GHZ\rangle = \frac{1}{\sqrt{2}}(|000\rangle + |111\rangle). \tag{1.15}$$

Figure 1. Pure states for a bipartite (left) and tripartite (right) system (Bengtsson & Życzkowski, 2006). The Bell states are correspond to two corners of a diagonal in the square (left). The $|GHZ_3\rangle = |000\rangle + |111\rangle$ is represented by two corners along the diagonal of the cube. The two parallel triangles in the cube correspond to two locally equivalent $|W_3\rangle = |100\rangle + |010\rangle + |001\rangle$ and $|\overline{W}_3\rangle = |011\rangle + |101\rangle + |110\rangle$ states.

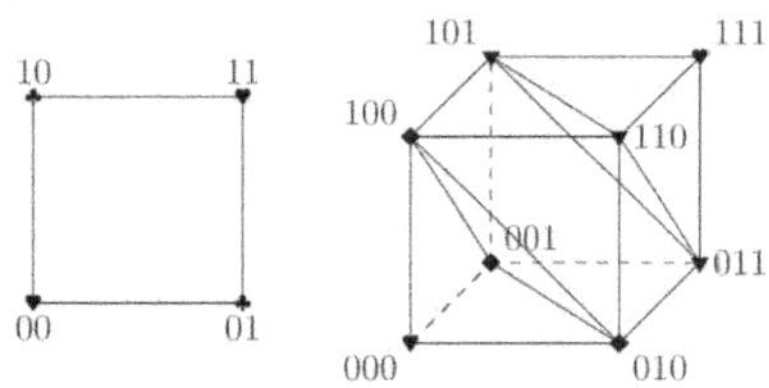

It is worth noting that if we trace out one subsystem of GHZ state, the result is a separable state. In this sense, the entanglement of GHZ is of a global nature, a property that holds for all Schmidt decomposable *(Thapliyal A. V., 1999).*

Another important entangled state is the W state which appear as triangle in *Figure 1.*

$$|W\rangle = \frac{1}{3}(|001\rangle + |010\rangle + |100\rangle). \tag{1.16}$$

The entanglement of W state is more robust than GHZ state in the sense that tracing out any subsystems leaves us with an entangled mixed state (see also Example 6). The complexity of *Figure 1* grows as the number of qubits is increased and this demands better organizing principles for a comprehensive analysis .

CREATION OF ENTANGLED STATES

Here we discuss some methods of creating entangled states. The central idea is that you have two particle in a well-defined state such that their individual states are indeterminate, but correlated and the correlation holds even when they are separated. Here are a few ways of creating entangled states. *(Orzel, 2017)*.

1. ***Entanglement from birth***. The most widely used entangled particles are photons, since it is relatively easy to entangle photons. This may be achieved by using a cascade transition, which involves exciting calcium atoms to higher energy levels from which a direct transition to the ground state is forbidden *(Freedman & Clauser, 1972)*. As a result, the atoms are de-excited via a metastable state emitting two photons, a few nano seconds apart. These photons are emitted in random directions, but when they are emitted in opposite directions, the conservation of angular momentum leads to the correlation in their polarization, the result is an entangled state.

Figure 2. Calcium cascade leading to the production of two entangled photons

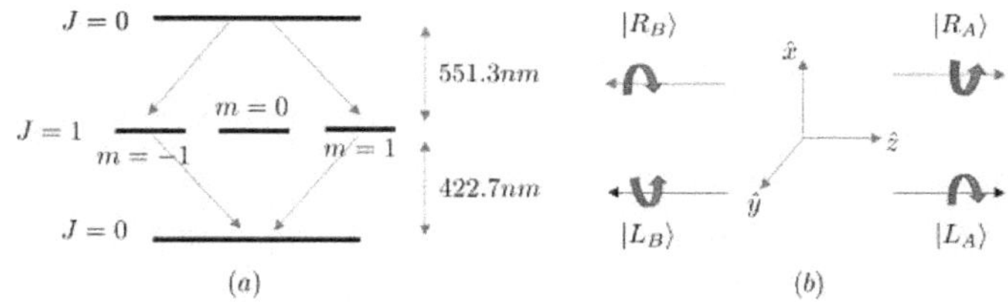

In the ground state, the calcium atom has two electron in outer most shell ($4s^2$), with opposite alignment of spin leading to total angular momentum $J = 0$. The transition from the excited state $4p^2$ to ground state $4s^2$(which also corresponds to $J = 0$) occurs via intermediate level $4s^1 4p^1$ (with $J = 1$). Two photons, emitted by calcium atoms, at wavelength $\lambda_A = 551.3\,nm$ and $\lambda_B = 422.7\,nm$. The two photon states can be realized in two ways: two left handed photons pass via intermediate level with $m = -1$, or two right handed photons via level with $m = +1$. In general, these are not emitted in the opposite direction, and once in a while, the two photon do come out in opposite directions, say $\pm z$-directions as depiction in *Figure 2*. The two possibilities are indistinguishable, leading to entangled state

$$|\psi\rangle = \frac{1}{\sqrt{2}}\left(|L_A\rangle \otimes |L_B\rangle + |R_A\rangle \otimes |R_B\rangle\right). \qquad (1.18)$$

The cascade process is slow since the photons are emitted randomly in all directions, and getting them in exactly opposite direction can take a while. A much efficient way is to use parametric down conversion sources, which use non-linear optical crystals (such as beta barium borate, BBO) to convert high energy photo into two daughter photons with half the initial energy.

2. ***Second generation entanglement*:** Though photons provide an easy way to demonstrate entanglement, they are hard to control as they are always moving with the speed of light. Thus it would be nice to entangle material particles instead. One way to achieve this is by using an entangled pair of photons and let them interact with and be absorbed by atoms. The state at the end of photon absorption will depend on the polarization of the photons. Thus the correlations from the polarization degree of freedom are transferred to atoms, leaving them entangled *(Hensen, et al., 2015)*.
3. ***Entanglement at a distance.*** This is kind of reverse of the earlier case, where you start with a pair of atoms at different locations that emit photons. By bringing these photons together in a right way, they can be entangled. This in turn, leads to entanglement of the original atoms. This was demonstrated by Chris Monroe's group at Maryland by using ytterbium ions held in separate ion traps *(Matsukevich, Maunz, Moehring, Olmschenk, & Monroe, 2008)*. The ions were excited to a state from which they could emit light with one of the two polarizations. The emitted photons were joined at a beam splitter, where 25 percent of the time they detected one photon at each output, confirming the entanglement of the source ions. The important thing to note is that the two ions are always separated and their entanglement is controlled by the light they emit.
4. ***Entanglement by interaction.*** This method involves bringing two particles together and let them interact in such a way such that the final states of the two particles depend on each other. The widely known example is "Rydberg blockade" scheme, which uses the fact that bringing two atoms close enough that exciting one to Rydberg state affects the energy level of the other such that this second atom cannot be excited by the same laser. Now trying to excite the second atom would lead to a superposition that is perfectly anti-correlated with the first atom, leaving the two atoms in entangled state *(Berman, Arimondo, & Lin, 2012)*.

Independent of the underlying physical mechanism, a compact way of describing the creation of entangled states is by using *quantum circuits*. A quantum circuit consists of a series of unitary transformation called *gates* sometimes supplemented with *measurements*. Let us consider the case when an arbitrary number of N qubits are converted into an N-partite entangled state. Assuming all the N-qubits are initialized in state $|0\rangle$. We need the following two gates to carry out this task. The Hadamard gate, represented by Hacting according to following:

$$H:\ |0\rangle \rightarrow \frac{|0\rangle + |1\rangle}{\sqrt{2}},\ |1\rangle \rightarrow \frac{|0\rangle - |1\rangle}{\sqrt{2}}$$

Another important gate that we need is the controlled-NOT (CNOT), described by the following operation:

$$CNOT:\ |00\rangle \rightarrow |00\rangle,\ |01\rangle \rightarrow |01\rangle, |10\rangle \rightarrow |11\rangle, |11\rangle \rightarrow |10\rangle.$$

Thus CNOT operation does not affect the first qubit (control qubit) but flips the second qubit (target qubit) if the control qubit is 1, and leaves the second qubit unchanged otherwise. For 2-qubit case, the Hadamard takes the input state $|0\rangle \otimes |0\rangle$ to $\left(\frac{|0\rangle + |1\rangle}{\sqrt{2}}\right)|0\rangle$, which is subsequently acted upon by CNOT and changed to $\frac{|00\rangle + |11\rangle}{\sqrt{2}}$. This procedure can be generalized as depicted in the circuit in *Figure 3*, resulting in N-qubit GHZ state (Braunstein & Loock, 2005).

Figure 3. Quantum circuit for generating N-qubit GHZ state using Hadamard and CNOT gates (Braunstein & Loock, 2005)

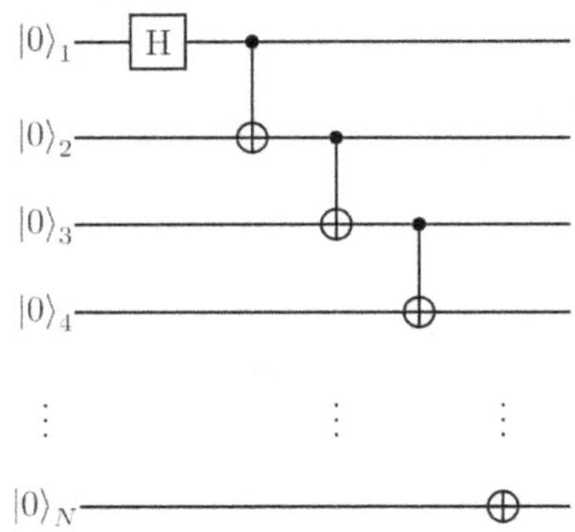

In general, two systems can be entangled if they interact in a controlled way. This raises the questions like given an interaction (a Hamiltonian), what is the best way of creating the entanglement? And how can one increase the efficiency of such entanglement generation processes? The entropy production is quantified in terms of the *state entanglement rate*$\Gamma(t) := \frac{d}{dt}E(|\psi(t)\rangle)$ where E is the von Neumann entropy of the reduced state $Tr_{A(B)}[|\psi(t)\rangle\langle\psi(t)|]$, where $|\psi(t)\rangle = \exp[-iHt]|\psi(0)\rangle$, with H being the Hamiltonian (Dür , Vidal , Cirac, Linden, & Popescu, 2001). The entanglement capacity of H is then defined as:

$$\mathscr{E}[H] := \max_{|\psi\rangle} \Gamma(t)|_{t\to 0}$$

In order to illustrate this concept, we take a simple example of a bipartite system subjected to Hamiltonian:

$$H = \omega_1 \sigma_1 \otimes \sigma_1 + \omega_2 \sigma_2 \otimes \sigma_2, \tag{1.17}$$

with $\omega_1, \omega_2 \in \mathbb{R}$, and $\sigma_1, \sigma_2$the Pauli matrices. The eigenvectors corresponding to this Hamiltonian are the four Bell states given in Eq. (1.5). We denote by $|\psi_E\rangle$ the initial state that leads to maximal entanglement rate Γ_E (say). For the Hamiltonian of form (1.10), such a state has the following form (Dür , Vidal , Cirac, Linden, & Popescu, 2001):

$$|\psi_E\rangle = \sqrt{p_0}|0\rangle \otimes |1\rangle + i\sqrt{1-p_0}|1\rangle \otimes |0\rangle$$

The time evolved state $\rho_E(t) = \exp[-iHt]|\psi_E\rangle\langle_E\psi|\exp[iHt]$, has the following matrix representation:

$$\rho_E(t) = \begin{pmatrix} 0 & 0 & 0 & 0 \\ 0 & x^2 & ixy^* & 0 \\ 0 & -ix^*y & y^2 & 0 \\ 0 & 0 & 0 & 0 \end{pmatrix}$$

where

$$x := \sqrt{p_0}\cos[(\omega_1+\omega_2)t] - \sqrt{1-p_0}\sin[(\omega_1+\omega_2)t],$$

and

$$y := \sqrt{p_0}\sin[(\omega_1+\omega_2)t] + \sqrt{1-p_0}\cos[(\omega_1+\omega_2)t].$$

The reduced state is given by

$$\rho_{1(2)} = \begin{pmatrix} x^2 & 0 \\ 0 & y^2 \end{pmatrix}$$

Therefore we have $\Gamma(t) = -\frac{d}{dt}[x^2\log_2 x^2 + y^2\log_2 y^2]$, which at $t = 0$ becomes:

$$\Gamma(0) := (\omega_1+\omega_2)2\sqrt{p_0(1-p_0)}$$

$$\log_2\left(\frac{p_0}{1-p_0}\right) = (\omega_1 + \omega_2)\kappa$$

where $\kappa = 2\sqrt{p_0(1-p_0)}\log_2\left(\frac{p_0}{1-p_0}\right)$. Thus the entanglement capability of the Hamiltonian H in Eq. (1.10) is given by:

$$E(H) = (\omega_1 + \omega_2)\kappa$$

ENTANGLEMENT DETECTION AND QUANTIFICATION

A general approach to characterize the entangled states is by using the notion of entanglement witness *(Chruściński & Sarbicki, 2014)*. A Hermitian operator W is an entanglement witness if for an entangled state ρ, $Tr[W\rho] < 0$ and $Tr[W\sigma] \geq 0$ for all separable states σ. Finding an entanglement witness for a given state is not an easy task and various numerical methods have been proposed which become inefficient with the increase in dimensionality of the system *(Brandão & Vianna, 2004), (Brandão, 2004), (Doherty, Parrilo, & Spedalieri, 2004)*. A famous example of entanglement witness is furnished by *Bell inequality,* which in *Clauser, Horne, Shimony, and Holt (CHSH)* form reads

$$Tr[B_{CHSH}\rho_{LHV}] \leq 2,$$

and is satisfied by all states ρ_{LHV} admitting a local hidden variable model. The CHSH operator is defined as

$$B_{CHSH} = A_1 \otimes (B_1 + B_2) + A_2 \otimes (B_1 - B_2) \tag{1.19}$$

with $A_j = a^j \cdot \sigma = \sum_{k=1}^{3} a^j{}_k \sigma_k$, $B_j = b^j \cdot \sigma = \sum_{k=1}^{3} b^j{}_k \sigma_k$, and $a^j, b^j \in \mathbb{R}^3$, and σ_k denote the Pauli matrices. One can now define the CHSH witness as

$$W_{CHSH} = 2\mathcal{I} \otimes \mathcal{I} - B_{CHSH}. \tag{1.20}$$

Then W_{CHSH} is a nonlocality witness i.e., if $Tr[\rho W_{CHSH}] < 0$, then ρ does not admit a LHV model. However, any such state is necessarily entangled and hence W_{CHSH} qualifies to be a legitimate entanglement witness.

At this point, we introduce reader to the following important observation. Given finite dimensional Hilbert spaces $\mathcal{H}_A$ and $\mathcal{H}_B$, and: $\mathcal{H} = \mathcal{H}_A \otimes \mathcal{H}_B$ denote their tensor product space. Let $|\psi\rangle$ be a unit vector in $\mathcal{H}$, and $\mathcal{O}$ be an observable acting in this space, with expectation value $\langle\psi|\mathcal{O}|\psi\rangle$. Then the following statements are equivalent for the pure states:

(i) With $|\psi_A\rangle \in \mathcal{H}_A$, and $|\psi_B\rangle \in \mathcal{H}_B$ being two normalized vector in the respective spaces, the state vector $|\psi\rangle = |\psi_A\rangle \otimes |\psi_B\rangle$ is *factorizable.*

(ii) The average of CHSH operator, defined in Eq. (1.11), with respect to $|\psi\rangle$, satisfies Bell inequality, i.e.,

$$\langle\psi|B_{CHSH}|\psi\rangle \leq 2.$$

(iii) The operators A and B acting in spaces $\mathcal{H}_A$ and $\mathcal{H}_B$, are statistically independent if:

$$\langle\psi|A \otimes B|\psi\rangle = \langle\psi|A \otimes \mathcal{I}_B|\psi\rangle\langle\psi|\mathcal{I}_A \otimes B|\psi\rangle$$

A direct way of telling whether a given quantum state is entangled is to see if it can be written as Eq. (1.6). However, this process is exceedingly complicated given that a state ρ can have infinitely many convex representations (Terhal, 2002). The Schmidt decomposition also provides a test for entanglement of bipartite *pure* states in terms of the existence of more than one Schmidt coefficients. For *mixed* states, the entanglement text goes by the name of positive partial transposition (PPT) criterion (Peres, 1996).

Definition 9 *A bipartite state* ρ_{AB} in a product basis $|ij\rangle = |i\rangle_A \otimes |j\rangle_B$ *i.e.,*

$$\rho_{AB} = \sum_{i,j,k,l} \rho_{ijkl}|ij\rangle\langle kl| \quad (1.21)$$

such that $\lambda_{ij,kl}$ are matrix elements of ρ_{AB} in this basis, we define partial transposition of ρ_{AB} as

$$\rho_{AB}^{T_B} = \sum_{ij,kl} \rho_{jikl}|ij\rangle\langle kl| \quad (1.22)$$

The PPT criterion then states that ρ_{AB} is entangled if $\rho_{AB}{}^{T_B}$ has a negative eigenvalue. We denote the set of states for which $\rho_{AB}{}^{T_B}$ has no negative eigenvalues by $\mathcal{S}^{PPT}$. It is worth mentioning here that $\rho \in \mathcal{S}^{PPT}$ does not always mean separability in higher dimensional systems.

The entangled states are indispensable for carrying out various quantum information tasks. Moreover, some entangled states are found more useful than other. Therefore, it is natural to ask how can one measure or quantify the degree of entanglement in a given state. For a state described by density matrix ρ, a good entanglement measure E must satisfy the following properties:

1. $0 \leq E(\rho) \geq 1$, with $E(\rho) = 0$ if and only if ρ is not entangled, and $E(\rho) = 1$ when ρ is maximally entangled state such as one of the Bell states.
2. $E(\rho)$ doesn't change under local unitary transformation, i.e., transformations of type $U_A \otimes U_B$.
3. Entanglement does not increase under local operations and classical communication
4. The entanglement measure of the full density operator cannot be greater than the weighted sum of the entanglement measures of its parts i.e., $E\left(\sum_j p_j \rho_j\right) \leq \sum_j p_j E(\rho_j)$.

Some well-known measures of quantum entanglement are:

1. **Entropy measure:** The *entropy of entanglement* of a pure state $|\psi\rangle$ is defined as

$$\begin{aligned} E(|\psi\rangle\langle\psi|) &:= S(Tr_A[|\psi\rangle\langle\psi|]) \\ &= S(Tr_B[|\psi\rangle\langle\psi|]) \end{aligned} \quad . \tag{1.23}$$

Here, $S(\rho) = -Tr[\rho \log \rho]$is the von-Neumann entropy and $Tr_{A(B)}$denotes the partial trance over system A(B).

2. **Entanglement of formation:** For a mixed state $\rho = \sum_j p_j |\psi_j\rangle\langle\psi_j|$, with:

$0 \leq p_j \leq 1$, $\sum_j p_j = 1$, the *entanglement of formation* is defined as

$$E_F(\rho) := \inf\left[\sum_j p_j E_F[|\psi_j\rangle\langle\psi_j|]\right]. \tag{1.24}$$

That is, this measure represents the minimal possible average entanglement over all pure state decompositions of ρ. Loosely speaking entanglement of formation tells us how many maximally entangled pairs are needed to prepare ρ. In other words, it quantifies of entanglement needed to synthesis ρ.

3. **Concurrence:** For a bipartite qubit states, the *concurrence* is defined as

$$C(\rho):=\max\left[0,\lambda_1-\lambda_2-\lambda_3-\lambda_4\right], \tag{1.25}$$

where λ_j (arranged in decreasing order) are the square roots of the eigenvalues of $\rho\left(\sigma_y\otimes\sigma_y\right)\rho^*\left(\sigma_y\otimes\sigma_y\right)$, where ρ^* is the complex conjugate of ρ. For general bi-partite state we have

$$E_F(\rho)=h\left(\frac{1+\sqrt{1-C^2(\rho)}}{2}\right), \tag{1.26}$$

with $h(x):=-x\log_2 x-(1-x)\log_2(1-x)$.

4. **Relative entropy:** The distinguishability between two quantum states $\rho$and σ can be quantified by *quantum relative entropy* defined as

$$S_R(\rho\ \|\ \sigma):=Tr[\rho\log\rho-\rho\log\sigma]. \tag{1.27}$$

The *relative entropy* is then defined with respect to a set X as

$$E_R{}^X(\rho):=\inf_{\sigma\in X}S_R(\rho\ \|\ \sigma). \tag{1.28}$$

Here, Xrepresents the set of separable states. Therefore, relative entropy of entanglement quantifies the distance of the entangled state $\rho$from the nearest separable state in the set X of separable states.

5. **Logarithmic Negativity:** This measure is based on partial transpose of a bipartite state ρ_{AB}, and is defined as

$$N(\rho):=\frac{\|\rho_{AB}{}^{T_B}\|-1}{2}, \tag{1.29}$$

where $\|x\| = Tr[\sqrt{x^{\dagger}x}]$is the trace norm. Thus this measures quantifies the degree to which positive partial transpose separability criterion is violated. The advantage of negativity is that it is easy to calculate, however, since PPT criterion is itself not able to detect all entangled states, consequently, there are entangled states with zero negativity.

MAXIMALLY ENTANGLED PURE AND MIXED STATES

The four Bell states introduced in Eq. (1.5) are examples of maximally entangled states in the sense that the reduced state left after tracing over one qubit is maximally mixed. The form of Bell states suggests a generalization to higher dimensions in two ways. First is to extend $|\Phi_{+}\rangle$to two component superposition with0's in one and 1's in the other component. Second option is to generalize $|\Psi_{+}\rangle$to an n-component superposition having 1in each qubit position. This leads to the two classes of maximally entangled sates

$$|GHZ\rangle : \frac{1}{\sqrt{2}}(|000\rangle + |111\rangle),$$

$$\frac{1}{\sqrt{2}}(|0000\rangle + |1101\rangle), \cdots, \frac{1}{\sqrt{2}}(|00\cdots0\rangle + |11\cdots1\rangle) \tag{1.30}$$

$$|W\rangle : \frac{1}{\sqrt{3}}(|100\rangle + |010\rangle + |001\rangle),$$

$$\frac{1}{\sqrt{4}}(|1000\rangle + |0100\rangle + |0010\rangle + |0001\rangle), \cdots ,$$

$$\frac{1}{\sqrt{N}}(|10\cdots0\rangle + |01\cdots0\rangle + \cdots + |00\cdots1\rangle) \tag{1.31}$$

Though both these classes represent pure maximally entangled states, but they differ fundamentally in the sense that they cannot be converted into one another using any unitary transformations. A natural question to ask is how can one generalize the concept of maximally entangled states to mixed states?

One option is to define a maximally entangled mixed state (MEMS) as a state that attains maximum entanglement for a give level of mixedness. However, one immediately runs into trouble with this definition, since the degree of mixedness and entanglement depends on the measures on chooses to quantify these notions. This problems seems to be fundamental and unavoidable, but as along as one sticks to a given measure of mixedness and of entanglement, certain interesting mixed entangled states emerge. For example, the following class of states MEMS was identified in (Wei, et al., 2003)

$$\rho_{MEMS} = \begin{pmatrix} \frac{1}{3} & 0 & 0 & \frac{b}{2} \\ 0 & \frac{1}{3} & 0 & 0 \\ 0 & 0 & 0 & 0 \\ \frac{b}{2} & 0 & 0 & \frac{1}{3} \end{pmatrix} \quad 0 \leq b \leq \frac{2}{3} \tag{1.32}$$

$$\begin{pmatrix} \frac{b}{2} & 0 & 0 & \frac{b}{2} \\ 0 & 1-b & 0 & 0 \\ 0 & 0 & 0 & 0 \\ \frac{b}{2} & 0 & 0 & \frac{b}{2} \end{pmatrix} \quad 0 \leq b \leq \frac{2}{3}$$

Using linear entropy as a measure of mixedness and concurrence as a measure of entanglement, one finds that for a given value of mixedness, these states show maximum entanglement.

DYNAMICS OF ENTANGLEMENT

From the discussion on the generation of entangled states in section IV, we note that two separated systems cannot be entangled unless they already share an entangled state. This is due to the fact that entangled states are created only via nonlocal operations . Therefore, in general, the creation of entanglement requires to put two systems in contact with some entangling device. Here, we discuss the opposite question, namely given that a system S is entangled with another system that we call R (reference system or reservoir or whatever you may like to think of it), what can one say about disentangling S from R? In particular, could one perform some quantum operation on system S to disentangle it from R when we don't know the exact nature of system R? It turns out that this is possible, by applying a special class of *quantum channels* known as *entanglement breaking channels.*

A *quantum channel* is a *linear* map $\mathscr{E}: \mathcal{S}(\mathscr{H}) \rightarrow \mathcal{S}(\mathscr{H})$ from the set of density operators $\mathcal{S}(\mathscr{H})$ acting in $\mathscr{H}$, satisfying the following properties:

(i) *Complete positivity (CP):* $\mathscr{E} \otimes \mathscr{I}_R$ on $\mathcal{S}(\mathscr{H} \otimes \mathscr{H}_R)$ is positive for all for all finite dimensional extension over a reference system with identity operator $\mathscr{I}_R$.
(ii) *Trace preserving (TP):* $Tr[\rho]= Tr[\mathscr{E}[\rho]]$, i.e., trace of input ρ equals the trace of the output $\mathscr{E}[\rho]$.

Keeping these in mind, we have the following definition of the entanglement breaking channel.

> **Definition 9.** A channel $\mathscr{E}:\mathcal{S}(\mathscr{H}) \rightarrow \mathcal{S}(\mathscr{H}_R)$ is called *entanglement breaking* if $\rho' = (\mathscr{E} \otimes \mathscr{I}_R)\rho$ is separable for all $\rho \in \mathcal{S}(\mathscr{H} \otimes \mathscr{H}_R)$ and for all reference systems $\mathscr{H}_R$.

Let $\mathscr{E}$ be entanglement breaking channel, then $\rho' = (\mathscr{E} \otimes \mathscr{I})\rho_{ME}$ where $\rho_{ME} = \frac{1}{d}\sum_{j,k}|j\rangle\langle k| \otimes |j\rangle\langle k|$, is the maximally entangled sate, must be separable. where d is the dimension of the space in which $\mathscr{E}$ acts. One can write

$$\rho' = (\mathscr{E} \otimes \mathscr{I})\frac{1}{d}\sum_{j,k}|j\rangle\langle k| \otimes |j\rangle\langle k|$$

$$= \frac{1}{d}\sum_{j,k}\mathscr{E}(|j\rangle\langle k|) \otimes |j\rangle\langle k| \tag{1.33}$$

Since $\mathscr{E}$ is entanglement braking, we have (see Eq. (1.8))

$$\rho' = \sum_n p_n|u_n\rangle\langle u_n| \otimes |v_n\rangle\langle v_n|, \tag{1.34}$$

with $0 \leq p_n \leq 1, \sum_n p_n = 1$. Consider a map $\mathscr{F}$ defined as

$$\mathscr{F}[\rho]= \sum_n(Tr[\rho F_n])\sigma_n, \tag{1.35}$$

with $\sigma_n = |v_n\rangle\langle v_n|, F_n = p_n d|u_n\rangle\langle u_n|$. Taking partial trace of Eq. (1.23), we have

$$Tr_B[(\mathscr{E} \otimes \mathscr{I})\rho_{ME}] = \frac{1}{d}\sum_{j,k}|j\rangle\langle k| \otimes Tr[|j\rangle\langle k|]$$

$$= \frac{\mathcal{I}}{d} = \sum_n p_n |u_n\rangle\langle u_n| \tag{1.36}$$

where the last equality holds because of Eq. (1.24). From this follows $\sum_k F_k = \mathcal{I}$. Now consider the following

$$(\mathcal{F} \otimes \mathcal{I})\rho_{ME} = \sum_{jkn} Tr\left[|j\rangle\langle k| p_n |u_n\rangle\langle u_n|\right] |v_n\rangle$$

$$\langle v_n| \otimes |j\rangle\langle k|$$

$$= \sum_{jkn} Tr\left[p_n \langle k||u_n\rangle\langle u_n||j\rangle\right] |v_n\rangle\langle v_n| \otimes |j\rangle\langle k|$$

$$= \sum_{jkn} p_n |v_n\rangle\langle v_n| \otimes |u_n\rangle\langle u_n| \tag{1.37}$$

But this is same as ρ', the output of channel $\mathcal{E}$, as defined in Eq. (1.24). Thus we conclude that the action of $\mathcal{E}$ and $\mathcal{F}$ is identical and write $\mathcal{E} = \mathcal{F}$. Therefore, if $\mathcal{E}$ is entanglement breaking, then it can be written in the form of Eq. (1.25), also known as *Holevo form*. In fact the converse also holds i.e., any channel of the form of (1.25) is entanglement breaking. The operator $(\mathcal{E} \otimes \mathcal{I})\rho_{ME}$ is known as the Choi–Jamiolkowski operator. A channel is entanglement braking if and only if the Choi–Jamiolkowski operator is separable. Therefore, the separability of the Choi–Jamiolkowski operator corresponding to a given channel, serves as a test for the channel to be entanglement breaking.

APPLICATIONS OF ENTANGLEMENT

Quantum entanglement is proving advantageous in various applications such as cryptography, superdense coding, and teleportation. The potential of quantum entanglement is attracting attention from various disciplines including industries, finance and banking looking to solve time and processing power-consuming problems with quantum computers.

Quantum cryptography. The traditional cryptography involves a sender encoding a message using a *key*, which is shared with the recipient who uses it to decode the message. This process has a risk of some third party learning the information about the key and hence compromising the security of cryptography. Quantum entanglement helps to create a secure channel for unbreakable cryptography between to parties. The central notion in quantum cryptography is the *quantum key distribution* (QKD), i.e., securing the distribution of encryption keys. The no-cloning principle further assures the impossibility of an unknown quantum state. The QKD involves sending information about the key using polarized photons. The recipient uses polarized filters to decipher the key with a chosen algorithm used to encrypt the message. The fact that disturbing the photons changes their state serves as a security check and allows the communicators to know about any breach. In 2004, the first QKD bank transfer was successfully made between Vienna City Hall and Bank Austria Creditanstalt.

Quantum teleportation. Quantum teleportation is a technique for transferring quantum states/information between two parties at different locations sharing an entangled state. This techniques works irrespective of the distance between the parties. Interestingly, the sender has no information of the state he is sending as well as the location of recipient. By sharing quantum states over large distance, scientist are hopping to create networks that can share information at very high speeds. However, due to the interaction with ambient environment, keeping this information transfer stable is an extremely difficult task. In 2020, a team involving NASA's jet propulsion laboratory successfully achieved sustained, high-fidelity (greater than 90%) quantum teleportation. The qubits were teleported across a distance of 44 kilometers (27 miles).

Superdense coding. The well-known *Holevo bound* dictates that *one qubit can carry at most only one bit of classical information* (Horodecki, Horodecki, Horodecki, & Horodecki, 2009). Superdense coding is a procedure of sending two classical bits of information using one *entangled* qubit, in contrast with teleportation which one qubit is transferred using two classical bits (Bennett & Wiesner, 1992). Thus superdense protocol empowers a channel to send, ahead of time, half of what will be needed to reconstruct a classical message. This protocol was experimentally demonstrated by experimentally in 1996 by Mattle, Weinfurter, Kwiat and Zeilinger using entangled photon pairs (Mattle, Weinfurter, Kwiat, & Zeilinger, 1996).

ILLUSTRATIVE EXAMPLES

In this section, we illustrate the concepts discussed so far by various examples (Nielsen & Chuang, 2010), (Braunstein & Loock, 2005).

Example 1. A symmetric matrix Mis defined as

$$\mathrm{M} = \begin{bmatrix} m_{11} & m_{12} & m_{13} & m_{14} \\ m_{12} & m_{22} & m_{23} & m_{24} \\ m_{13} & m_{23} & m_{33} & m_{34} \\ m_{14} & m_{24} & m_{34} & m_{44} \end{bmatrix}$$

where $m_{jk} \in \mathbb{R}$, and the orthonormal Bell basis defined as

$$|\Phi_{\pm}\rangle = \frac{1}{\sqrt{2}}\begin{bmatrix} 1 \\ 0 \\ 0 \\ \pm 1 \end{bmatrix},$$

$$|\Psi_{\pm}\rangle = \frac{1}{\sqrt{2}}\begin{bmatrix} 0 \\ 1 \\ \pm 1 \\ 0 \end{bmatrix}$$

(1.38)

Let $\widetilde{\mathrm{M}}$ denote the matrix Min Bell basis. Extract the conditions on matrix elements m_{jk}, such that $\widetilde{\mathrm{M}}$ is diagonal.

Solution. The matrix elements in Bell basis are given by

$$\widetilde{m}_{11} = \langle \Phi_{+}|\mathrm{M}|\Phi_{+}\rangle = \tfrac{1}{2}(m_{11} + 2m_{14} + m_{44}),$$

$$\widetilde{m}_{12} = \langle \Phi_{+}|\mathrm{M}|\Phi_{-}\rangle = \tfrac{1}{2}(m_{11} - m_{44}),$$

and so on. Putting the off-diagonal elements to zero, one obtains

$$m_{11} = m_{44},\ m_{22} = m_{33},$$

$$m_{12} = m_{13} = m_{24} = m_{34} = 0$$

with m_{14}, m_{23} arbitrary. Thus the matrix Mwill be diagonal in Bell basis if it is of the following form

$$\mathrm{M} = \begin{bmatrix} m_{11} & 0 & 0 & m_{14} \\ 0 & m_{22} & m_{23} & 0 \\ 0 & m_{23} & m_{22} & 0 \\ m_{14} & 0 & 0 & m_{11} \end{bmatrix}.$$

Example 2. Prove that the condition for separability of $|\psi\rangle \in \mathbb{C}^4$, where $|\psi\rangle = (a \quad b \quad c \quad d)^T$is $ad = bc$. Here T denotes the transposition operation.
Solution. Let us calculate the *entropy of entanglement* defined in Eq. (1.14) for the given state. A straightforward calculation shows

$$E(|\psi\rangle\langle\psi|) = S(Tr_A[|\psi\rangle\langle\psi|]) = S(Tr_B[|\psi\rangle\langle\psi|])$$

$$= -e_+ \log_2 e_+ - e_- \log_2 e_-$$

where $e_\pm = \frac{1}{2}[1 \pm \sqrt{1 - 4|ad - bc|^2}]$.
The separability demands that $E(|\psi\rangle\langle\psi|) = 0$, which occurs at $ad = bc$.

Example 3. Let $|0\rangle = (1 \quad 0)^T$, where T denotes the transpose, denote the eigenvector of Pauli matrix σ_3. With $a \in \mathbb{R}$, what can we say about the separability of $\exp[ia(\sigma_2 \otimes \sigma_2)]|0\rangle \otimes |0\rangle$?
Solution. Using Euler matrix identity, one can write

$$\exp[ia(\sigma_2 \otimes \sigma_2)] = \cos(a)\mathcal{I}_4 + i(\sigma_2 \otimes \sigma_2)\sin(a)$$

this lead to

$$\exp[ia(\sigma_2 \otimes \sigma_2)]|0\rangle \otimes |0\rangle$$

$$= \exp[ia(\sigma_2 \otimes \sigma_2)]\begin{pmatrix} 1 \\ 0 \\ 0 \\ 0 \end{pmatrix} = \begin{pmatrix} \cos(a) \\ 0 \\ 0 \\ -i\sin(a) \end{pmatrix}$$

Using the condition derived in previous problem, the state is separable if $\sin(a)\cos(a) = 0$, or $a = n\pi/2$, with $n = 0, 1, 2 \cdots$.

Example 4. Check the concurrence measure of entanglement, given in Eq. (1.16), for the Werner state defined as

$$\rho_W = p|\Phi_+\rangle\langle\Phi_+| + \frac{(1-p)}{4}\mathcal{I}_4$$

Where $0 \le p \le 1$, and $|\Phi_+\rangle$is the Bell state defined in Eq. (1.21).

Solution. In matrix representation, the Werner state is given by

$$\rho_W = \frac{1}{4}\begin{pmatrix} 1+p & 0 & 0 & 2p \\ 0 & 1-p & 0 & 0 \\ 0 & 0 & 1-p & 0 \\ 2p & 0 & 0 & 1+p \end{pmatrix}$$

with eigenvalues $\lambda_1 = (1+3p)/4$, $\lambda_2 = \lambda_3 = \lambda_4 = (1-p)/4$. This gives

$$\lambda_1 - \lambda_2 - \lambda_3 - \lambda_4 = (3p-1)/2,$$

and hence

$$C(\rho_W) = \max[(3p-1)/2, 0].$$

Corresponding to $p = 0, \frac{1}{2}, 1$, we have $C(\rho_W) = 0, \frac{1}{4}, 1$, respectively.

Example 5. Check the entanglement of the eigenvectors of the following Hamiltonian

$$H = a_1\sigma_1 \otimes \sigma_1 + a_2\sigma_2 \otimes \sigma_2 \tag{1.39}$$

Solution. The matrix representation of the given Hamiltonian is

$$H = \begin{pmatrix} 0 & 0 & 0 & a_1 - a_2 \\ 0 & 0 & a_1 + a_2 & 0 \\ 0 & a_1 + a_2 & 0 & 0 \\ a_1 - a_2 & 0 & 0 & 0 \end{pmatrix}.$$

The eigenvectors corresponding to eigenvalues $a_1 - a_2$, $a_2 - a_1$, $a_1 + a_2$, and $-a_1 - a_2$ are $\frac{1}{\sqrt{2}}(1 \quad 0 \quad 0 \quad 1)^T$, $\frac{1}{\sqrt{2}}(1 \quad 0 \quad 0 \quad -1)^T$, $\frac{1}{\sqrt{2}}(0 \quad 1 \quad 1 \quad 0)^T$, and $\frac{1}{\sqrt{2}}(0 \quad 1 \quad -1 \quad 0)^T$, respectively. They are precisely the Bell states introduced Eq. (1.5), and correspond to the maximally entangled states in $\mathbb{C}^4$.

Example 6. The GHZ and W states defined in Eqs. (1.8) and (1.9), respectively, are examples of genuinely (or fully) tripartite entangled states, in the sense that entanglement of these three qubit states is not just present between two parties while the remaining party is separated. Show that the although W state is genuinely entangled, it is also "readily bipartite", in the sense that tracing out one party leads to a bipartite entangled state. Compare this with GHZ state.
Solution. The partial trace of W state leads to the reduced state:

$$Tr_A[|W\rangle\langle W|] = \frac{1}{3}(|00\rangle\langle 00| + |10\rangle\langle 10| + |01\rangle\langle 01| + |01\rangle\langle 10 + |10\rangle\langle 01||)$$

This state is inseparable, as can be seen by using the PPT criterion, see Eq. (1.13). The eigenvalues of the partial transpose of above reduced state are 1/3,1/3, $(1 + \sqrt{5})/6$, and $(1 - \sqrt{5})/6$. The last eigenvalue being negative, confirms the inseparability of the reduced state $Tr_A[|W\rangle\langle W|]$.

The reduced after tracing over one party in GHZ state is given by:

$$Tr_A[|GHZ\rangle\langle GHZ|] = \frac{1}{2}(|00\rangle\langle 00| + |11\rangle\langle 11|)$$

$$= \frac{1}{2}(|0\rangle\langle 0| \otimes |0\rangle\langle 0| + |1\rangle\langle 1| \otimes |1\rangle\langle 1|)$$

which is separable according to our definition of separability of density operators in Eq. (1.6). The above reduced state of GHZ state should not be confused with maximally mixed state $\mathcal{I}^{\otimes 2}/4$. However, tracing out two parties does indeed leave us with maximally mixed state, i.e.,

$$Tr_{AB}[|GHZ\rangle\langle GHZ|] = \frac{1}{2}(|0\rangle\langle 0| + |1\rangle\langle 1|).$$

Further, a local measurement on one party in the basis $|\pm\rangle = \frac{1}{\sqrt{2}}(|0\rangle \pm |1\rangle)$, leads to:

$$\frac{\frac{1}{2}|\pm_A\rangle\langle_A\pm||GHZ\rangle}{\left\|\frac{1}{2}|\pm_A\rangle\langle_A\pm||GHZ\rangle\right\|} = \frac{1}{\sqrt{2}}|\pm_A\rangle \otimes |\Phi_\pm\rangle,$$

where $|\Phi_\pm\rangle$ are the maximally entangled states (Bell states) defined in Eq. (1.5).

Example 7. Consider the state:

$$|\psi\rangle = \cos\eta|0\rangle \otimes |0\rangle + \sin\eta|1\rangle \otimes |1\rangle,$$

where $\eta$is called as *Schmidt angle* with $0 < \eta < \pi/4$.

(i) Show that the eigenvalues of $|\psi\rangle\langle\psi|$are independent of η.
(ii) Find the reduced density matrix corresponding to $|\psi\rangle\langle\psi|$.
(iii) Show that the eigenvalues of the reduced density matrix are $\cos^2\eta$and $\sin^2\eta$.
(iv) How does the von Neumann entropy of the reduced density matrix vary with Schmidt angle η?

Solution. (i) The density matrix corresponding to $|\psi\rangle$turns out to be

$$|\psi\rangle\langle\psi| = \begin{pmatrix} \cos^2\eta & 0 & 0 & \frac{1}{2}\sin 2\eta \\ 0 & 0 & 0 & 0 \\ 0 & 0 & 0 & 0 \\ \frac{1}{2}\sin 2\eta & 0 & 0 & \sin^2\eta \end{pmatrix},$$

with eigenvalues 0,0,0,1. Thus the eigenvalues are independent of the η.

(ii) The reduced density matrix turns out to be

$$Tr_B[|\psi\rangle\langle\psi|] = \begin{pmatrix} \cos^2\eta & 0 \\ 0 & \sin^2\eta \end{pmatrix},$$

with eigenvalues $e_1 = \cos^2\eta$and $e_2 = \sin^2\eta$.

(iii) The von Neumann entropy of the reduced state is given by

$$S_A = S[Tr_B[|\psi\rangle\langle\psi|]] = -e_1 \log_2 e_1 - e_2 \log_2 e_2, \quad (1.40)$$

which is monotonically increasing function of ηin the range $0 < \eta < \pi/4$, as depicted in Figure 4.

Figure 4. Depicting von Neumann entropy as defined in Eq. (1.40)

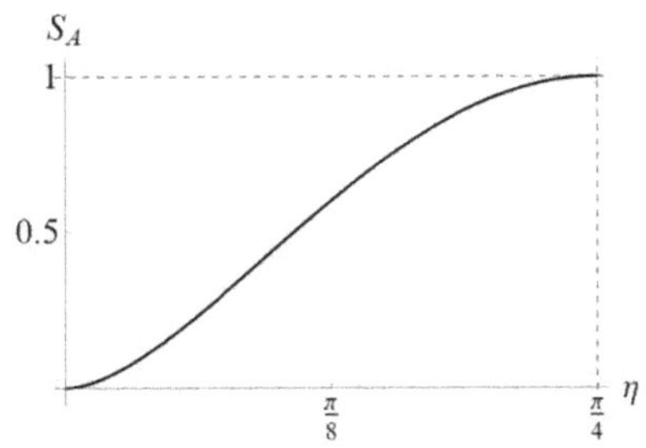

Example 8. The CHSH operator introduced in Eq. (1.11) is rewritten here:

$$B_{CHSH} = A_1 \otimes (B_1 + B_2) + A_2 \otimes (B_1 - B_2),$$

where A_i and B_i are dichotomic operators by which we mean that they have only two eigenvalues given by ± 1. From this information, what can be said about the eigenvalues of B_{CHSH} and $(B_{CHAH})^2$?

Solution. Since A_i and B_i are dichotomic operators, we have $A_i^2 = \mathcal{I}_A$, and $B_i^2 = \mathcal{I}_B$, where $\mathcal{I}_{A(B)}$ is the identity operator associated with space $\mathcal{H}_{A(B)}$. Using this fact, we can write:

$$(B_{CHSH})^2 = 4\mathcal{I}_A \otimes \mathcal{I}_B + (i[A_1, A_2]) \otimes (i[B_1, B_2]),$$

where $[x, y]$ denotes the commutator $xy - yx$. Note that $i[x, y]$ is a Hermitian operator and has real eigenvalues. Thus one can diagonalize the two commutators appearing in above equation by choosing unitary operators U and V as:

$$U(i[A_1,A_2])\,U^* \;=\; diag.[u_1,u_2,\cdots\;,u_{d_A}],$$

$$V(i[A_1,A_2])\,V^* \;=\; diag.[v_1,v_2,\cdots\;,v_{d_B}].$$

Here $u_i, v_i \in \mathbb{R}$ are the eigenvalues. We have:

$$(U\otimes V)(B_{CHSH})^2(U^*\otimes V^*)$$

$$= 4\mathcal{I}_A\otimes\mathcal{I}_B + diag.[u_1,u_2,\cdots\;,u_{d_A}]$$

$$\otimes\, diag.[v_1,v_2,\cdots\;,v_{d_B}]$$

With the form of A_i and B_i given below Eq. (1.11), one can check that the eigenvalues of the commutator $i[A_1,A_2]$ and $i[B_1,B_2]$ are individually bounded by $-2 \le u_j \le 2$ and $-2 \le v_k \le 2$, leading to $-4 \le u_j v_k \le 4$, with $j = 1,2,\cdots\;,d_A$, and $k = 1,2,\cdots\;,d_B$. Therefore, the eigenvalues of $(B_{CHSH})^2$ are given by

$$|w_{jk}| \;=\; \sqrt{4+u_j v_k}.$$

REFERENCES

Banerjee, S. (2019). *Open Quantum Systems: Dynamics of Nonclassical Evolution*. Springer. 10.1007/978-981-13-3182-4

Bengtsson, I., & Życzkowski, K. (2006). *Geometry of Quantum States: An Introduction to Quantum Entanglement*. Cambridge University Press. 10.1017/CBO9780511535048

Bennett, C., & Brassard, G. (1984). Quantum cryptography: Public key distribution and coin tossing. *Proceedings of IEEE International Conference on Computers, Systems, and Signal Processing*, (pp. 175-179). IEEE.

Bennett, C., Brassard, G., Popescu, S., Schumacher, B., Smolin, J., & Wootters, W. (1996). Purification of Noisy Entanglement and Faithful Teleportation via Noisy Channels. *Physical Review Letters*, 76(5), 722–725. 10.1103/PhysRevLett.76.72210061534

Bennett, C., & Wiesner, S. (1992). Communication via one- and two-particle operators on Einstein-Podolsky-Rosen states. *Physical Review Letters*, 69(20), 2881–2884. 10.1103/PhysRevLett.69.288110046665

Bennett, C. H., Brassard, G., Crépeau, C., Jozsa, R., Peres, A., & Wootters, W. K. (1993). Teleporting an unknown quantum state via dual classical and Einstein-Podolsky-Rosen channels. *Physical Review Letters*, 70(13), 1895–1899. 10.1103/PhysRevLett.70.189510053414

Berman, P., Arimondo, E., & Lin, C. (2012). *Advances in Atomic, Molecular, and Optical Physics*. Elsevier. Retrieved from https://www.elsevier.com/books/advances-in-atomic-molecular-and-optical-physics/berman/978-0-12-396482-3

Brandão, F. G., & Vianna, R. O. (2004a). Robust semidefinite programming approach to the separability problem. *Physical Review A*, 70(6), 062309. 10.1103/PhysRevA.70.062309

Brandão, F. G., & Vianna, R. O. (2004b). Separable Multipartite Mixed States: Operational Asymptotically Necessary and Sufficient Conditions. *Physical Review Letters*, 93(22), 220503. 10.1103/PhysRevLett.93.22050315601074

Braunstein, S. L., & Loock, P. (2005). Quantum information with continuous variables. *Reviews of Modern Physics*, 77(2), 513–577. 10.1103/RevModPhys.77.513

Chruściński, D., & Sarbicki, G. (2014). Entanglement witnesses: Construction, analysis and classification. *Journal of Physics. A, Mathematical and Theoretical*, 47(48), 483001. 10.1088/1751-8113/47/48/483001

Dagmar, B., & Gerd, L. (2019). *Quantum Information: From Foundations to Quantum Technology Applications*. John Wiley & Sons.

Doherty, A. C., Parrilo, P. A., & Spedalieri, F. M. (2004). Complete family of separability criteria. *Physical Review A*, 69(2), 022308. 10.1103/PhysRevA.69.022308

Dür , Vidal , Cirac, Linden, & Popescu. (2001). Entanglement Capabilities of Nonlocal Hamiltonians. *Physical Review Letters, 87*, 137901. Retrieved from https://link.aps.org/doi/10.1103/PhysRevLett.87.137901

Einstein, A., Podolsky, B., & Rosen, N. (1937). Can Quantum-Mechanical Description of Physical Reality Be Considered Complete? *Physical Review*, 47(10), 777–780. 10.1103/PhysRev.47.777

Ekert, A. K. (1991). Quantum cryptography based on Bell's theorem. *Physical Review Letters*, 67(6), 661–663. 10.1103/PhysRevLett.67.66110044956

Freedman, S. J., & Clauser, J. F. (1972). Experimental Test of Local Hidden-Variable Theories. *Physical Review Letters*, 28(14), 938–941. 10.1103/PhysRevLett.28.938

Heathcote, A. (1990). Unbounded Operators and the Incompleteness of Quantum Mechanics. *Philosophy of Science*, 57(3), 523–534. 10.1086/289572

Hensen, B., Bernien, H., Dréau, A. E., Reiserer, A., Kalb, N., Blok, M. S., Ruitenberg, J., Vermeulen, R. F. L., Schouten, R. N., Abellán, C., Amaya, W., Pruneri, V., Mitchell, M. W., Markham, M., Twitchen, D. J., Elkouss, D., Wehner, S., Taminiau, T. H., & Hanson, R. (2015). Loophole-free Bell inequality violation using electron spins separated by 1.3 kilometres. *Nature*, 526(7575), 682–686. 10.1038/nature1575926503041

Horodecki, R., Horodecki, P., Horodecki, M., & Horodecki, K. (2009). Quantum entanglement. *Reviews of Modern Physics*, 81(2), 865–942. 10.1103/RevModPhys.81.865

Kitaev, A., & Preskill, J. (2006). Topological Entanglement Entropy. *Physical Review Letters*, 96(11), 110404. 10.1103/PhysRevLett.96.11040416605802

Matsukevich, D. N., Maunz, P., Moehring, D. L., Olmschenk, S., & Monroe, C. (2008). Bell Inequality Violation with Two Remote Atomic Qubits. *Physical Review Letters*, 100(15), 150404. 10.1103/PhysRevLett.100.15040418518088

Mattle, K., Weinfurter, H., Kwiat, P. G., & Zeilinger, A. (1996). Dense Coding in Experimental Quantum Communication. *Physical Review Letters*, 76(25), 4656–4659. 10.1103/PhysRevLett.76.465610061348

Naikoo, J., Dutta, S., & Banerjee, S. (2019). Facets of quantum information under non-Markovian evolution. *Physical Review. A*, 99(4), 042128. 10.1103/PhysRevA.99.042128

Nielsen, M. A., & Chuang, I. L. (2010). *Quantum Computation and Quantum Information*. Cambridge University Press. 10.1017/CBO9780511976667

Orzel, C. (2017). *How Do You Create Quantum?* Forbes. Retrieved from https://www.forbes.com/sites/chadorzel/2017/02/28/how-do-you-create-quantum-entanglement/?sh=308175441732

Osterloh, A., Amico, L., Falci, G., & Fazio, R. (2002). Scaling of entanglement close to a quantum phase transition. *Nature*, 608-610(6881), 608–610. Advance online publication. 10.1038/416608a11948343

Peres, A. (1996). Separability Criterion for Density Matrices. *Physical Review Letters*, 77(8), 1413–1415. 10.1103/PhysRevLett.77.141310063072

Shi, Y. (2004). Quantum entanglement in second-quantized condensed matter systems. *Journal of Physics. A, Mathematical and General*, 37(26), 6807–6822. 10.1088/0305-4470/37/26/014

Terhal, B. M. (2002). Detecting quantum entanglement. *Theoretical Computer Science*, 287(1), 313–335. 10.1016/S0304-3975(02)00139-1

Thapliyal, A. V. (1999). Multipartite pure-state entanglement. *Physical Review A*, 59(5), 3336–3342. 10.1103/PhysRevA.59.3336

Thapliyal, K., Pathak, A., & Banerjee, S. (2017). Quantum cryptography over non-Markovian channels. *Quantum Information Processing*, 16(5), 115. 10.1007/s11128-017-1567-1

Vidal, G., Latorre, J. I., Rico, E., & Kitaev, A. (2003). Entanglement in Quantum Critical Phenomena. *Physical Review Letters*, 90(22), 227902. 10.1103/PhysRevLett.90.22790212857342

Wei, T.-C., Nemoto, K., Goldbart, P. M., Kwiat, P. G., Munro, W. J., & Verstraete, F. (2003). Maximal entanglement versus entropy for mixed quantum states. *Physical Review A*, 67(2), 022110. 10.1103/PhysRevA.67.022110

Chapter 5
The Potential of Quantum Cryptography in Securing Future Communication Channels

Shyam Sihare
https://orcid.org/0000-0003-2096-8273
Dr. A.P.J. Abdul Kalam Government College, India

ABSTRACT

This chapter discusses the significance of quantum cryptography in securing communication channels for the future. It highlights the challenges posed by quantum computing to traditional cryptographic systems and the potential solutions offered by quantum-resistant protocols. The chapter emphasizes the transition from classical to quantum-resistant cryptography, highlighting hybrid cryptosystems, algorithm agility, and standards development. It discusses the vulnerability of classical cryptographic systems to quantum algorithms, such as Shor's and Grover's algorithms. It also explains the concept of hybrid cryptosystems, which combine classical algorithms with post-quantum key exchange protocols.

INTRODUCTION TO QUANTUM CRYPTOGRAPHY

Cryptography, the ancient art and science of securing communication and information, boasts a history stretching back millennia (Kahn, 1968). Early civilizations utilized basic encryption methods like substitution ciphers, including the renowned Caesar cipher. The middle ages saw advancements with polyalphabetic substitution techniques, such as the Vigenère cipher. The Renaissance ushered in further

DOI: 10.4018/978-1-7998-9522-0.ch005

innovations, including Leon Battista Alberti's pioneering cipher disk and Johannes Trithemius' influential work “Polygraphia” (Kahn, 1968).

The 19th century witnessed a surge in cryptographic techniques and the rise of dedicated codebreaking organizations. The Playfair cipher, introduced in 1855, aimed to bolster encryption security (Oppliger, 2021). Auguste Kerckhoff's principle, emphasizing key management, significantly impacted the development of secure telegraph and electronic communication systems (Singh, 2000).

During both world wars, cryptography played a pivotal role in military strategies. The Enigma machine, a German encryption device, was famously cracked by Alan Turing and other codebreakers at Bletchley Park, significantly contributing to Allied victory (Singh, 2000).

The advent of public-key cryptography, pioneered by Whitfield Diffie and Martin Hellman in 1976, revolutionized secure communication in the digital age (Diffie & Hellman, 2022). The RSA algorithm, based on the challenging task of factoring large integers, became a cornerstone of secure online transactions and communications (Diffie & Hellman, 2022; Rivevst et al., 1978). However, the ever-evolving landscape of computing technology has thrown a new challenge at the doorstep of classical cryptography: the emergence of quantum computing (Bernstein, 2009; Van et al., 2014).

Quantum computers leverage the principles of quantum mechanics to perform calculations at speeds that dwarf those of classical computers (Nielsen & Chuang, 2001; Martınez, 2014). This poses a significant threat to existing cryptographic protocols, as powerful quantum algorithms like Shor's algorithm can efficiently factor large numbers, rendering RSA and similar schemes vulnerable (Shor, 1999). Additionally, Grover's algorithm can accelerate the search for solutions within exponentially large datasets, potentially compromising cryptographic hash functions and other widely used primitives (Grover, 1996).

The implications for secure communication are significant. Widespread adoption of quantum computing could render currently employed cryptographic protocols obsolete, jeopardizing the security of online transactions, sensitive communications, and critical infrastructure (Albert et al., 2022; Lo & Chau, 1999; Huang et al., 2022; Easttom, 2019; Sihare, S. R (2023 (b))). Transitioning to quantum-resistant cryptography will necessitate significant research, development, and infrastructure updates to ensure the continued protection of our digital world (Alagic et al., 2019; Sihare, S. R. (2022 (a))).

Quantum cryptography, emerging as a revolutionary paradigm within the cybersecurity landscape, harnesses the principles of quantum mechanics like superposition and entanglement to redefine secure communication (Ekert, 1991; Sihare, S. R. (2022 (b))). These unique properties of quantum particles, their ability to exist in multiple states simultaneously, unlock unprecedented advantages in terms of security and privacy, far exceeding the capabilities of traditional cryptography (Gisin & Thew, 2007; Sihare, S. R. (2022 (c))). At the heart of this paradigm shift lies quantum key distribution (QKD), a groundbreaking approach to establishing secure cryptographic keys between communicating parties (Bennett & Brassard, 2014; Sihare, S. R. (2022 (d))).

Pioneering the field, Charles Bennett and Gilles Brassard's revolutionary BB84 protocol, devised in 1984, served as the cornerstone for demonstrating the feasibility of secure QKD without compromising confidentiality (Bennett & Brassard, 2014). The protocol's elegance lies in its ability to detect eavesdropping attempts through statistical inconsistencies, thereby providing an inherent layer of security (Bennett & Brassard, 2014; Sihare, S. R. (2023 (a))). The crucial role of quantum entanglement, intricately intertwined with the EPR paradox, becomes vividly apparent within many QKD protocols (Horodecki et al., 2009). Entangled particles, often photon pairs, are generated and shared between communicating parties (Alice and Bob), guaranteeing instantaneous correlation of their states regardless of physical separation (Horodecki et al., 2009; Sihare, S., & Khang, A. (2023)). This entanglement-based verification serves as a powerful tool for exposing any interference by a third party, ensuring the tamper-proof nature of the key exchange process (Gisin & Thew, 2007).

While still in its nascent stages, the potential of quantum cryptography to revolutionize secure communication is undeniable (Noor-ul-Ain et al., 2016). With continuous advancements in technology and protocols, its widespread adoption holds the promise of safeguarding our increasingly digital world from the vulnerabilities of classical cryptography in the face of evolving computational threats (Pan et al., 2023).

Transcending theoretical bounds, quantum cryptography presents compelling applications in tackling real-world challenges within the burgeoning landscape of information security (Orieux & Diamanti, 2016). A prominent application lies in building secure communication networks (Pan et al., 2023). These networks seamlessly integrate with existing infrastructures, facilitating secure communication between geographically dispersed entities (Pan et al., 2023). This holds immense significance for the financial sector, where quantum key exchange safeguards the confidentiality and integrity of sensitive financial data (Noor-ul-Ain et al., 2016). Recognizing its immense potential, government and defense sectors have spearheaded the adoption of quantum cryptography to secure critical communications (Nahar et al., 2023).

As the specter of quantum computing looms large, necessitating secure solutions beyond the vulnerabilities of classical cryptography, the development and implementation of quantum-safe cryptographic algorithms gain momentum (Alagic et al., 2019). However, several challenges and hurdles remain (Albert et al., 2022). These include the practicalities of implementing and maintaining stable, high-performance quantum communication systems (Orieux & Diamanti, 2016). Optimizing the speed and efficiency of QKD protocols and the ongoing pursuit of global standardization present further hurdles on the path to widespread adoption (Alagic et al., 2019). It is imperative for organizations still reliant on classical cryptographic systems to proactively transition to quantum-safe alternatives while the window of opportunity remains open (Alagic et al., 2019).

The paper presents a clear and well-structured introduction and literature review, followed by an exploration of quantum key distribution, entanglement-based cryptography, and post-quantum cryptography. The subsequent sections cover quantum-safe cryptographic protocols, the cryptographic implications of quantum computing, and the challenges and opportunities in quantum cryptography. Quantum cryptography in practice examines real-world applications and case studies. The chapter concludes with insights into the future landscape and a final outlook on the subject.

ANALYSIS OF QUANTUM GATES

Quantum gates are fundamental building blocks of quantum circuits, used to manipulate qubits for quantum computations. This chapter has introduced various quantum gates, but a deeper analysis is required to fully understand their characteristics, limitations, and potential for optimization.

1. Single-Qubit Gates

Hadamard Gate (H)

The Hadamard gate, denoted as *H*, is a single-qubit quantum gate that is fundamental to quantum computing algorithms. It creates superpositions and is the quantum equivalent of a classical coin flip. The Hadamard gate transforms the basis states $|0\rangle$ and $|1\rangle$ into equal superpositions.

The action of the Hadamard gate on a qubit can be represented as follows:

$$H|0 = \frac{1}{\sqrt{2}}(|0 + |1)$$

$$H|1 = \frac{1}{\sqrt{2}}(|0 - |1)$$

In matrix form, the Hadamard gate can be expressed as:

$$H = \frac{1}{\sqrt{2}}\begin{bmatrix} 1 & 1 \\ 1 & -1 \end{bmatrix}$$

The truth table for the Hadamard gate is as follows:

Table 1. Truth Table

Input	*Output*
$\lvert 0\rangle$	$\frac{1}{\sqrt{2}}(\lvert 0\rangle + \lvert 1\rangle)$
$\lvert 1\rangle$	$\frac{1}{\sqrt{2}}(\lvert 0\rangle - \lvert 1\rangle)$

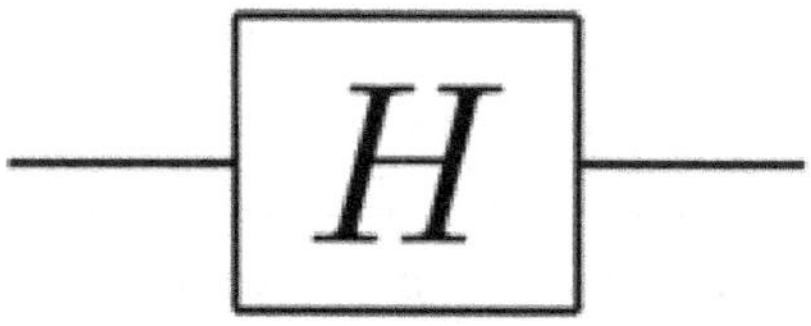

When applied to the $|0\rangle$ state, it creates an equal superposition, and when applied to the $|1\rangle$ state, it creates a superposition with a relative phase. The Hadamard gate is a crucial element in numerous quantum algorithms as it generates quantum interference patterns and enables quantum parallelism.

Pauli Gates (X, Y, Z)

The Pauli-X gate is often referred to as the NOT gate in classical computing. It flips the state of a qubit from $|0\rangle$ to $|1\rangle$ and vice versa.

Table 2. Truth Table

<table>
<tr><th>Expression</th><th>Matrix Representation</th><th>Circuit Diagram</th><th>Truth Table</th></tr>
<tr><td>$X|0\rangle=|1\rangle$
$X|1\rangle=|0\rangle$</td><td>$X = \begin{bmatrix} 0 & 1 \\ 1 & 0 \end{bmatrix}$</td><td>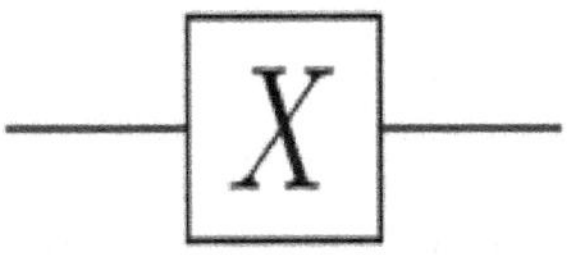</td><td><table><tr><th>Input</th><th>Output</th></tr><tr><td>$|0\rangle$</td><td>$|1\rangle$</td></tr><tr><td>$|1\rangle$</td><td>$|0\rangle$</td></tr></table></td></tr>
</table>

The Pauli-Y gate introduces a phase shift and performs a bit-flip on the qubit.

Table 3. Truth Table

Expression	Matrix·Representation	Circuit·Diagram	Truth·Table
$Y\|0\rangle=i\|1\rangle$ $Y\|1\rangle=-i\|0\rangle$	$X=\begin{bmatrix}0 & -i\\ i & 0\end{bmatrix}$	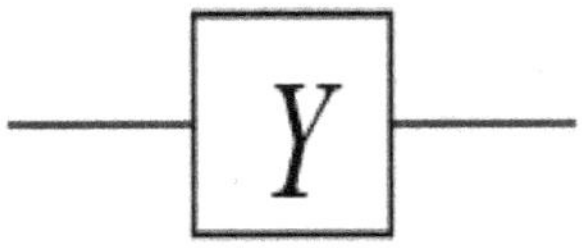	Input: $\|0\rangle$ → Output: $i\|1\rangle$ Input: $\|1\rangle$ → Output: $-i\|0\rangle$

The Pauli-Z gate leaves the $|0\rangle$ state unchanged and flips the sign of the $|1\rangle$ state.

Table 4. Truth Table

Expression	Matrix·Representation	Circuit·Diagram	Truth·Table
$Z\|0\rangle=\|0\rangle$ $Z\|1\rangle=-\|1\rangle$	$X=\begin{bmatrix}1 & 0\\ 0 & -1\end{bmatrix}$	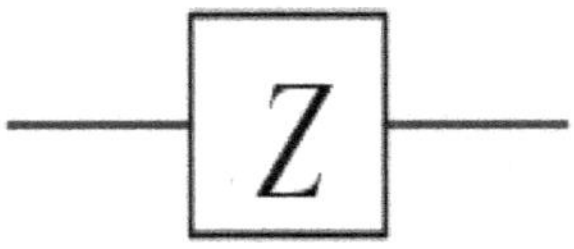	Input: $\|0\rangle$ → Output: $\|0\rangle$ Input: $\|1\rangle$ → Output: $-\|1\rangle$

QUANTUM KEY DISTRIBUTION

QKD is a fundamental aspect of quantum cryptography, providing a secure communication method based on the principles of quantum mechanics. Its protocols, security proofs, and transformative impact are significant.

$$|\psi = \alpha|0 + \beta|1 \qquad (1)$$

Here, the basis states $|0$ and $|1$ represent the classical bits 0 and 1, respectively. The complex probability amplitudes α and β satisfy the normalization condition $|\alpha|^2 + |\beta|^2 = 1$. Superposition allows for the encoding of information in quantum states, forming the basis of QKD protocols.

Let $\mathscr{H}_A \otimes \mathscr{H}_B$ be the Hilbert space of the bipartite system of qubits A and B, where $\mathscr{H}_A$ and $\mathscr{H}_B$ are the individual Hilbert spaces of each qubit (typically $\mathbb{C}^2$). The Pauli operators are defined as follows:

$$\sigma_x = \begin{pmatrix}0 & 1\\ 1 & 0\end{pmatrix} \sigma_y = \begin{pmatrix}0 & -i\\ i & 0\end{pmatrix}$$

$$\sigma_z = \begin{pmatrix} 1 & 0 \\ 0 & -1 \end{pmatrix} \tag{2}$$

Introduce the computational basis states:

$$|0_A, |1_A \in \mathscr{H}_A$$

$$|0_B, |1_B \in \mathscr{H}_B \tag{3}$$

The entangled Bell state is represented as:

$$|\phi^- = \frac{1}{\sqrt{2}}(|0_A \otimes |1_B - |1_A \otimes |0_B) \tag{4}$$

The tensor product is denoted as ⊗, acts on vectors in different Hilbert spaces:

$$|v_A \otimes |w_B = \begin{pmatrix} v_0 \otimes \mathrm{w}_0 \\ v_0 \otimes \mathrm{w}_1 \\ v_1 \otimes \mathrm{w}_0 \\ v_1 \otimes \mathrm{w}_1 \end{pmatrix} \tag{5}$$

The Bell state is maximally entangled, indicated by its to be expressed as a product of individual qubit states:

$$|\phi^- \neq |a_A \otimes |b_B \forall |a \in \mathscr{H}_A, |b \in \mathscr{H}_B \tag{6}$$

Entangled states like the Bell state enable secure key exchange in QKD protocols. Correlations in measurements of entangled qubits, performed in different bases, reveal any eavesdropping attempt by introducing detectable errors in the correlations. Denoted as $\mathscr{H}$, representing the Hilbert space of a single qubit. Therefore, rectilinear basis state defines as:

$$|R_0\rangle = |0\rangle = \begin{bmatrix} 1 \\ 0 \end{bmatrix}^T \in \mathscr{H}$$

$$|R_1\rangle = |1\rangle = \begin{bmatrix} 0 \\ 1 \end{bmatrix}^T \in \mathscr{H} \tag{7}$$

And diagonal basis states:

$$\left|D_0\right\rangle = |+\rangle = \frac{1}{\sqrt{2}}(|0 + |1) = \begin{bmatrix} \frac{1}{\sqrt{2}} \\ \frac{1}{\sqrt{2}} \end{bmatrix}^T \in \mathscr{H} \tag{8}$$

$$\left|D_1\right\rangle = |-\rangle = \frac{1}{\sqrt{2}}(|0 - |1) = \begin{bmatrix} \frac{1}{\sqrt{2}} \\ -\frac{1}{\sqrt{2}} \end{bmatrix}^T \in \mathscr{H} \tag{9}$$

Alice selects basis $b \in \{R,D\}$ for each qubit. The qubit is prepared in the superposition state $|\psi = \alpha|R_0 + \beta|R_1$, where $\alpha,\beta \in \mathbb{C}$ and $|\alpha|^2 + |\beta|^2 = 1$. Alice sends the qubit $|\psi$ to via a quantum channel.

Bob randomly chooses a measurement basis $b' \in \{R,D\}$. The probability of correctly measuring the basis is given by $P(CorrectBasis) = |\psi| b_0'|^2 + |\psi| b_1'|^2$.

The inner product $\psi|\phi$ between states $|\psi$ and $|\phi$ is a complex number representing the projection of one state onto the other. Alice and Bob publicly compare their basis choices. Qubits with different bases are discarded. The remaining qubits with the same basis form the raw key. Error correction and privacy amplification are performed for a secure key.

Two non-orthogonal bases are defined:

Rectilinear Basis (R): $\left\{\left|R_0\right\rangle, \left|R_1\right\rangle\right\}$

Diagonal Basis (D):

$$\left\{\left|D_0\right\rangle, \left|D_1\right\rangle\right\} \tag{10}$$

Consider an entangled state shared by Alice and Bob, denoted as $|\Psi \in \mathscr{H} \otimes \mathscr{H}$. Measurement operators for each basis are defined:

R basis:

$$\{\hat{A}_0, \hat{A}_1\}$$

basis:

$$\{\widehat{D}_0, \widehat{D}_1\} \tag{11}$$

The Bell parameter S is defined as:

$$S = \sum_{i,j} \lambda_{i,j} P(A_i \cap B_j) \tag{12}$$

where $\lambda_{i,j}$ are real coefficients depending on the chosen Bell inequality, and $P(A_i \cap B_j)$ is the joint probability of measuring outcomes A_i and B_j for Alice and Bob, respectively.

$$|\Psi \neq |\Phi \tag{13}$$

distinguishes between entangled state $|\Psi$ and separable state $|\Phi$. Entangled states cannot be expressed as a tensor product of individual qubit states. Alice and Bob choose random bases for each entangled qubit. They measure and publicly communicate their bases. Qubits with different bases are discarded. Classical error correction and privacy amplification are applied to obtain a secure key. The impossibility of perfectly copying an unknown quantum state is mathematically expressed as:

$$|\psi \otimes |C \otimes |C'\rangle = |\psi \otimes |C \otimes |C'\rangle \tag{14}$$

The BB92 security proof ensures the security of the BB84 protocol against specific eavesdropping strategies, relying on the inconsistency between different bases used by legitimate parties.

$|\alpha|\beta + |\beta|\alpha$and $|\alpha'|\beta + |\beta'|\alpha$ where $|\alpha$ and $|\beta$ are basis states, and $|\alpha'\rangle$ and $|\beta'\rangle$ are other possible states. This inconsistency exposes the presence of the eavesdropper. Furthermore, Heisenberg's uncertainty principle in quantum mechanics states that the precision of one property is directly proportional to the precision of another complementary property.

In quantum mechanics, quantum measurement introduces inherent uncertainties due to Heisenberg's uncertainty principle. This principle states that the more precisely one property is measured, the less precisely another complementary property can be known.

$$\Delta x \cdot \Delta p \geq \frac{\hbar}{2} \tag{15}$$

where Δx is the uncertainty in position, Δp is the uncertainty in momentum, and $\hbar$ is the reduced Planck constant ($\approx 1.054 \times 10^{-34} Js$). This inequality highlights the inherent trade-off between measuring position and momentum simultaneously. The six-state protocol employs non-orthogonal states represented as:

$$|\Psi_j = \frac{1}{\sqrt{3}}\left(e^{i\frac{2\pi j}{3}}|0 + e^{i\frac{4\pi j}{3}}|1 + |2\rangle\right) \tag{16}$$

where $j = 0, 1, ..., 5$. The six states form three mutual unbiased bases (MUBs):

$$\left\{|\Psi_0\rangle, |\Psi_1, |\Psi_2\right\}, \{|\Psi_3, |\Psi_4, |\Psi_5\}$$

,

$$\left\{|\Psi_0\rangle + |\Psi_3, |\Psi_1 + |\Psi_4, |\Psi_2 + |\Psi_5\right\} \tag{17}$$

Satisfying the property:

$$|\psi_i^m | \psi_j^n|^2 = \frac{1}{3} \text{ for all } i \neq j, m, n \tag{18}$$

Mutual information $I(A;B)$ quantifies shared information between Alice and Bob's measurements and is calculated as:

$$I(A;B) = H(A) + H(B) - H(AB) \tag{19}$$

where $H(A)$ and $H(B)$ are the Shannon entropies of Alice and Bob's individual measurements, and $H(AB)$ is the conditional entropy of Alice's measurements given Bob's.

The optimization problem involves determining the optimal measurement settings to maximize key rate by maximizing mutual information based on the specific properties of the six-state protocol (see Figure 1).

Quantum key decryption is employed in quantum communication networks to ensure secure communication among multiple parties. Bell inequalities, which are derived from physicist John Bell, quantify the correlations between entangled particles, ensuring the absence of malicious eavesdroppers and establishing shared key security.

QKD protocols provide secure communication in a range of domains, including secure communication networks, government and military communications, financial transactions, healthcare information exchange, satellite communication, critical infrastructure protection, data centres and cloud computing, quantum internet, smart cities, IoT security, and authentication and secure access. QKD is a highly secure method suitable for scenarios where traditional cryptographic methods may be vulnerable to quantum attacks. As quantum technologies continue to advance, the integration of QKD into real-world applications is expected to increase, thereby contributing to a more secure and resilient communication infrastructure.

Figure 1. (a) Bit error rate vs. distance, (b) Secret key rate vs. distance, and (c) Bit error rate vs. secret key rate. These plots showcase the relationship between distance, bit error rate, and secret key rate in realistic QKD settings.

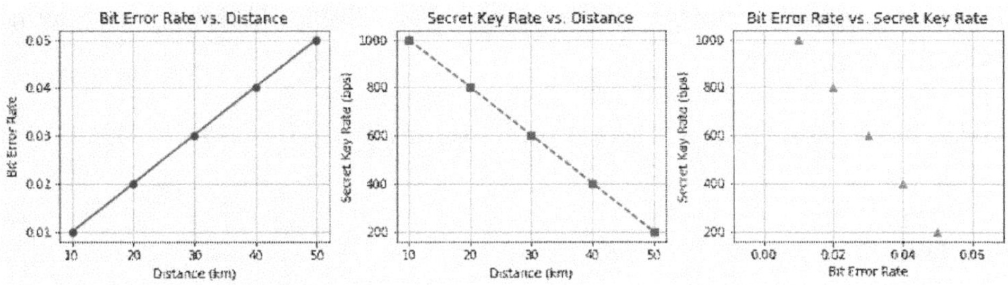

ENTANGLEMENT-BASED CRYPTOGRAPHY

Entanglement is a crucial phenomenon in quantum cryptography and QKD in quantum mechanics. Computational basis is:

$$\left\{ |0\rangle, |1 \right\} \in \mathscr{H} |\phi^{+} = \frac{1}{\sqrt{2}} \left(|0_{A} \otimes |1_{B} + |1_{A} \otimes |0_{B} \right) \tag{20}$$

This represents a maximally entangled state with perfectly correlated measurement outcomes.

$$|\Psi = \alpha|00 + \beta|01 + \gamma|10 + \delta|11 \tag{21}$$

Represents a general entangled state for two qubits, with coefficients satisfying the normalization condition,

$$|\alpha|^2 + |\beta|^2 + |\gamma|^2 + |\delta|^2 = 1 \tag{22}$$

Entangled states represented using the density matrix formalism:

$$\rho = |\Psi\rangle\langle\Psi| \tag{23}$$

Where ρ is a Hermitian, positive semi-definite matrix with trace 1. Measurement operators for the computational basis:

$$\{\hat{J}_0, \hat{J}_1\} \tag{24}$$

Where $\hat{J}_0$ and $\hat{J}_1$ are projectors onto the $|0$ and $|1$ states, respectively. Quantitative measures like Von Neumann entropy, entanglement fidelity, and Bell parameters measure entanglement. Entanglement is crucial in quantum cryptography and QKD protocols, ensuring security in key exchange. For Alice's and Bob's bases, define projection operators:

Alice's basis $\{A_a\}:\{A_0, A_1\}$ where $A_a|\psi\rangle = |a\rangle\langle a||\psi\rangle$

Bob' basis

$\{B_b\}:\{B_0, B_1\}$

where

$$B_b|\psi\rangle = |b\rangle\langle b||\psi\rangle \tag{25}$$

QKD protocols use Bell inequalities for secure key exchange, with Alice preparing a qubit in two bases and Bob measuring it in a randomly chosen base. The probability of joint measurement outcomes (a, b) for Alice and Bob is defined as:

$$P(a,b) = |\psi|A_a \otimes B_b|\psi|^2 \quad (26)$$

This expression utilizes the shared state $|\psi\rangle$ and projection operators for their chosen bases. Entanglement in QKD ensures security by violating Bell inequalities, constraints impossible with classical communication, and revealing correlations unrelated to local hidden variables, deterring eavesdroppers. The GHZ state, a maximally entangled state for three qubits is defined as:

$$|GHZ = \frac{1}{\sqrt{2}}(|000 + |111) \quad (27)$$

This state demonstrates multi-partite entanglement, where qubits are correlated regardless of distance. Quantum coin flipping uses entangled states like EPR pairs for fairness and unbiased outcomes, preventing potential adversaries from security analysis.

$$|EPR = \frac{1}{\sqrt{2}}(|0_A \otimes |1_B - |1_A \otimes |0_B) \quad (28)$$

Alice and Bob share a maximally entangled Bell state, measuring qubits in rectilinear or diagonal bases. The coin flip outcome depends on their measurement correlations, with heads indicating the same outcome (both R or D) and tails indicating different outcomes (R and D).

$$P(Heads) = |EPR|(R_A \otimes R_B \, or \, D_A \otimes D_B)|EPR|^2 \, P$$
$$(Tails) = |EPR|(R_A \otimes D_B \, or \, D_A \otimes R_B)|EPR|^2 \quad (29)$$

Both probabilities sum to 1, ensuring a fair and unbiased outcome. Entanglement prevents adversaries from manipulating randomness, while eavesdropping attempts disrupt the state and introduce detectable errors, ensuring communication security. Fidelity measures the stability of entangled states by comparing the actual state to the ideal state.

$$F(\rho, |\psi\rangle) = \psi|\rho|\psi \quad (30)$$

Here, F is the fidelity, ρ is the actual state, and $|\psi$ is the ideal entangled state. Quantum repeaters enhance secure quantum communication by creating entanglement over shorter segments and using entanglement swapping to connect them, ensuring scalability and distributing entangled pairs.

$$|\psi = |\phi_1 \otimes |\phi_2 \otimes \ldots \otimes |\phi_n \quad (31)$$

Quantum-safe algorithms, such as entanglement and Bell inequalities, protect quantum information from errors, enhancing secure communication, secret sharing, and coin flipping. This is particularly important due to the threat posed by quantum computers.

The system consists of n qubits, represented by the overall Hilbert space $\mathscr{H}_n$. The code is defined by k independent stabilizer operators $S_1, S_2, ..., S_k$, all acting on the n-qubit system. The codewords are simultaneous +1 eigenstates of all stabilizers. Mathematically, a codeword $|\psi$ satisfies:

$$S_i|\psi = |\psi$$

for all

$$i = 1, 2, ..., k \tag{32}$$

Each stabilizer S_i is expressed as:

$$S_i = \alpha_i I + \beta_i P_i \tag{33}$$

Where, I is the identity operator, P_i is a Pauli operator (X, Y, or Z) acting on specific qubits within the system and α_i and β_i are complex numbers. The code encodes k logical qubits into n physical qubits, with a minimum distance of d and maximum error correction capability of t. It meets the Knill-Laflamme condition if $d \geq 2t + 1$, ensuring a minimum number of independent one-qubit errors needed to transform one codeword into another (see Figure 2).

Alice and Bob have a shared entanglement pair $|\phi^+$. Alice performs a joint Bell measurement on the qubit $|\psi$ she wishes to teleport and her own qubit A. The Bell measurement involves the projection operators. In the Bell base measurement, the projection operators are involved.

$$|\phi^+\phi^+| = |0000| + |1111|$$

$$|\phi^-\phi^-| = |0101| + |1010| \tag{34}$$

$$|\psi^+\psi^+| = \frac{1}{\sqrt{2}}(|0001| + |0100| + |1110| + |1011|)$$

$$|\psi^-\psi^-| = \frac{1}{\sqrt{2}}(|0010| + |0111| + |1100| + |1001|) \tag{35}$$

Quantum cryptography is informed by information theory, which quantifies information using principles such as entropy and mutual information to analyze security and efficiency.

$$H(X) = -\sum_i P(x_i)\log_2 P(x_i) \tag{36}$$

X is a discrete random variable, where $P(x_i)$ is the probability of the i-th outcome, summed over all possible outcomes. Density matrix of entangled state is defined as:

$$\rho_{AB} = |\phi^+\rangle\langle\phi^+| = \frac{1}{2}|00\rangle\langle 00| + \frac{1}{2}|11\rangle\langle 11| + \frac{1}{2}|01\rangle\langle 10| + \frac{1}{2}|10\rangle\langle 01| \tag{37}$$

This represents the mixed state of the qubits after sharing the $|\phi^+$ entanglement. Hence, reduced density matrix of Alice's qubit is defined as:

$$\rho_A = tr_B(\rho_{AB}) = \frac{1}{2}|00| + \frac{1}{2}|11| \tag{38}$$

Demonstrates Alice's qubit in a completely mixed state, indicating the non-local nature of entanglement. Bell state projector probabilities is:

$$P(\phi^+) = P(\phi^-) = \frac{1}{2}\ P(\psi^+) = P(\psi^-) = \frac{1}{2} \tag{39}$$

Equal likelihood of obtaining any Bell state outcome in a perfect Bell measurement.

$$I(A{:}B) = H(A) - H(AB) = H(B) - H(BA) \tag{40}$$

Measures shared information between Alice and Bob's qubits, maximal for a maximally entangled state like $|\phi^+$. Density operator evolution under unitary transformations is defined as:

$$\rho_A^{'} = U_A \rho_A U_A^{\dagger} \tag{41}$$

Describes how Alice's qubit state is transformed by a unitary operation U_A. Therefore, classical channel capacity is defined as:

$$C = \max_{p(x)} I(X;Y) \tag{42}$$

Represents the maximum achievable transmission rate through a noisy channel, where X is the input, Y is the output, and $I(X; Y)$ is mutual information. Similarly, quantum channel capacity is:

$$C = \max_{\rho_{in}} I(A;B) \tag{43}$$

Defines the maximum quantum information rate through a quantum channel, where A and B represent input and output qubits, and ρ_{in} is the input state.

$$F(\rho,\sigma) = tr\left(\sqrt{\sqrt{\rho}\,\sigma\sqrt{\rho}\,\sigma}\right) \tag{44}$$

Measures the similarity between two quantum states, where ρ and σ are density operators.

$$E_i{:}\mathscr{H}_a \rightarrow \mathscr{H}_b \tag{45}$$

Describe the error process transforming an input state in Hilbert space $\mathscr{H}_a$ to an output state in $\mathscr{H}_b$.

$$|\psi_E = \sum_i \alpha_i |\psi_i \tag{46}$$

Where $|\psi_i$ are encoded basis states and α_i are coefficients.

$$D{:}\mathscr{H}_b \rightarrow \mathscr{H}_a \tag{47}$$

Attempts to recover the original state from the error-affected output. A set of encodings, error detection operators, and decoder maps that can correct or detect errors with certain probabilities.

$$H_{Eve} \leq S(\rho_E) - S(\rho_{AB}) \tag{48}$$

Limits the eavesdropper's extractable information, where ρ_E is the eavesdropper's state, ρ_{AB} is the joint state of Alice and Bob, and S denotes von Neumann entropy. The study employs classical randomness to minimize eavesdropper information, ensuring secure key generation, and analyzes complex QKD protocols by creating individual security proofs of sub-protocols.

$$S_E(\rho_{AB}) = -tr(\rho_A \log_2 \rho_A) \tag{49}$$

Measures the entanglement between subsystems, where ρ_A is the reduced density matrix of subsystem A. Hence, CHSH inequality is defined as:

$$|E(a,b) + E(a',b') - E(a,b') - E(a',b)| \leq 2 \quad (50)$$

Where $E(a,b)$ is the expectation value of measurement outcome a by Alice and b by Bob in specific bases. Violated by maximally entangled states like the EPR pair.

$$S(\rho_{AB}) + S(\rho_{A'}) + S(\rho_{B'}) \leq 2\log_2(d) \quad (51)$$

Restricts achievable correlations, where d is the dimension of the Hilbert space, and ρ_{AB}, $\rho_{A''}$, and $\rho_{B'}$ are density operators. Quantifying security with mutual information by chain rule is defined as:

$$I(A{:}B{:}E) = I(A{:}B) + I(A{:}EB) \quad (52)$$

Represents mutual information between three parties, aiding security analysis against joint attacks by Eve. Moreover, quantum Rényi entropy is written as:

$$Sn(\rho) = \frac{1}{1-n}\log_2 tr(\rho^n) \quad (53)$$

Generalizes von Neumann entropy, providing more information about the state's structure. Game Theory with Quantum Strategies uses matrix representations, non-commutative strategies, Nash equilibrium, and Grover's search algorithm. Entanglement, fairness, and Bell inequality violations ensure fairness in coin flipping games.

$$D_{q,d,s}(a) = Pr\left[a \overset{\$}{\leftarrow} \mathbb{Z}_q^n : a^T \cdot s = b + e mod q\right] \quad (54)$$

Where q is a large prime, d is the dimension of the vectors, s is the secret vector, b is a noise term, and e is sampled from a discrete error distribution. Given samples (a_i, c_i) from $D_{q,d,s}$, distinguish them from uniformly random samples in $\mathbb{Z}_q^n \times \mathbb{Z}_q$. Given samples (a_i, c_i) from $D_{q,d,s}$, find the secret vector s. Given a lattice L and a target length t, find a non-zero vector in L with a norm less than t (see Figure 3).

Quantum cryptography, based on entanglement, has revolutionized secure communication by offering unconditional security against technological advancements. Despite challenges like photon loss and noise, quantum technologies are improving scalability and developing quantum networks. Future directions include bridging the gap between classical and quantum cryptographic techniques, exploring other cryptographic primitives, and integrating quantum cryptography into real-world applications like financial transactions, healthcare, and government communications. However, practical considerations like experimental imperfections and specialized hardware pose challenges. The future of cryptography lies at the intersection of classical and quantum methods.

Figure 2. (a) The real part of different Bell states as distance increases. Each state is represented by a different colour, highlighting its changes over the given distance range. (b) The entropy function in classical systems is related to the probability of outcomes. (c) Quantum entropy is measured against the density operator. (d) The mutual quantum information between two quantum systems as a function of distance, indicating the level of correlation and information exchange. (e) The Bell inequality equation as a function of distance. Bell inequalities quantify the correlations between entangled particles. (f) The probability that Alice wins in a quantum coin flip game.

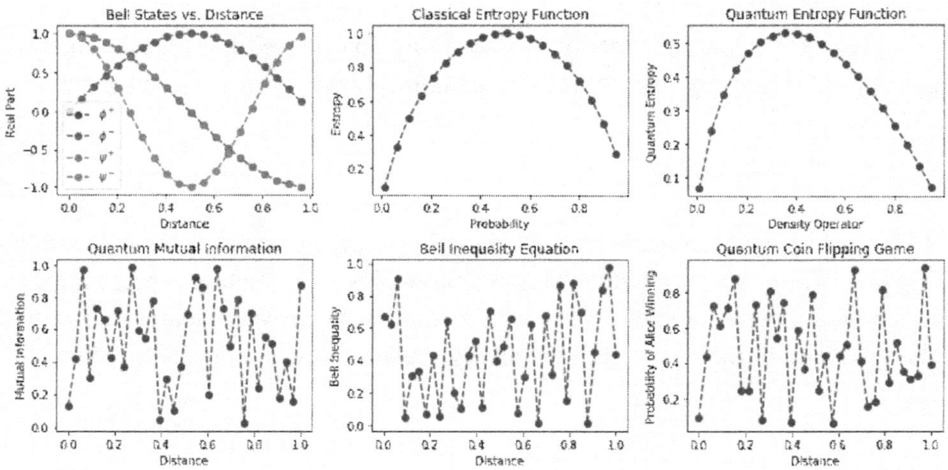

Figure 3. (a) Sinusoidal relationship between non-locality and shared randomness through Bell inequality violation. (b) Security degrades as entanglement distance increases, as quantified by QBER. (c) Efficiency of key generation, measured in secret key rate (bps). (d) Correlation between entanglement distance and quantum state fidelity, which represents the success rate of quantum communication.

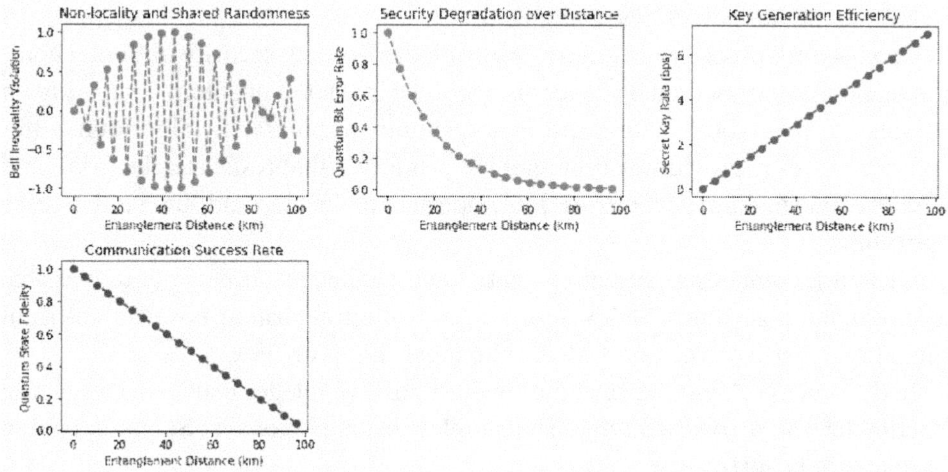

POST-QUANTUM CRYPTOGRAPHY

The emergence of quantum computing has had a significant impact on the security of classical cryptographic algorithms, which have been the foundation of secure communication in the classical computing era. These cryptographic systems rely on the computational complexity of certain mathematical problems, and are now vulnerable to being compromised by the potential capabilities of quantum computers. Shor's algorithm, a quantum algorithm developed by mathematician Peter Shor in 1994, poses a significant threat to classical cryptographic systems, particularly those based on the difficulty of integer factorization.

Another class of cryptographic algorithms widely used in secure communication is elliptic curve cryptography (ECC), which provides strong security with shorter key lengths compared to traditional public-key cryptosystems. However, quantum computers can perform a quadratic speedup in searching an unstructured database, which can halve the effective key length of symmetric-key algorithms in the presence of a quantum adversary. Therefore, it is necessary to carefully examine cryptographic infrastructures.

The cryptographic community is actively developing and standardizing post-quantum cryptographic algorithms to withstand potential threats posed by quantum computers. The following cryptographic algorithms are included: lattice-based cryptography, code-based cryptography, hash-based cryptography, and multivariate polynomial cryptography. The transition to PQC is a complex process due to various factors, such as interoperability and standardization, key management, quantum-safe cryptographic infrastructures, and QKD.

The need for proactive measures to ensure the ongoing security of digital communication is underscored by concerns about the impact of quantum computing on classical cryptographic algorithms. Post-quantum cryptographic algorithms offer a path forward by providing cryptographic primitives that resist attacks from both classical and quantum adversaries. It is important to develop and standardize these algorithms.

Quantum computing presents potential and challenges, especially to classical cryptographic algorithms. Shor's algorithm, a quantum algorithm, poses a significant threat to integer factorization and discrete logarithm problems.

In the order problem, we seek the smallest positive integer r such that a specific function $f(x) = a^x modNequals1$. This problem is foundational in Shor's algorithm for quantum factorization.

$$f(x) = a^x modN \tag{55}$$

The quantum Fourier transform (QFT) is applied to superposed states and maps a quantum state $|x$ to a superposition of states.

$$QFT|x = \frac{1}{\sqrt{N}} \sum_i e^{\left(\frac{2\pi i \cdot x \cdot i}{N}\right)} |i \tag{56}$$

Quantum order-finding subroutine employs QFT and its inverse to determine the order of a with high probability, especially when r is large and classical factoring of N is challenging.

$$gcd(r, N) = 1, then\, a^{\left(\frac{r}{2}\right)} modNhasorder2 \tag{57}$$

$gcd(a^{\left(\frac{r}{2}\right)} - 1, N)$ is non-trivial, revealing a factor of $N$with high probability. Modular exponentiation represents $a^x modN$ as a product of prime powers:

$$a^x modN = \prod_j (a^{p_j^{e_j}} modN) \tag{58}$$

When $gcd(r,N) = 1$, $a^{(\frac{r}{2})} modN$ exhibits an order of 2. If $gcd(a^{(\frac{r}{2})} - 1, N)$ is non-trivial, it reveals a factor of N with high probability. The general number field sieve (GNFS) is a prominent classical factoring algorithm, with complexity $O(e^{(c\,(logN)^{\frac{1}{3}})})$. Shor's algorithm demonstrates a significantly faster complexity for large N:

$$O(\log^2 NloglogNlogloglogN) \tag{59}$$

In RSA, a public key (N, e) is generated by the user, where N the product of two is large primes, and e is a public exponent. The private key (d) is derived from the factors of N using integer factorization, ensuring that e and d are inverses modulo $\phi(N)$. RSA key generation and security for public key is:

$$N = p \times q(largeprimes), e(publicexponent, coprimeto(p-1)(q-1)) \tag{60}$$

And for private key:

$$d(modularinverseofemodulo(p-1)(q-1)) \tag{61}$$

The encryption function involves raising the message m to the power of the public exponent e modulo N.

$$c = m^e modN \tag{62}$$

Decryption is achieved by raising the ciphertext c to the power of the private key d modulo N.

$$m = c^d modN \tag{63}$$

The security of RSA relies on the difficulty of factoring N into its prime components p and q (RSA problem). DSA key generation and security for public key is:

$(p, q, g, y), wherepisaprimenumber, qisalargeprimedividingp - 1,\ gisageneratorof$-$subgroupoforderqin\ \mathbb{Z}_p^*$,

$$andy = g^x modp(xisprivatekey) \tag{64}$$

A digital signature is created using a random value r and involves modular arithmetic with a hash function H.

$$(r,s), where r is random in \mathbb{Z}_q^*$$

$$and s = (k^{-1})(H(m) + xr)$$

$$mod q\left(k is another random in \mathbb{Z}_q^*, H is a hash function\right) \tag{65}$$

Verification checks if a certain equation involving g, y, r, s, and m holds true.

$$g^{ks^{-1}} \equiv y^r \times m^q mod p \tag{66}$$

The security of DSA relies on the difficulty of finding the private key x given the public key parameters (p,q,g,y) (Discrete Logarithm Problem). Diffie-Hellman key exchange and security for public key is:

$$(p,g), where p is a prime, g is a generator of subgroup of order p - 1 in \mathbb{Z}_p^* \tag{67}$$

And for private keys:

$$a and b (chosen randomly by Alice and Bob) \tag{68}$$

The shared secret K is computed independently by both parties using their private keys and public values.

$$K = g^{ab} mod p \tag{69}$$

The security of Diffie-Hellman key exchange relies on the difficulty of computing either a or b given certain public values (Discrete Logarithm Problem). Shor's algorithm efficiently finds p and q, breaking RSA encryption and signatures. Shor's algorithm efficiently finds x in DSA and a/b in DH, compromising key exchange security. Development of new cryptosystems resistant to quantum attacks, such as lattice-based, code-based, and multivariate cryptography.

The Discrete Logarithm problem involves finding x in $g^x \equiv h mod p$, foundational for Diffie-Hellman (DH) key exchange.

$$g^x \equiv h mod p \tag{70}$$

In DH key exchange, Alice and Bob select private keys a and b, compute public values $g^a mod p$ and $g^b mod p$, with shared secret key $K = g^{ab} mod p$.

$$A = g^{a} modp, B = g^{b} modp, K = g^{ab} modp \quad (71)$$

The computational difficulty of solving Discrete Logarithm for x, especially with large prime p, secures DH against unauthorized access. Shor's Algorithm efficiently finds the order r of $hmodg$ with a subroutine crucial for DL solution.

$$ristheorderofhmod \quad (72)$$

Shor's quantum algorithm efficiently solves DL by computing order r and using modular arithmetic, threatening DH on quantum computers.

$$g, h, pgiven, recoverxusingorderr \quad (73)$$

Symmetric key encryption with block cipher like AES encrypts plaintext M with secret key K.

$$C = E_K(M) \quad (74)$$

In brute-force attack, all possible keys are tried until correct ciphertext C is found, revealing plaintext M. Try all possible keys until $C = E_K(M)$.

Grover's algorithm, a quantum algorithm, provides quadratic speedup over classical brute-force search, threatening security for smaller key sizes. Grover's Algorithm Complexity is $O(N/2)$ for small N. Grover's algorithm poses a threat to symmetric key encryption with smaller key sizes (e.g., 64-bit keys). Threat to $O(N)$ classical search for smaller key sizes (see Figure 4).

Figure 4. (a) Growth of Shor's algorithm efficiency concerning the size of RSA keys. (b) Increasing size of RSA keys required to maintain security against quantum threats. (c) Compares classical and quantum factorization times. (c) Impact of Shor's algorithm on RSA security levels. (e) Compares the key sizes of RSA with quantum-resistant alternatives, Dilithium and Kyber. (f) Relationship between RSA key size and the cost of breaking it using quantum computers.

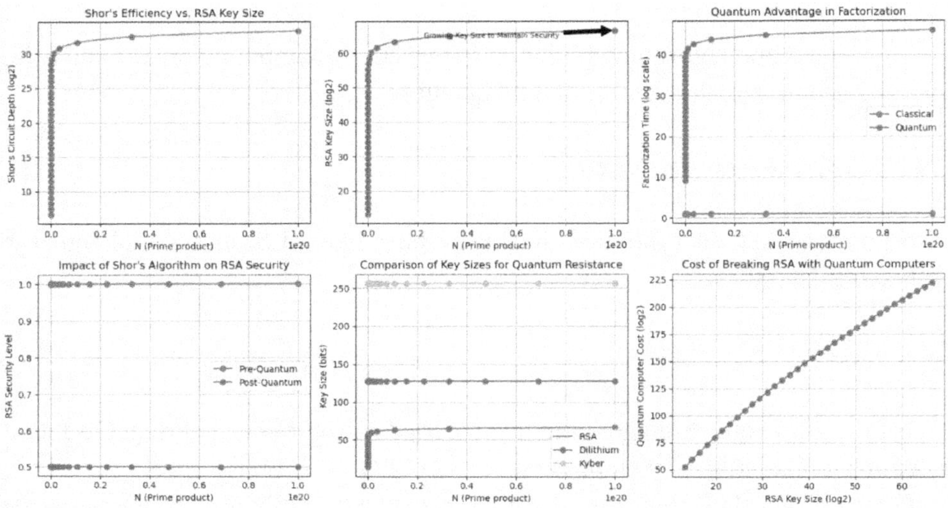

Classical brute-force search attempts all possible key combinations (2^n) for a key length of n. However, Grover's algorithm achieves a quadratic speedup by utilizing superposition and oracle queries, requiring only $\sqrt{2^n} = 2^{n/2}$ queries. This results in an effective key length reduction by $n/2$ bits, which impacts the security strength of symmetric key encryption schemes.

While classical search has exponential complexity $O(2^n)$ with key length, Grover's algorithm reduces it to $O(2^{n/2})$. This reduction means that 128-bit keys may be vulnerable to quantum computers. The probability of success for Grover's algorithm is $1 - 2^{-2r^2}$. The LWE distribution, denoted as $D_q, d, s(a)$, involves the probability,

$$Pr\left[a \leftarrow \mathbb{Z}_q^n : a^T \cdot s = b + e \bmod q\right]$$

, (75)

where q is a large prime, d is the dimension of vectors, s is the secret vector, b is noise, and e is sampled from a discrete error distribution. Decisional LWE (DLWE) distinguishes samples from D_q, d, s, while search LWE (SLWE) aims to find the secret vector s. LWE-based cryptosystems demonstrate their strength by reducing security risks associated with lattice problems. These cryptographic methods rely on error-correcting codes and utilize the one-wayness and collision resistance properties of hash functions to exploit lattice problems. Lattice L is defined as $\{\lambda_v \mid v \in \mathbb{Z}_n\}$ for a vector $\lambda \in \mathbb{R}^n$. Basis B is a set of linearly independent vectors $B = \{b_1, \ldots, b_n\} \in \mathbb{R}^n$ that spans L. The norm v for integer vectors v is given by the maximum absolute value of its components is $v = \max_i |v_i|$. The distance $dist(u, v)$ between lattice points u and v is calculated as $u - v$. The LWE distribution $D_q, d, s(a)$ is defined as the probability $Pr\left[a \leftarrow \mathbb{Z}_q^n : a^T \cdot s\right.$

$\left. = b + emodq\right]$, where q is a large prime, d is the dimension of the vectors, s is the secret vector, b is a noise term, and e is sampled from a discrete error distribution.

Decisional LWE (DLWE) involves distinguishing given samples (a_i, c_i) from D_q $, d, s$ against uniformly random samples in$\mathbb{Z}_q^n \times \mathbb{Z}_q$. Search LWE (SLWE) aims to find the secret vector s given samples (a_i, c_i) from D_q, d, s (see **Figure 5**).

Figure 5. (a) Effect of Grover's algorithm on the key search complexity. As the key length increases, the time complexity for key search is compared between Grover's quantum algorithm and classical methods. (b) Relationship between the LWE problem difficulty, represented on a logarithmic scale, and the lattice dimension. (c) Ciphertext overhead of lattice-based cryptography. (d) Effect of noise variance on the decryption success rate. (e) Key exchange latency in milliseconds as a function of the lattice dimension. (f) Compares the performance of different PQC approaches.

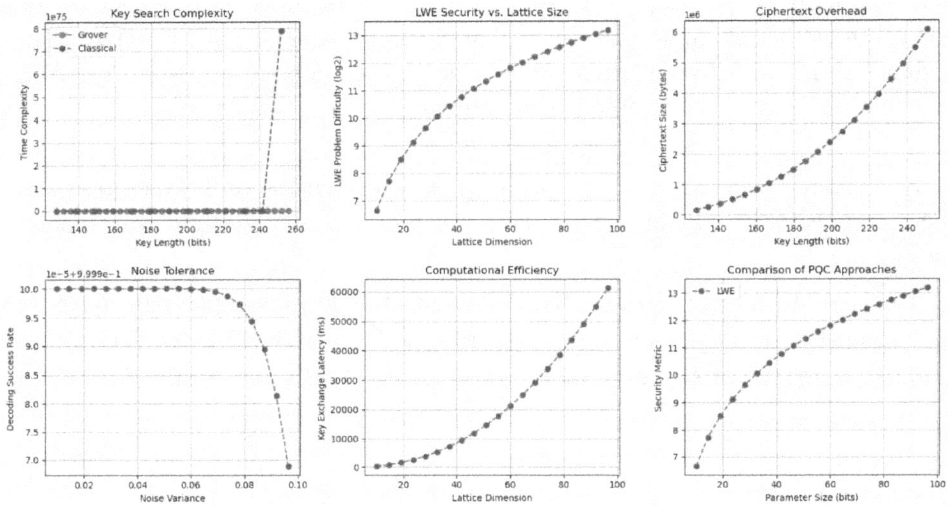

The McEliece cryptosystem encrypts by multiplying plaintext by a generator matrix and adding an error vector; decryption uses error-correcting codes. In classic McEliece, $c = mg + e(mod 2^t)$, where c is the ciphertext, m is the plaintext, g is the generator matrix, e is the error vector, and t is the code length. Ring LWE operates on polynomials instead of vectors, and its equation $c = as + e(modq)$ includes ciphertext c, public polynomial a, secret polynomial s, error polynomial e, and modulus q. Regev's LWE is a specific LWE scheme using binary vectors and specific noise distributions; it is represented as $c = as + e(mod2)$, including ciphertext c, public vector a, secret vector s, error vector e, and modulo operation 2. Collision resistant hash functions aim to prevent finding different x_1 and x_2 with the same hash value. Merkle-Damgård construction uses hash value recursion, PQC adapts key lengths to resist Grover's algorithm, and standardization efforts aim to select and standardize quantum-safe algorithms.

Integrating PQC into existing systems requires careful planning, diverse environments, and security maintenance. Hybrid cryptosystems offer a solution by integrating classical and post-quantum algorithms.

In hybrid encryption, the classical encryption scheme $E_{classical}\left(m, k_{classical}\right) = c_{classical}$ operates on plaintext m using the classical key $k_{classical}$ to produce ciphertext $c_{classical}$. Simultaneously, the post-quantum encryption scheme $E_{post-quantum}\left(m, k_{post-quantum}\right) = c_{post-quantum}$ encrypts m with the post-quantum key $k_{post-quantum}$ to yield ciphertext $c_{post-quantum}$. The hybrid encryption scheme $E_{hybrid}\left(m, k_{classical}, k_{post-quantum}\right) = E_{classical}\left(E_{post-quantum}\left(m, k_{post-quantum}\right), k_{classical}\right)$ combines both, ensuring security dependent on chosen parameters and implementation details.

In BB84 QKD, qubit states $|0\rangle$, $|1\rangle$, and $|\psi^{+}\rangle/\psi^{-}\rangle$ are used with Alice randomly selecting rectilinear or diagonal bases for qubit transmission. Bob, in turn, randomly measures received qubits, leading to error correction and key agreement based on public comparison, discarding mismatched bases, and using remaining qubits for reconciliation.

For ring-LWE signature scheme, public parameters like modulus q, polynomial ring R_q, and error distribution χ are employed. The signing key involves a secret polynomial s, while the verification key comprises a public polynomial $h = as\text{-}modq$. Signing generates a polynomial t from χ, computes a message representative u, and creates a signature $\sigma = (u, v)$, where $v = t + usmodq$. Verification checks if $h(u) = vmodq$.

Additional advanced equations include the security dependency of E_{hybrid} on both classical and post-quantum encryption, as well as the reliance of BB84 error correction on classical error-correcting codes like Cascade. The security of Ring-LWE is rooted in the hardness of the polynomial-based Short Integer Solution Problem (SIS).

In post-quantum signature schemes, public key generation involves sampling a from a secure parameter space using a probability distribution D (e.g., lattices, multivariate quadratic equations). Signature generation, specific to the chosen scheme (e.g., Ring-LWE), is denoted as $c = Sign(m, y, s)$, where y is a random element from the same space as a, s is the secret key, and c is the signature.

Verification, represented as $Verify(c, a, m)$, outputs True if c is a valid signature for message m with public key a, otherwise False. Verification checks specific mathematical properties based on the scheme (e.g., Ring-LWE verifies $h(u) = vmodq$ for public key h and signature (u, v)).

Encoding maps the message to a binary vector of appropriate length. Goppa code construction defines a Goppa code with a generator matrix G and error locator polynomial g. Hashing involves $c = m \cdot G$, multiplying the message with the generator matrix. Permutation and syndrome application are denoted as $h = P \cdot c$, applying a public permutation P and extracting the syndrome. The output $H(m) = Decode(h, g)$ decodes the syndrome using knowledge of g to obtain the hash value (see Figure 6).

Figure 6. (a) Relationship between entanglement fidelity and key generation rate, emphasising the impact of entanglement quality on secure key generation. (b) Relationship between channel noise and QBER, where QBER is simulated based on channel noise and entanglement fidelity. (c) Relationship between key length and security level as simulated based on key length and entanglement fidelity. (d) Correlation between distribution distance and QBER increase. The simulation shows that the QBER increases with the distribution distance, indicating the limitations of quantum communication over longer distances. (e) Correlation between the efficiency of decoy states and the detection rate of eavesdropping. The detection rate is simulated based on the efficiency of decoy states, demonstrating their role in enhancing the security of quantum communication. (f) Compares different methods of entanglement distribution, focusing on performance metrics.

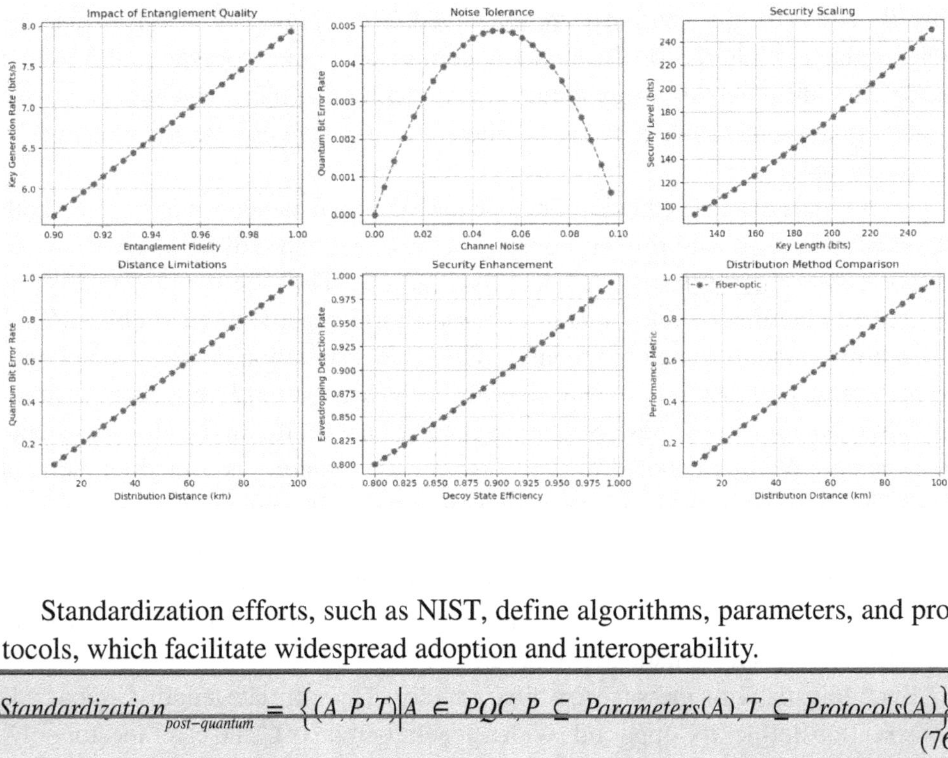

Standardization efforts, such as NIST, define algorithms, parameters, and protocols, which facilitate widespread adoption and interoperability.

$$Standardization_{post-quantum} = \left\{(A,P,T) \middle| A \in PQC, P \subseteq Parameters(A), T \subseteq Protocols(A)\right\} \quad (76)$$

$Standardization_{post-quantum}$ encompasses the collection of standardized PQC schemes. The symbol A denotes an individual post-quantum cryptographic algorithm. The set PQC represents the entirety of candidate post-quantum cryptographic algorithms. The expression $P \subseteq Parameters(A)$ signifies that P is a subset of the parameters specific to algorithm A, implying that P comprises valid parameter choices for A. Similarly, $T \subseteq Protocols(A)$ conveys that T is a subset of the protocols compatible with algorithm A.

$$Certify(M,S,R) = \{True, False\}$$

$$Test(M,S) \wedge Verify(M,R) \rightarrow Certify(M,S,R) \quad (77)$$

The cryptographic module (M) is evaluated based on compliance with applicable standards such as FIPS 140-2 and ISO/IEC 19790. The evaluation considers the fulfilment of security requirements (R), including aspects such as confidentiality, integrity and authenticity. The test process ($Test(M,S)$) verifies that the module implementation conforms to the specified standards, ensuring conformance. In addition, the verification process ($Verify(M,R)$) rigorously assesses whether the cryptographic module meets the defined security requirements, using comprehensive testing and analysis methodologies. Transition strategies incorporate PQC, which employs classical and quantum-resistant cryptographic principles. Let $C_{classical}$ represent the ciphertext generated by a classical encryption scheme:

$$C_{classical} = E_{classical}\left(M, K_{classical}\right) \quad (78)$$

Here, M is the plaintext, and $K_{classical}$ is the classical encryption key. Similarly, let $C_{post-quantum}$ represent the ciphertext generated by a post-quantum encryption scheme:

$$C_{post-quantum} = E_{post-quantum}\left(M, K_{post-quantum}\right) \quad (79)$$

The hybrid encryption process involves combining both ciphertexts:

$$C_{hybrid} = E_{classical}\left(C_{post-quantum}, K_{classical}\right) \quad (80)$$

Hybrid cryptosystem security depends on classical and post-quantum encryption schemes. It transitions to quantum-resistant key exchange protocols, such as the BB84 QKD protocol, which involves qubit preparation and measurement. In the signature generation process, let s represent the private key, a represent the public key, and y represent a random polynomial. To verify the signature, check if the received signature c is valid for the given message and public key. The signature generation and verification processes can be represented as follows:

$$SignatureGeneration{:}y \leftarrow Sample(\mathcal{D}), c \leftarrow Sign(y,s)$$

$$SignatureVerification\text{:}\,Verify(c, a, Message) \tag{81}$$

The security of a digital signature scheme depends on the computational hardness of the lattice structure, which ensures resistance against quantum attacks. To achieve quantum-resistant hash functions, linear code properties are used, such as in McEliece code-based cryptography. The hash value is obtained by applying a public permutation matrix P to the codeword c, where G is the generator matrix of the code and m is the input message.

$$c = m \cdot G$$

$$\mathscr{H}(m) = P \cdot (m \cdot G) \tag{82}$$

PQC offers quantum era communication security, but challenges include practical, and strategic aspects (see Figure 7 and Figure 8).

Figure 7. (a) Relationship between QKD distance and the resulting secure key rate, considering the influence of noise in the communication channel. (b) Analyses the computational efficiency of lattice-based cryptography by correlating lattice dimension with signature verification time. (c) Security of code-based hash functions by analysing the collision resistance against varying code lengths. (d) Performance comparison of key exchange algorithms. It includes the approximate key exchange time for classical algorithms, the time for post-quantum algorithms, and the hybrid approach. (e) Trade-offs between computational cost and security level for different hybrid cryptographic algorithms. Scheme 1 and Scheme 2 are represented by the red and blue dots, respectively, demonstrating the trade-off between computational cost and security level.

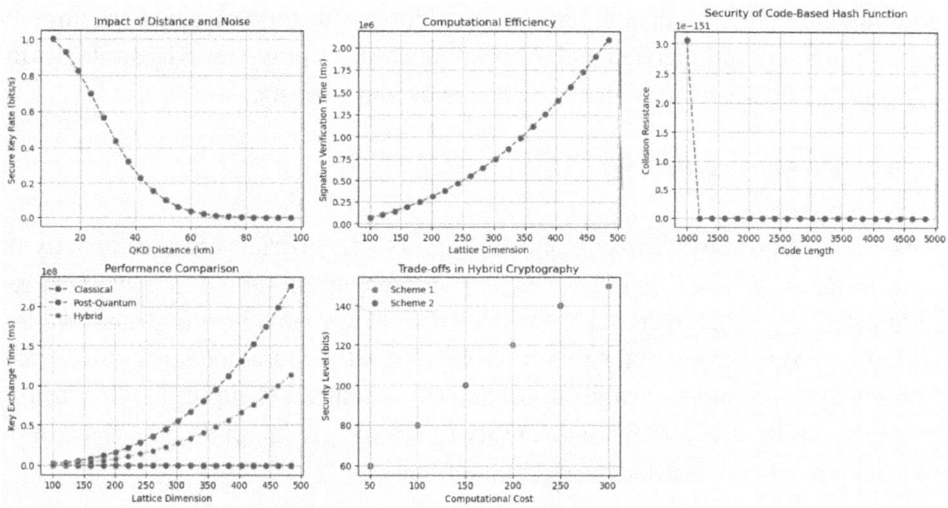

Attention is required for quantum-resistant key management, QKD, and entanglement and superposition issues.

$$K_{QR} = f(d, N, R) \tag{83}$$

The variable K_{QR} represents the objective of achieving quantum-resistant key management. The challenges associated with d, noise level (N), and device reliability (R) are encapsulated in the function $f(d, N, R)$. The function suggests that achieving quantum-resistant key management depends on effectively addressing certain challenges. These challenges include distance, which can be expressed as a numerical value or set, noise level, which can be either a numerical or qualitative variable, or device reliability, which can be measured as a metric or categorical variable. In order to address the challenges posed by post-quantum cryptographic algorithms, it is necessary to develop innovations in quantum communication protocols, error correction, and robust key distribution technologies. The concept of post-quantum cryptographic computational overhead (C_{PQ}) involves optimizing various factors, including metrics such as execution time, energy consumption, and memory usage. The optimization process, denoted as $Optimize(Parameters, Techniques, AlgorithmVariants)$, aims to identify the optimal configuration of parameters (P), techniques (T), and algorithm variants (A) to minimize C_{PQ} while ensuring the desired security levels. Parameters are specific settings that influence the behaviour of an algorithm, such as key sizes and security parameters. The techniques used in post-quantum cryptosystems include algorithmic choices and implementation strategies, represented as a set T. The algorithm variants consist of diverse post-quantum algorithms with varying security and efficiency profiles, denoted as a set A. This optimization framework aims to address the computational demands of post-quantum cryptosystems by strategically adjusting key elements for enhanced efficiency and security.

$$C_{PQ} = \min(Optimize(P, T, A)) \tag{84}$$

The concept of quantum-safe protocol agility (A_{PQ}) pertains to the capacity of a cryptographic protocol to effectively adjust to emerging threats and evolving requirements. A_{PQ} is characterized by the design's inherent flexibility and adaptability. Flexibility (F) encompasses the protocol's capability to accommodate modifications to parameters, techniques, or algorithm choices without necessitating a full redesign, while adaptability (Ad) refers to its ability to seamlessly integrate new algorithms or techniques as they are discovered in the future.

$$A_{PQ} = Design(F, Ad) \tag{85}$$

The integration of PQC into existing systems is a critical phase in the evolution of cryptography, addressing complexities, practical challenges and strategic considerations. Hybrid cryptosystems facilitate coexistence. They ensure reliability and interoperability.

Figure 8. (a) Relationship between lattice dimension and quantum circuit depth for the short integer solution (SIS) problem. (b) Impact of McEliece cryptosystem field size on security. (c) Performance of code-based cryptographic algorithms in terms of codeword length and error correction.

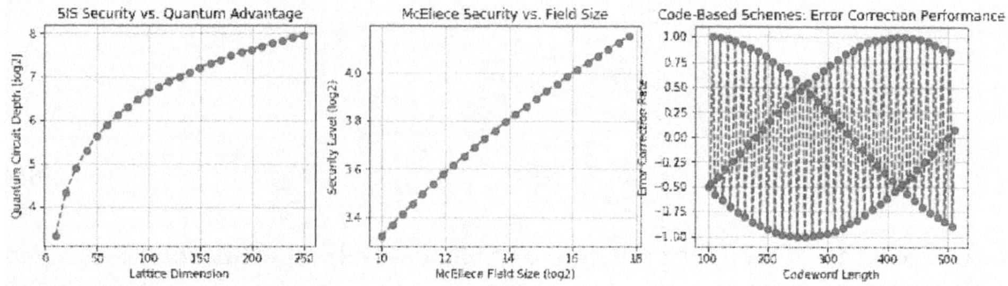

QUANTUM-SAFE CRYPTOGRAPHIC PROTOCOLS

Quantum computing poses a threat to classical cryptographic protocols, which has led to the development of quantum-safe protocols. These protocols, such as QKD, rely on mathematical problems that are resistant to quantum algorithms. An example of such a protocol is the BB84 protocol, which uses principles of quantum mechanics to establish secure keys.

Consider the preparation of qubits in the rectilinear basis (|0, |1) and the diagonal basis (|+,·−). The qubits are denoted as $|\psi_0, |\psi_1, |\psi_+, |\psi_-$:

$$|\psi_0 = |0, |\psi_1 = |1, |\psi_+ = \frac{1}{\sqrt{2}}(|0 + |1), |\psi_- = \frac{1}{\sqrt{2}}(|0 - |1) \quad (86)$$

Quantum gates and operators are utilized in transmission and measurement processes. Qubit states form the foundation for secure key generation, supported by quantum entanglement principles.

$$|\psi_0, |\psi_1, |\psi_+, |\psi_- \rightarrow QuantumGatesandOperators \rightarrow SecureKeyGeneration \quad (87)$$

QKD security is based on the no-cloning theorem, which prevents eavesdroppers from measuring quantum states. Code-based cryptography, such as McEliece, provides post-quantum security by decoding error-correcting codes.

The McEliece cryptosystem utilises a generator matrix G to represent a systematic linear code. The public key is generated as G, while the private key involves a random invertible matrix S used for decoding. The encryption process is represented as $c = m \cdot G + e$, where m is the message and e is an error vector. The decryption process involves finding m by applying the inverse of S.

$$c = m \cdot G + e$$

$$S^{-1} \cdot c = m + S^{-1} \cdot e \tag{88}$$

The security of the McEliece cryptosystem depends on the decoding of linear codes, which is a challenging task even for quantum computers. Hash-based cryptographic protocols that use the Merkle-Damgård construction rely on the hardness of certain functions.

The Merkle-Damgård construction applies a compression function f iteratively to blocks of input data, incorporating the hash of the previous block. In this context, $\mathscr{H}$ represents the hash function, M_i represents the i-th block of the message, and $\mathscr{H}_i$ represents the intermediate hash value after processing M_i.

$$\mathscr{H}_0 = InitialValue$$

$$\mathscr{H}_i = f(\mathscr{H}_i - 1, M_i) \tag{89}$$

The final hash value is $\mathscr{H}_n$, where n represents the number of blocks in the message.

$$\mathscr{H}_n = f(\mathscr{H}_n - 1, M_n) \tag{90}$$

Due to the one-way nature of the underlying hash functions, hash-based cryptographic protocols are designed to resist quantum attacks. Quantum-safe protocols, such as QKD, are based on mathematical principles, including the no-cloning theorem. This theorem states that it is impossible to create a perfect copy of an unknown quantum state.

If the quantum state is represented by $|\psi$, there is no unitary operator U that can satisfy the equation $U|\psi \otimes |0 = |\psi \otimes |\psi$. Hash-based cryptographic protocols focus on the computational challenge of finding a pre-image for a given hash value, exploiting the one-way property of hash functions for security. The hash function is represented by $\mathscr{H}$ and the message is represented by M. The resulting hash value is denoted as $\mathscr{H}(M)$, and the security of the system relies on the challenge of finding M given $\mathscr{H}(M)$.

$$\mathscr{H}(M)\xrightarrow{Hard}M \tag{91}$$

Quantum-safe protocols, such as the one-way property of hash functions, are essential for secure communication. It presents two case studies that explore the development and implementation of quantum-safe protocols. The first case study focuses on the implementation of QKD in quantum communication networks. QKD uses quantum mechanics principles to establish secure keys between communicating parties. It discusses key components required for secure communication, including QKD nodes, quantum channels, classical communication, and key generation and privacy amplification techniques.

Additionally, it presents a case study on the deployment of code-based cryptography, specifically the McEliece cryptosystem, in secure communication scenarios. Key generation involves mathematical operations on matrices and the selection of appropriate linear codes. The security of the system relies on the difficulty of decoding the chosen linear code. Encryption and decryption processes involve encoding the message using a public key and decoding the received ciphertext using a private key. The security of code-based cryptography relies on the difficulty of decoding linear codes.

Parameter selection is crucial for the security and efficiency of code-based cryptography, as it involves balancing security requirements with computational feasibility. This case study demonstrates the practical considerations involved in deploying code-based cryptography for secure communication. It highlights the importance of mathematical foundations in parameter selection and key operations.

With NIST bodies defining specifications, algorithms, security levels and protocols, standardisation of quantum-safe cryptographic algorithms ensures consistency, interoperability and widespread adoption. PQC algorithms are a set of approved or recommended cryptographic schemes, represented as A_{QS}. Parameters, which influence the security and efficiency of PQC algorithms, are a function associated with each algorithm in A_{QS}. Protocols, specific cryptographic schemes, are represented as $Prot$ in A_{QS}.

$$S_{QS} = \{(a, P(a), p) | a \in A_{QS}, p \in Prot\} \tag{92}$$

Collaboration among researchers, industry experts, and cryptographic authorities is essential to establish a shared framework for quantum-safe cryptographic algorithms. The proposed quantum-safe cryptography framework involves three key metrics. Firstly, consistency (C_{QS}), represented as a binary variable, evaluates the adherence to standardized algorithms, parameters, and protocols across diverse systems. It assumes a value of 1 when consistency is achieved and 0 otherwise. Secondly, exchangeability (E_{QS}), another binary variable, assesses the seamless exchange of cryptographic data among compliant systems, taking a value of 1 for successful exchangeability and 0 otherwise. Lastly, utilization (U_{QS}), expressed as a variable ranging from 0 to 1, measures the effective and widespread application of standardized quantum-safe cryptography across various use cases.

$$I_{QS} = C_{QS} \cdot E_{QS} \cdot U_{QS} \tag{93}$$

Interoperability standards aim to establish a unified ecosystem for quantum-safe cryptographic algorithms. In the context of quantum-safe cryptography, the concept of backward compatibility (BC_{QS}), represented as a binary variable, focuses on ensuring compatibility with existing cryptographic systems during the transition. BC_{QS} assumes a value of 1 if backward compatibility is achieved and 0 otherwise. In addition, gradual adoption (GA_{QS}), represented as a variable with a possible range of 0 to 1, refers to the gradual integration of quantum-resistant cryptography, reflecting the pace or progress of adoption over time. Finally, hybrid cryptosystems (HC_{QS}), represented by a variable with a possible range of 0 to 1, refers to systems that seamlessly combine traditional and quantum-resistant cryptography, allowing a smoother transition between cryptographic paradigms.

$$M_{QS} = BC_{QS} \cdot GA_{QS} \cdot HC_{QS} \tag{94}$$

Quantum-safe cryptographic protocols, such as QKD, code-based cryptography, and hash-based cryptography, are intended to withstand attacks from quantum computers. These protocols rely on mathematical principles, including quantum mechanics, code-based cryptography, and hash-based protocols. Case studies on the implementation of QKD in quantum communication networks and the deployment of code-based cryptography in secure communication highlight the practical aspects of quantum-safe protocols. Future perspectives on quantum-safe protocols involve advancements in mathematical foundations, protocol design, and integration strategies. Advancements in mathematical structures, protocol design innovations, and hybrid cryptosystems are crucial for ensuring the coexistence of classical and quantum-safe algorithms. As quantum communication networks continue to evolve, the role of quantum-safe protocols becomes increasingly significant. Future perspectives involve developing lightweight quantum-safe protocols that are suitable for resource-constrained IoT devices, addressing energy efficiency, secure key exchange, and resistance against quantum attacks (see Figure 9).

Figure 9. The effect of different parameters on the design and performance of post-quantum cryptographic algorithms. Each subplot explores different aspects, including the relationship between code dimension and minimum distance, the trade-off between security and efficiency in weight distributions, the influence of decoding algorithms on key generation time, the correlation between standardization level and algorithm performance, challenges in formal verification based on protocol complexity, and trade-offs in migration considering backward compatibility level and security degradation.

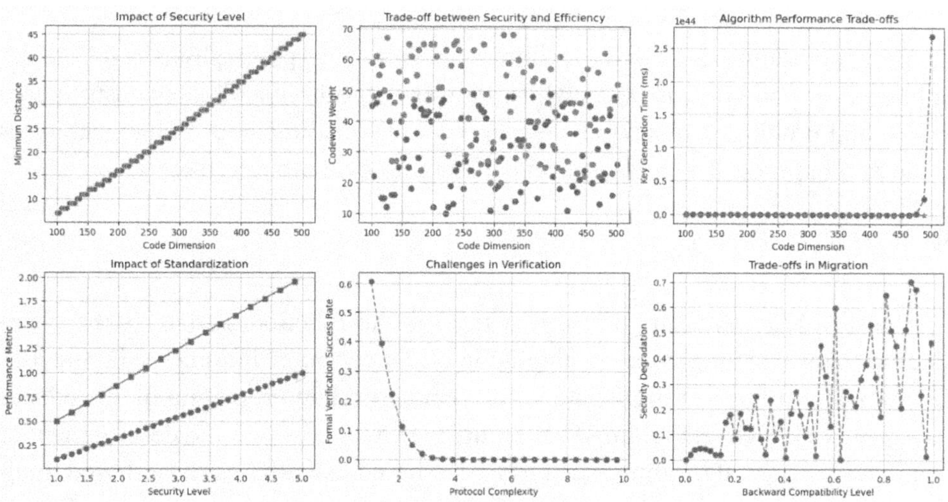

QUANTUM COMPUTING AND ITS CRYPTOGRAPHIC IMPLICATIONS

Quantum computing utilizes qubits to exist in multiple states simultaneously, enabling superposition and entanglement. Cryptographic practices must adapt to quantum computing, taking into account quantum algorithms that could compromise current encryption methods. The expression of the evolution of a quantum state $|\psi(t)\rangle$ over time t is:

$$|\psi(t)\rangle = U(t)|\psi(0)\rangle \tag{95}$$

where *U(t)* is a unitary operator. The potential impact of quantum computing on classical cryptographic systems, including RSA encryption, is significant due to its ability to efficiently solve mathematical problems such as integer factorization and discrete logarithm problems. Shor's algorithm comprises the following steps:

1. **Initialization**: Choose a random integer $a < N$.
2. **QFT**: Apply a QFT to determine the period r of the function $f(x) = a^x modN$.
3. **Classical Post-Processing**: Use classical algorithms to find the factors of N based on the obtained period r.

Shor's algorithm has a significant impact on classical cryptography as it undermines public-key cryptosystems. Meanwhile, Grover's algorithm addresses unstructured search and provides a quadratic speedup over classical methods, which enhances solution amplitude. The quantum state undergoes evolution as follows:

$$|\psi_k = U_s U_f(2|\psi_{k-1} - |\psi_0) \tag{96}$$

Here, U_f represents the oracle that reflects the function, and U_s is the diffusion operator that amplifies the amplitude of the correct solution (see Figure 10). Grover's algorithm has implications for symmetric key cryptography. It reduces the effective key length required for security. PQC focuses on the development of quantum-resistant algorithms, such as lattice-based, code-based, hash-based, multivariate polynomial-based, and code-based signature schemes. The aim of these algorithms is to provide secure cryptographic primitives that are able to withstand the computational power of quantum algorithms. QKD is a protocol that uses quantum mechanical principles for secure key exchange and is resistant to quantum attacks.

Figure 10. The performance characteristics of quantum algorithms, specifically Shor's algorithm for integer factorisation and Grover's algorithm for database search. (a) Success rate of Shor's algorithm as a function of the size of the modulus N. (b) Runtime of Shor's algorithm as a function of the period r. (c) Speedup achieved by Grover's algorithm when searching a database of varying sizes. (d) Effect of the initial choice of integer on the running time of Shor's algorithm. (e) Convergence of Grover's algorithm by showing the probability of finding the correct solution over iterations. (f) Comparison of factoring time between Shor's and Grover's algorithms for selected modulus sizes.

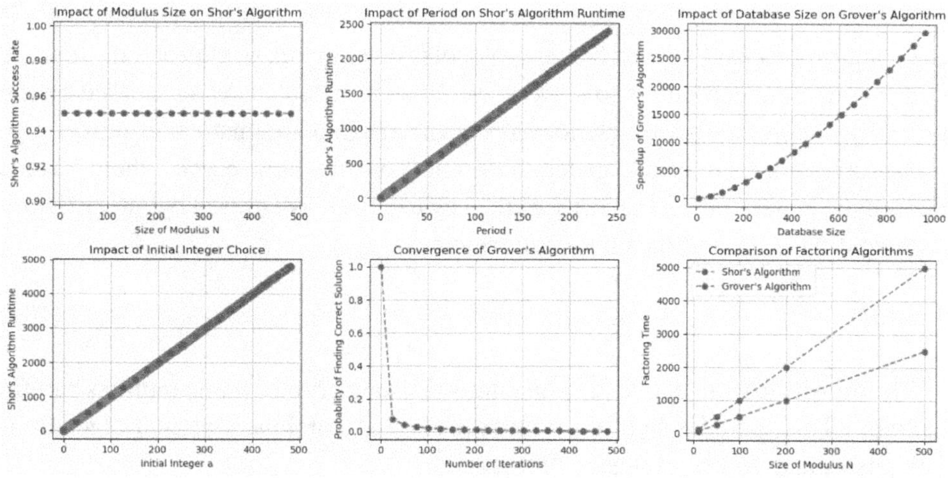

Hash-based cryptography, which utilizes the one-way property of hash functions, provides a secure method of communication that is resistant to quantum attacks through the Merkle-Damgård construction. The hash function is represented by $\mathscr{H}$, the *i*-th block of the message is represented by M_i, and the intermediate hash value after processing M_i is represented by $\mathscr{H}_i$.

$$\mathscr{H}_0 = InitialValue$$

$$\mathscr{H}_i = f(\mathscr{H}_i - 1, M_i) \tag{97}$$

The final hash value is $\mathscr{H}_n$, where *n* represents the number of blocks in the message.

$$\mathscr{H}_n = f(\mathscr{H}_n - 1, M_n) \tag{98}$$

Hash-based cryptographic protocols provide a quantum-resistant solution for digital signatures and secure communication. Quantum-resistant key exchange protocols adapt to quantum computing by establishing a shared secret key. lattice-based cryptography, code-based cryptography, hash-based cryptography, and quantum-resistant key exchange protocols.

$$Given\left(A, b = A_s + e\right), finds where A is a matrix, b is a vector, e is noise,$$

$$and s is a short vector in the lattice generated by A \quad (99)$$

Quantum-resistant key exchange protocols use mathematical problems that are hard for quantum computers to solve to establish a shared secret key, based on the difficulty of computing discrete logarithms in certain mathematical groups.

A secret scalar a is used to generate a point Q on an elliptic curve, where P is the point of the generator. The security of the key exchange is ensured by the difficulty of computing P, Q, and the curve parameters.

$$Q = aP \quad (100)$$

The transition from classical to quantum-resistant cryptography requires careful planning. This includes hybrid cryptosystems that combine classical algorithms with post-quantum key exchange protocols.

$$HybridEncryption\left(M, C_C, C_Q, A_C, A_Q\right) = Enc_C(M, K_C) \| K_C + K_Q \| Dec_Q\left(K_C, A_Q\right) \| Enc_Q \left(M, K_Q\right) \quad (101)$$

M for the plaintext message, C_C and C_Q for the classical and quantum ciphertexts, A_C and A_Q for the classical and quantum decryption algorithms, Enc_C and Enc_Q for the classical and quantum encryption functions, and K_C and K_Q for the classical and quantum encryption keys, respectively. Eq. (101) describes the hybrid encryption process that combines classical and quantum encryption techniques. The symbol | | signifies concatenation.

Algorithm agility enables organizations to transition to quantum-resistant algorithms without having to completely overhaul their existing cryptographic systems. This ensures compatibility and security through mathematical considerations such as parameter negotiation and key sizes.

$$A_A = \{(p, k, prim) | p \in P, k \in K, prim \in Prim \quad (102)$$

Algorithm agility (A_A) involves adapting based on parameters (P), representing settings affecting the algorithm's behaviour; key sizes (K), determining key length and security; and primitives ($Prim$), the fundamental building blocks. Organisations are adapting to advances in quantum computing and cryptographic standards through algorithmic agility, standardisation and interoperability. This ensures a cohesive and interoperable cryptographic landscape.

$$S_{QR} = \{(a, P(a), p) | a \in A_{QR}, p \in Prot_{QR}\}$$

$$S_{QR} \Rightarrow Interoperability(S_{QR}) \tag{103}$$

Quantum-resistant cryptographic standards (S_{QR}) comprise three main components: PQC (A_{QR}), which represents approved or recommended PQC algorithms; Parameters ($P(a)$), which denote specific settings that impact the security and efficiency of each algorithm a in the set A_{QR}; and protocols ($Prot_{QR}$), which comprise approved or recommended cryptographic schemes that utilize PQC algorithms.

Interoperability is a critical aspect of quantum-resistant standards, ensuring seamless operation across different systems, platforms and applications.

$$I_{QR} = C_{QR} \cdot E_{QR} \cdot U_{QR} \tag{104}$$

Consistency, denoted as a binary variable C_{QR} (1 for achieved, 0 otherwise), refers to the use of standardized algorithms, parameters, and protocols across diverse systems. Exchangeability, represented by a binary variable E_{QR} (1 for achieved, 0 otherwise), signifies the seamless interchange of cryptographic data among compliant systems. The variable U_{QR} measures the extent to which standardized quantum-resistant cryptography is effectively and widely used across different contexts, with values ranging from 0 to 1.

Interoperability standards are intended to create a unified ecosystem for quantum-resistant cryptographic algorithms, but concerns have been raised about the potential impact of these standards on classical encryption and public key cryptography.

The impact on RSA encryption is the factorization of the public modulus N. The private exponent d is computed as the modular inverse of the public exponent $emodulo\phi(N)$, where $\phi(N)$is Euler's totient function.

$$d \equiv e^{-1}(mod\phi(N)) \tag{105}$$

Shor's algorithm reduces the computational complexity of RSA and ECC, while Grover's algorithm provides a quadratic speed-up in the search for secret keys in unstructured databases. The transition from classical to PQC requires careful planning, backward compatibility, gradual adoption, and the development of hybrid cryptosystems to ensure security.

$$BC_{PQC} = C_{PQC} \cdot I_{PQC} \cdot Comm_{PQC} \tag{106}$$

Eq. (106) shows that achieving good backward compatibility in PQC transition requires ensuring coexistence (C_{PQC}), interoperability (I_{PQC}), and secure communication ($Comm_{PQC}$) between classical and PQC systems.

Backward compatibility in cryptography requires designing protocols for both classical and post-quantum cryptographic entities. This allows for a gradual adoption (G_A) process, enabling the incremental replacement of classical algorithms with post-quantum counterparts.

$$G_A = Integrate(Phase1, Phase2, \ldots, PhaseN) \tag{107}$$

Hybrid cryptosystems combine classical and post-quantum algorithms. They integrate PQC into existing systems, ensuring a seamless transition.

$$HC = Enc_C(M, K_C) \| K_C + K_Q \| Dec_Q(K_C, A_Q) \| Enc_Q(M, K_Q) \tag{108}$$

The message to be encrypted is denoted by M. C_C and C_Q represent the classical and PQC components, respectively. The encryption functions for the classical and PQC components are symbolized by Enc_C and Enc_Q, while the secret keys generated by the classical and PQC components are represented by K_C and K_Q. The symbol | | denotes the concatenation of distinct outputs. Dec_Q represents the decryption function for the PQC component, utilizing the public key A_Q.

Hybrid cryptosystems combine classical and post-quantum algorithms to enable a secure and gradual transition. This requires careful planning, algorithm agility, and a phased approach. The challenges posed by quantum computing to classical cryptographic algorithms necessitate the implementation of quantum-safe protocols with careful planning.

CHALLENGES AND OPPORTUNITIES IN QUANTUM CRYPTOGRAPHY

Quantum cryptography presents challenges for large-scale implementation due to mathematical intricacies and technological constraints. These challenges include QKD and integrating quantum-safe protocols into existing communication infrastructures. QKD is the foundation of quantum cryptography, enabling secure key exchange. However, the practical implementation of QKD protocols can affect their scalability and performance. The transmission of quantum states over long distances presents challenges, such as signal degradation and loss. The exponential decay of quantum states describes this phenomenon.

$$P(Success) = e^{-\alpha L} \tag{109}$$

The probability of successful transmission ($P(Success)$) is determined by the attenuation coefficient (α) and the distance traveled (L). Quantum channels are susceptible to noise and errors and these can affect fidelity. The challenge for QKD protocols is to minimize QBER and ensure key security.

$$F(\rho_1,\rho_2) = f(Q_C, P_E(e)) \tag{110}$$

Eq. (110) describes the relationship between the fidelity of the transmitted qubit (F), the characteristics of the quantum channel (Q_C), and the probability distribution of errors (P_E). To minimize QBER, the equation should be manipulated to optimize F with respect to Q_C and P_E.

Technological constraints in quantum cryptography involve the generation, transmission, and detection of quantum states. These constraints pose challenges such as decoherence and environmental interactions, which require the use of quantum repeaters. The effectiveness of quantum repeaters (E) can be quantified by the rate of entanglement generation (R), F, and d.

$$E = R \cdot F \cdot d \tag{111}$$

Developing efficient quantum repeater protocols that balance these factors is a challenge. Retrieving quantum information while preserving it remains a technological challenge. Quantum memory is essential for quantum repeaters and long-distance communication, as it determines coherence time (T_2) and F.

$$QuantumMemoryEfficiency = T_2 \cdot F \tag{112}$$

The development of hybrid cryptosystems is necessary to address the challenges of compatibility, interoperability, and coexistence that arise from integrating quantum-safe protocols into classical communication systems. The integration is a combination of classical public key algorithms ($PK_{Classical}$) and post-quantum key exchange protocols ($KE_{Quantum-Safe}$).

$$HybridCryptosystems = PK_{Classical} + KE_{Quantum-Safe} \tag{113}$$

The challenge is to ensure the security and functionality of the hybrid approach. Algorithm agility is a key consideration for a smooth transition from classical to quantum-safe cryptography.

$$AlgorithmAgility = Select\left(CA_{Classical}, CA_{Quantum-Safe}\right) \tag{114}$$

CA stands for cryptographic algorithms. Quantum cryptography offers opportunities for technological advancements, protocol design, and QKD, enabling secure communication over longer distances and overcoming limitations in classical algorithms. The scalability of QKD networks can be defined by three parameters: the number of users (U), key generation rate (R), and security parameters (P).

$$S = U \cdot R \cdot P \tag{115}$$

Improvements in quantum repeaters and efficient error correction techniques can extend the range of QKD, providing opportunities in network topologies and protocols. The rate of entanglement generation (R), F, and error correction efficiency (E) are important factors to consider.

$$QKDRange = R \cdot F \cdot E \tag{116}$$

Breakthroughs in quantum repeater technologies could enhance scalability of quantum communication. Quantum entanglement offers secure communication beyond key distribution, and quantum teleportation for information transfer. Quantum teleportation involves the use of entangled states ($|\phi^{+}$) and quantum operations ($\mathcal{U}$).

$$|\psi_1 \otimes |\phi^{+}\mathcal{U}|\psi_2 \otimes |\phi^{+} \tag{117}$$

Quantum teleportation technologies and quantum-safe cryptographic protocols provide secure communication in quantum computing. A formal proof (P) establishes the security (S) of a cryptographic algorithm.

$$P \Rightarrow S \tag{118}$$

Interdisciplinary collaboration can be defined as the fusion of knowledge (*K*) and expertise (*E*).

$$InterdisciplinaryCollaboration = K \cdot E \tag{119}$$

Quantum cryptography encounters challenges in large-scale distribution, technological constraints, and integration with classical communication systems. Opportunities for advancements include QKD protocols and secure communication (see Figure 11).

Figure 11. Key parameters in QKD systems, illustrating their impact on various performance metrics. (a) Effect of optical attenuation on the QKD success probability as a function of distance. (b) Relationship between quantum memory efficiency and fidelity, highlighting the effect of coherence time on memory performance. (c) Scalability of the QKD network as a function of the number of users, showing the influence of the key rate on the network capacity. (d) Effect of fidelity on QKD range, assuming a constant error correction rate.

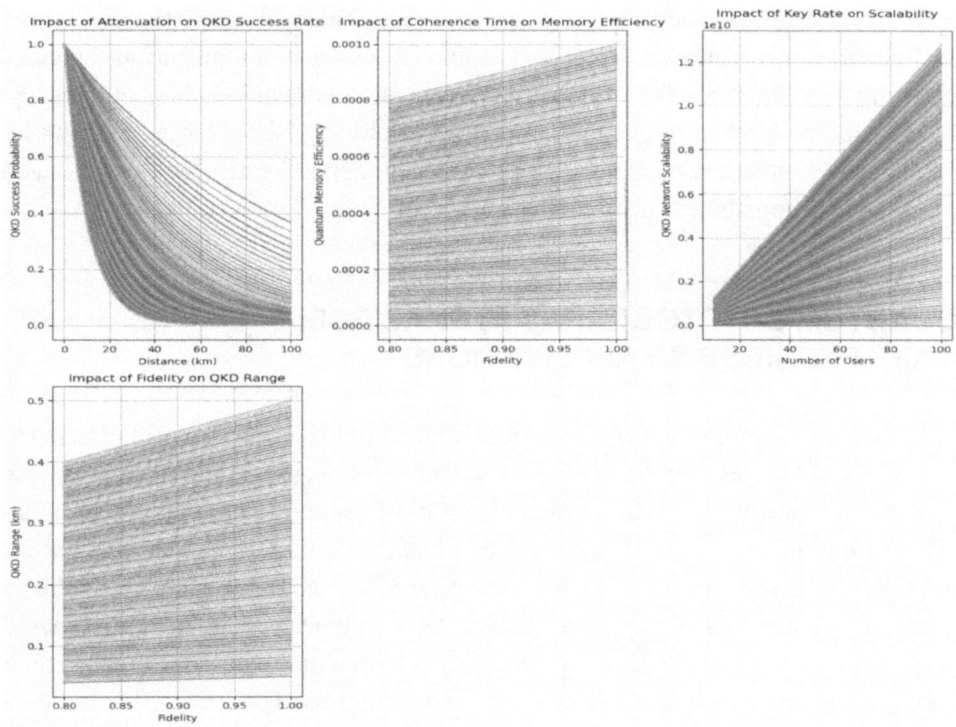

The integration of quantum computing concepts with existing cryptographic systems is a major shift in secure communications. Quantum-safe cryptographic algorithms are being developed to withstand attacks from quantum computers, and the transition from classical to post-quantum cryptography requires careful planning and backward compatibility. Hybrid cryptosystems, which combine classical and post-quantum encryption techniques, play a key role in ensuring a seamless transition by integrating post-quantum cryptography into existing frameworks.

Algorithmic agility is critical for organisations to adapt to evolving cryptographic standards without overhauling existing systems. This includes dynamic adaptation of parameters, key sizes and cryptographic primitives. Backward compatibility in the post-quantum transition requires coexistence, interoperability and secure communication between classical and post-quantum systems.

Quantum computing poses specific challenges to classical encryption schemes, such as Shor's algorithm and Grover's algorithm, which may affect the security of RSA encryption. Hybrid cryptosystems provide security by integrating both classical and post-quantum algorithms, allowing organisations to gradually adopt quantum-resistant cryptography while maintaining the integrity of their existing systems.

Real-world implementations of quantum cryptography demonstrate its practical applications in industries such as finance, government and healthcare, securing communication channels and protecting sensitive data. Telecommunications networks use quantum cryptographic protocols to establish secure communication channels, ensuring the confidentiality of transmitted data and maintaining channel integrity.

The future landscape of quantum cryptography envisions large-scale implementations of practical quantum-safe protocols, integrated with cloud services and machine learning algorithms for enhanced security.

QUANTUM CRYPTOGRAPHY IN PRACTICE: CASE STUDIES AND APPLICATIONS

Quantum cryptography is a practical application that has been implemented in various industries, demonstrating its potential for secure communication. Case studies have shown successful implementations of quantum cryptography in financial institutions, government agencies, and healthcare data. Financial institutions use QKD to secure communication between branches and data centers, providing a quantum-resistant key exchange mechanism. Government agencies use lattice-based cryptography, which is believed to be resistant to quantum attacks, to secure their communication infrastructure. Healthcare institutions use QKD to establish secure communication channels for transmitting patient records and medical information.

The security of these communications relies on the hardness of mathematical problems, such as lattice problems, which are assumed to withstand quantum attacks.

The adoption of quantum cryptography in industry varies. Banking and finance have adopted QKD to secure financial transactions and communications, while government agencies and defense organizations use quantum-safe communication to protect classified information. They employ post-quantum cryptographic algorithms such as lattice-based cryptography to provide a robust defense against potential quantum threats. These industries demonstrate the significance of quantum cryptography in protecting confidential patient data and ensuring data privacy.

Quantum cryptography is a secure solution for communication in various industries, such as healthcare and telecommunications. Telecommunication networks use quantum cryptographic protocols to establish secure communication channels. The security of these communications is vital due to the confidentiality of transmitted data and the integrity of communication channels.

Real-world implementations of quantum cryptography include proactive adoption for long-term security, seamless integration with existing infrastructure, collaboration and knowledge sharing, continuous evaluation of security posture, training and skill development, addressing challenges related to distance limitations, noise and error rates, authentication, and balancing quantum and classical technologies.

The integration of quantum computing concepts with existing cryptographic systems is a major shift in secure communications. Quantum-safe cryptographic algorithms are being developed to withstand attacks from quantum computers, and the transition from classical to post-quantum cryptography requires careful planning and backward compatibility. Hybrid cryptosystems, which combine classical and post-quantum encryption techniques, play a key role in ensuring a seamless transition by integrating post-quantum cryptography into existing frameworks.

Algorithmic agility is critical to enable organisations to adapt to evolving cryptographic standards without overhauling existing systems. This includes the ability to dynamically negotiate parameters, key sizes and cryptographic primitives. Backward compatibility in the post-quantum transition requires coexistence, interoperability and secure communication between classical and post-quantum systems.

Quantum computing poses specific challenges to classical encryption schemes, such as Shor's algorithm and Grover's algorithm, which affect the security of RSA encryption. Hybrid cryptosystems provide security by integrating both classical and post-quantum algorithms, allowing organisations to gradually adopt quantum-resistant cryptography while maintaining the integrity of their existing systems.

Real-world implementations of quantum cryptography demonstrate its practical applications in diverse industries such as finance, government and healthcare to secure communication channels and protect sensitive data. Quantum gates play a key role in quantum cryptography, contributing to the development of quantum-safe cryptographic algorithms that can withstand attacks from quantum computers. Fault-tolerant quantum computing ensures the reliability of quantum cryptographic protocols in the presence of faults, while quantum-safe cryptographic algorithms resist attacks from quantum computers.

Companies offering quantum cryptography solutions often incorporate quantum gates into their commercial products, often performing tasks such as quantum key generation, distribution and secure communication. Overall, the integration of quantum computing concepts with current cryptographic systems presents both challenges and opportunities for secure communications.

THE FUTURE LANDSCAPE OF QUANTUM CRYPTOGRAPHY

The future of quantum cryptography appears promising for transformative developments and advancements that may reshape the landscape of secure communication. Emerging technologies and trends, such as quantum hardware developments, quantum-safe cryptography research, and global collaborations, will shape the trajectory of quantum communication.

The development of a quantum internet, which utilizes principles of quantum entanglement and superposition, is highly anticipated in the field of quantum cryptography. Quantum-safe cryptographic protocols are likely to be integrated with cloud services to protect against potential threats from quantum adversaries. The integration of quantum-safe cryptographic techniques into machine learning algorithms could enhance data processing and model training for quantum machine learning security.

In the future, PQC standardization is likely to provide a foundation for interoperability and ensure a consistent and secure cryptographic landscape. Large-scale implementation of practical quantum-safe protocols is possible, including the deployment of quantum-resistant cryptographic algorithms such as lattice-based, code-based, and hash-based algorithms in real-world applications.

Advancements in QKD technologies are expected to lead to significant improvements, expanding secure QKD networks. Emerging technologies and trends in quantum computing include quantum hardware developments, quantum-safe cryptography research, integration with classical cybersecurity, global collaboration in quantum research, and the establishment of quantum-secure communication standards.

FUTURE RESEARCH AND DEVELOPMENT DIRECTIONS

Quantum cryptography encounters various challenges, such as technological limitations in generating, transmitting, and detecting quantum states. To overcome these challenges, future research should concentrate on developing more efficient quantum repeater protocols that balance factors such as entanglement generation, fidelity, and distance. Improving quantum memory is essential for quantum repeaters and long-distance communication. Researchers should explore ways to enhance efficiency by extending coherence time and fidelity. It is crucial to maintain algorithm agility as quantum-resistant cryptographic standards evolve. The large-scale implementation of quantum-safe protocols, including lattice-based, code-based, and hash-based algorithms, is necessary for securing communication in various industries. Advancements in quantum communication networks, such as QKD networks, can benefit from improvements in quantum repeaters and error correction techniques. Integrating quantum-safe cryptographic techniques into machine learning algorithms presents an exciting area for research. Global collaboration and interdisciplinary research are essential for addressing complex challenges in quantum communication. The development of a quantum internet, which leverages principles such as quantum entanglement and superposition, is a transformative goal. The evaluation of security in quantum cryptographic systems should be an ongoing process, with a focus on continuous improvement, training, and skill development to stay ahead of potential threats from quantum adversaries.

CONCLUSION AND OUTLOOK

The research chapter examines the transformative impact of quantum cryptography on secure communication. It highlights the potential threats to classical cryptographic systems and the development of quantum-safe cryptographic solutions. It discusses the vulnerabilities of classical cryptographic systems to quantum algorithms and the importance of post-quantum cryptographic algorithms such as lattice-based, code-based, and hash-based schemes. It also mentions the integration challenges addressed by hybrid cryptosystems. In the quantum era, quantum-safe protocols like QKD and post-quantum cryptographic algorithms are essential components for secure communication. The chapter highlights the significance of taking into account algorithm agility, standards, and interoperability during the shift from classical to quantum-resistant cryptography.

Recommendations for future research include exploring algorithm agility and standards, implementing quantum cryptography in various industries, integrating quantum-safe cryptographic protocols with cloud services and machine learning algorithms, deploying quantum-safe protocols on a large scale, and collaborating globally in quantum research. The findings provide a basis for a quantum-secure future. Further research will be crucial in fully realizing the potential of quantum cryptographic solutions.

REFERENCES

Alagic, G., Alagic, G., Alperin-Sheriff, J., Apon, D., Cooper, D., Dang, Q., ... Smith-Tone, D. (2019). *Status report on the first round of the NIST post-quantum cryptography standardization process*. Academic Press.

Bennett, C. H., & Brassard, G. (2014). Quantum cryptography: Public key distribution and coin tossing. *Theoretical Computer Science*, 560, 7–11. 10.1016/j.tcs.2014.05.025

Bernstein, D. J. (2009). Introduction to post-quantum cryptography. In *Post-quantum cryptography* (pp. 1–14). Springer Berlin Heidelberg. 10.1007/978-3-540-88702-7_1

Diffie, W., & Hellman, M. E. (2022). New directions in cryptography. In *Democratizing Cryptography: The Work of Whitfield Diffie and Martin Hellman* (pp. 365-390). 10.1145/3549993.3550007

Easttom, C. (2019). *Computer security fundamentals*. Pearson IT certification.

Ekert, A. K. (1991). Quantum cryptography based on Bell's theorem. *Physical Review Letters*, 67(6), 661–663. 10.1103/PhysRevLett.67.66110044956

Gisin, N., & Thew, R. (2007). Quantum communication. *Nature Photonics*, 1(3), 165–171. 10.1038/nphoton.2007.22

Grover, L. K. (1996, July). A fast quantum mechanical algorithm for database search. In *Proceedings of the twenty-eighth annual ACM symposium on Theory of computing* (pp. 212-219). 10.1145/237814.237866

Horodecki, R., Horodecki, P., Horodecki, M., & Horodecki, K. (2009). Quantum entanglement. *Reviews of Modern Physics*, 81(2), 865–942. 10.1103/RevModPhys.81.865

Huang, Z., Joshi, S. K., Aktas, D., Lupo, C., Quintavalle, A. O., Venkatachalam, N., ... & Rarity, J. G. (2022). Experimental implementation of secure anonymous protocols on an eight-user quantum key distribution network. *NPJ Quantum Information, 8*(1), 25.

Kahn, D. (1968). *The Story of Secret Writing*. Weidenfeld & Nicolson.

Lo, H. K., & Chau, H. F. (1999). Unconditional security of quantum key distribution over arbitrarily long distances. *Science, 283*(5410), 2050-2056.

Martınez, I. L. (2014). *Real world quantum cryptography* (Doctoral dissertation, University of Calgary).

Nahar, S., Pithawa, D., Bhardwaj, V., Rawat, R., Rawat, A., & Pachlasiya, K. (2023). Quantum Technology for Military Applications. *Quantum Computing in Cybersecurity*, 313-334.

Nielsen, M. A., & Chuang, I. L. (2001). Quantum computation and quantum information. *Physics Today*, 54(2), 60.

Noor-ul-Ain, W., Atta-ur-Rahman, M., Nadeem, M., & Abbasi, A. G. (2016). Quantum cryptography trends: a milestone in information security. In *Hybrid Intelligent Systems: 15th International Conference HIS 2015 on Hybrid Intelligent Systems, Seoul, South Korea, November 16-18, 201515* (pp. 25-39). Springer International Publishing. 10.1007/978-3-319-27221-4_3

Oppliger, R. (2021). *Cryptography 101: From Theory to Practice*. Artech House.

Orieux, A., & Diamanti, E. (2016). Recent advances on integrated quantum communications. *Journal of Optics*, 18(8), 083002. 10.1088/2040-8978/18/8/083002

Pan, D., Song, X. T., & Long, G. L. (2023). Free-space quantum secure direct communication: basics, progress, and outlook. *Advanced Devices & Instrumentation, 4*, 0004.

Rivest, R. L., Shamir, A., & Adleman, L. (1978). A method for obtaining digital signatures and public-key cryptosystems. *Communications of the ACM*, 21(2), 120–126. 10.1145/359340.359342

Shor, P. W. (1999). Polynomial-time algorithms for prime factorization and discrete logarithms on a quantum computer. *SIAM Review*, 41(2), 303–332. 10.1137/S0036144598347011

Sihare, S., & Khang, A. (2023). Effects of Quantum Technology on the Metaverse. In *Handbook of Research on AI-Based Technologies and Applications in the Era of the Metaverse* (pp. 174–203). IGI Global. 10.4018/978-1-6684-8851-5.ch009

Sihare, S. R. (2022a). Dynamic multi-party quantum key agreement protocol based on commutative encryption. *International Journal of Theoretical Physics*, 61(9), 242. 10.1007/s10773-022-05203-w

Sihare, S. R. (2022b). Multi-party quantum key agreement protocol for detection of collusive attacks in each sub-circle segment by headers. *International Journal of Theoretical Physics*, 61(7), 208. 10.1007/s10773-022-05184-w

Sihare, S. R. (2022c). Transformation of Classical to Quantum Image, Representation, Processing and Noise Mitigation. *International Journal of Image. Graphics and Signal Processing*, 12(5), 10. 10.5815/ijigsp.2022.05.02

Sihare, S. R. (2022d). Qubit and bit-based quantum hybrid secret key generation. *The European Physical Journal D*, 76(11), 222. 10.1140/epjd/s10053-022-00532-1

Sihare, S. R. (2023a). Potential of quantum computing to effectively comprehend the complexity of brain. *Applied Intelligence*, 53(22), 1–24. 10.1007/s10489-023-04857-1

Sihare, S. R. (2023b). Multi-Party Quantum Key Distribution Using Variational Quantum Eigensolvers. *Advanced Quantum Technologies*, 2300270.

Singh, S. (2000). *The code book: the science of secrecy from ancient Egypt to quantum cryptography*. Anchor.

Van Tilborg, H. C., & Jajodia, S. (Eds.). (2014). *Encyclopedia of cryptography and security*. Springer Science & Business Media.

KEY TERMS AND DEFINITIONS

Algorithm Agility: The ability of organizations to adapt cryptographic algorithms in response to advances in quantum computing.

Backward Compatibility: Designing protocols for both classical and post-quantum cryptographic entities to enable a gradual adoption process.

Gradual Adoption: Incremental replacement of classical algorithms with post-quantum counterparts.

Hybrid Cryptosystems: The combination of classical and post-quantum algorithms for secure transitions.

Interoperability (Quantum-Resistant): Ensuring consistency, exchangeability, and utilization of quantum-resistant cryptographic algorithms across different systems.

Quantum Cryptography: The overarching theme, referring to cryptographic systems based on principles of quantum mechanics.

Quantum Memory Efficiency: The efficiency of quantum memory, determining coherence time and fidelity.

Quantum Repeater: Technology addressing constraints in quantum cryptography involving the generation, transmission, and detection of quantum states.

Signal Degradation: Challenges associated with the transmission of quantum states over long distances.

Standards (Quantum-Resistant): Specifications defining post-quantum cryptographic parameters and protocols.

Chapter 6
Quantum Program:
A Sequence of Quantum Circuits Using Qiskit

Amlan Sengupta
Jadavpur University, India

Debotosh Bhattacharjee
https://orcid.org/0000-0002-1163-6413
Jadavpur University, India

ABSTRACT

The rapid evolution of quantum theory and technology has improved a lot in diverse fields. Quantum computing develops quantum-mechanical effects to execute a computation efficiently, and its benefits reduce both the execution duration and energy consumption compared to conventional computing. Recently, Google declared that quantum supremacy reached a maximum reach, and the quantum computer can effectuate an intractable calculation on a supercomputer. The different quantum algorithms implemented in quantum computers enhance efficiency and speed up the process with classical algorithms. The quantum software Qiskit is used to write quantum computing codes with different stages including building and execution stages. The single Qubit gates controlled two-bit gates and multi-controlled gates help identify the rotations of different dimensions of the plans. The three phenomena of quantum computing will be explained in detail on superposition, quantum measurement, and entanglement to evaluate its functioning.

DOI: 10.4018/978-1-7998-9522-0.ch006

INTRODUCTION

In today's world, fast computing plays a significant role in every individual's personal and public life. However, this leads to a lot of power consumption. The Complementary Metal-Oxide Semiconductor (CMOS) transistor shown by progressions in transistor technology solved this problem. Yet, various applications of computing technologies such as trade, medicine, administration, and logistics using CMOS still demand an even higher computing speed. Therefore, the current computation methods require a more efficient replacement. Quantum computing is considered as one such notable alternative (Wille et al. 2019). Quantum computation is defined as the data processing technique using the physical properties of quantum states on a quantum computer (Black et al., 2002). It is the juncture of fields like mathematics, physics, and computer science. The initial point of development of quantum computers can be traced to the 1980s when physicists questioned whether a universal device could simulate quantum mechanical systems (Hassija et al., 2020). Quantum computers perform computations based on the fundamental properties of quantum mechanics like a superposition of states, interference, and entanglement. In addition, they possess the computational benefit of having critical properties of reversibility. These properties enable quantum computers with the prospect of widespread abilities of quantum data processing in the areas like sensing and communication (Wille et al. 2019). Since quantum computing can process information exponentially, quicker than any supercomputer, many private sector companies have started participating in quantum computing-based R&D to avail of maximum opportunities. Numerous experts in this area consider that the improvement of quantum computing technologies may not follow the standard smooth curve of evolution. It may take several years before it can be employed to solve real-life tasks. In this initial development stage, corporations that have started financing and developing plans to integrate their trade structure with quantum supremacy have far superior chances to take advantage of the upcoming market(Hassija et al., 2020).

There are four groups of problems where a quantum computer can stand considerably beneficial over a classical computer. They cover most of the applications established by many industries to produce new commercial prospects and ensure a competitive advantage.

- Combinatorial optimization is the method of detecting the maxima or minima of an objective function, for example, judging the shortest distance among a specified set of points.
- Linear algebra problem: It functions as an essential pillar to machine learning, which has a noticeable impact in several uses across industries.

- Differential equation-related problem: It can be employed to model complex systems' performance involving basic physics laws. Various applications built on simulation belong to this category.
- Factorization: Currently, computer security and cryptography are greatly dependent on classical computers' infeasibility in factoring the product of two prime numbers. Quantum computing can be of great help in this scenario (Hassija et al., 2020).

The most valued quality of a quantum computer is its capability to execute extensive simulations. This property guarantees its prospect for a significant number of uses in numerous types of industries (Hassija et al., 2020). Considering the considerable influence quantum computers can have in the future, many governmental organizations have also financed R&D in quantum computing technologies. Despite having several market-ready and established technologies such as artificial intelligence and blockchain, quantum computing has received massive attention worldwide. With the increased research work, the number of patents based on quantum computing is also growing. Likewise, the publications on quantum computing in journals such as IEEE have also increased in number (Hassija et al., 2020).

Superposition, quantum measurement, and entanglement are three phenomena that are central to quantum computing.

BACKGROUND

A Qubit is the fundamental unit of information in quantum computing. A qubit uses the quantum mechanical phenomena of superposition to achieve a linear combination of two states. A classical binary bit can only represent a single binary value, such as 0 or 1, meaning it can only be in two possible states. A qubit, however, can represent a 0, a 1, or any proportion of 0 and 1 in a superposition of both states, with a certain probability of being a 0 and a certain probability of being a 1.

A Qubit is a quantum mechanical state vector and is essentially indeterminate, being neither of the two-bit states of $|0\rangle$ or $|1\rangle$ but rather their quantum mechanical superposition.

Like classical computation, in quantum computation also the input string(s) and output string(s) are both determinate string(s). Once the input string(s) is specified, the quantum computation process is carried out through a series of mappings called Quantum Gates, using the laws of Quantum Mechanics.

Entanglement

The most exciting phenomenon of quantum mechanics is the ability of two or more quantum particles to become entangled with each other. When particles become entangled, they form a single system. The quantum state of any one particle cannot be described independently of the quantum state of the other particles. This means that whatever operation or process you apply to one particle correlates to the other particles as well.

The quantum entanglement performs an operation that a classical system cannot achieve. The entanglement is possible to be - transformed, refined, and measured in the process. The probability distribution measures in an entangled state are applied in the quantum communication channel to effectuate cryptographic and computing challenging tasks in traditional systems.

The quantum information includes the characteristics, implications, & benefits of quantum entanglement. The states of entanglement are engrossing because of the exposition of correlation with no rules of interpretation.

Non-Maximal Entanglement

The maximal entanglement states in quantum theory predicted the violation of Bell inequality. However, there is no Bell inequality violation for general states due to the local variables and procreate in the quantum theory prediction for a single qubit in the pure state. The Bell inequality violations calculated maximally tangled state in quantum theory, and it is not very confident in the general states that stated below:

$$|\varphi i = \alpha|00i + \beta|11i$$

Here, α and β are called non-negative - real.

Entanglement Benefits

The entanglements benefits are stated below,

Superdense Coding

The superdense coding will start with proper communication. For instance, Patty is sending a text to Joey. The communication will become effective for both parties when connected to the quantum channel.

No-Cloning

The no-cloning concept is used to prevent the duplication of an arbitrary unidentified quantum state in quantum physics. This no-cloning plays a vital role in quantum cryptography to avoid specific cyber breaches and creating duplication of data and a cryptographic key. The quantum entanglement helps to prevent superluminal communication.

Quantum Teleportation

This quantum teleportation approach helps transfer the quantum data from one place(location) to another place(far away). In this case, the sender will send the data or information via a specific quantum state without knowing the sender's location. Figure 1 demonstrates the transfer of information through classical bits and entangled qubits to prevent third-party intruders' privacy.

Figure 1. Quantum Teleportation

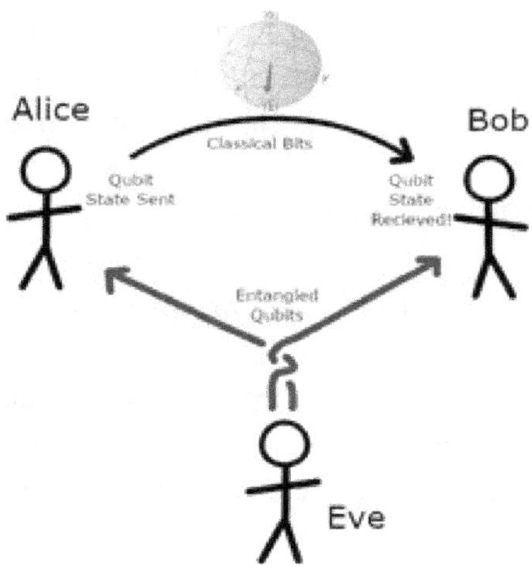

REAL-TIME APPLICATIONS OF QUANTUM TELEPORTATION

The traditional network still exists in the internet world. The quantum internet techniques, implemented to save the information passed through the channels, protecting online transactions. Quantum computing is also used in several organizations to resolve complex supply chain operations and enhance the network's speed faster than traditional network usage.

The Improved microscopy(the entanglement-enhanced microscope) was invented at Hokkaido University to increase the sharpness and the information collected through the samples. Quantum Entanglement created a new path in the technology world that was thought impossible once in the scientific experiments.

Quantum Interference

The quantum interference - a superposition by-product permits to bias the accurate quantum of the qubit to a set of required states. The following formula - used to fix ideas using qubit in superposition(ψ).

$$|\psi\rangle = \alpha|0\rangle + \beta|1\rangle$$

Kravchuk Transform

This Kravchuk transform(KT) reawakened in computer science, showing the best outcome in sound & image processing. Moreover, many algorithms should implement for this KT which results in accurate and effective outcomes.

The Principle of Quantum Interference

This section will describe the principle of quantum interference in quantum computing.

Quantum interference can be a single particle or composite system or collection of fragments with no binding or interaction. The first two categories are QTS - tight quantum system, and the rest known as LQS - Loose quantum system.

- The interference of the waves allows the LQS or QTS to interfere selves. In the case of representation of waves as a whole quantum system, it will enable to separate more than one path and finally connect up to make the occurrence of a quantum system.
- The quantum system occurs during the concurrence of waves happens in the same place and at the same time. The final quantum system results from the total summary of sub-waves.
- The exact time and the space are needed to evaluate the waves in interference. The addition of identical particles in the system also plays a prominent role during the evaluation.

- The waves on the quantum system are calculated with the mass motion number, isospin, polarization, spin, and other construction particles of quantum number.

Dirac's path integral formula to calculate the quantum interference that stated below:

ψ(x1, . . ., xn; s1, s2, . . ., sn, t)
= A Σ - all paths e i/~S[x,s,x0, s0]ψ(x 0 1, . . ., x0 n ; s 0 1, s0 2, . . ., s0 n, t0),

Relevance of Quantum Gates

The most popular model of computation for quantum computers is the quantum circuit model. A quantum circuit is a model for quantum computation in quantum information theory, similar to classic digital circuits. An analysis is a sequence of quantum gates, measurements, initializations of qubits to known values, and possibly other actions.

Single Qubit Gates

The Bloch Sphere represents the qubit states in the geometrical form to describe the quantum information clearly. The single-qubit gates resemble rotation of the axis - XY-plan can be executed using RF pulses. The states of single-qubit gates represented as

|q⟩ =cosθ2|0⟩ +eiϕsinθ2|1⟩

ϕ & θ - real numbers, In Figure 2, the Bloch Sphere diagram depicted and the Bloch Sphere's main qualities are orthogonality of opposite points and rotation operators, which help the bloch vector rotate and estimate the exponentials as represented.

Rx(θ) e−iθX/2 Ry(θ) e−iθY /2 Rz(θ) e−iθZ/2

Figure 2. Bloch Sphere

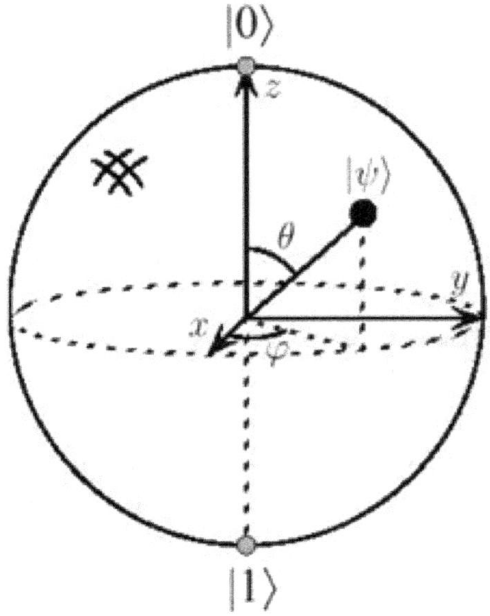

Operators Functions

This technique helps measure the exponential of paul's matrices with the property to determine the rotation operators. The measurement will be done with an X-measurement circuit, Y-measurement circuit, and z-measurement circuit. The z-components - estimated with the Z-measurement circuit. In recent papers, it has been explained that we can re-normalize the noise using the quantum state tomography and suggested that the researchers use it.

Rotation Gates

There are many possible ways available to construct single-qubit gates. The three essential gates used in single-qubit gates are Identity gate, Hadamard gate, and Pauli gates known as (I), (H), and (X, Y, Z gates). The transformations of the single-qubit gates are depicted as rotations encircling to the axis (x,y,z) known as Bloch Sphere.

The three rotations - R1x, R1y and R1z with Pauli Gates (X, Y, Z) represented as below:

R1x = [cos(φ2)−sin(φ2)i−sin(φ2)icos(φ2)] R1y = [cos(φ2)−sin(φ2)sin(φ2) icos(φ2)] R1z

$$= [ei\varphi200ei\varphi2]$$

The rotations using Pauli's gates are apparent and represent the required rotations needed for the experiments.

Digression - (X, Y, and Z) Bases

The appearance of Z-gate will not have any effect on the qubit. This is because of the eigenstates $|0\rangle$ and $|1\rangle$. The other vectors $|+\rangle$ & $|-\rangle$ by eigenstates of X-gates based on X-Basis.

$$|+\rangle = 1\sqrt{2}(|0\rangle + |1\rangle) = 1\sqrt{2} \quad [11]$$

$$|-\rangle = 1\sqrt{2}(|0\rangle - |1\rangle) = 1\sqrt{2}[1-1]$$

The Hadamard Gate

It is also called an H-gate, a primary quantum gate, and it allows the creation of a superposition and departs with poles of the BP - Bloch sphere. The matrix has represented below:

$$H1 = 1\sqrt{2}[111-1]$$

Where $H1|0\rangle = |+\rangle$, $H1|1\rangle = |-\rangle$

P-Gate

The p-gate will perform a Z-axis rotation, and it is parameterized, so it should be mentioned about the activities that take place. The matrix form is represented below with the actual number ($\phi1$).

$$P(\phi1) = [1\ 0,\ 0\ ei\ \phi1]$$

The P-Gates are available in the reference gates, including - I-Gates, S- Gates, T-Gates, and U-Gates with different notations in quantum computing. The I - Gates and S- Gates notation, included below:

I - Gates: I = XX

S - Gates: SS|q⟩ =Z|q⟩

Controlled Two-Bit Gates

The controlled two-bit gate is also known as CNOT gate, a prototype model of two-bit gates. The CNOT states as controlled not. The main concentration of the two-bit gates is the control bit and target bit. Figure 3 represents the controlled-NOT gate.

Figure 3. controlled-NOT gate

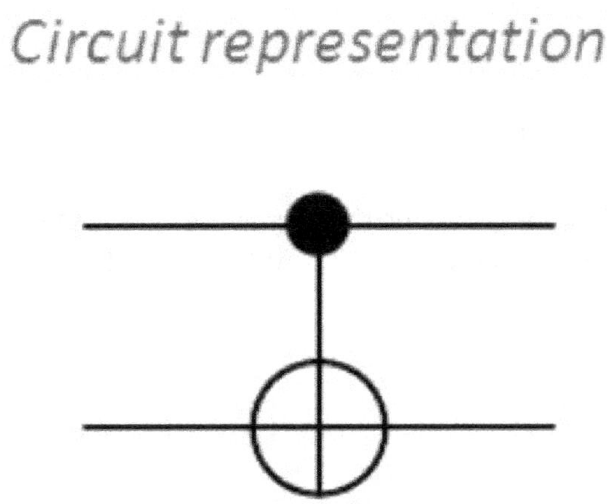

Multi-Controlled Gates

The Multi-controlled gates have four possible states – 00, 01, 10, and 11. To explain the two qubits, it is required that the state have four complex amplitudes, and the amplitudes collected should be stored in the 4D vector. Ψ represents the Quantum Logic gates and Multiple Qubits.

$$|\psi\rangle = \begin{bmatrix} a \\ b \end{bmatrix} \otimes \begin{bmatrix} c \\ d \end{bmatrix} = \begin{bmatrix} ac \\ ad \\ bc \\ bd \end{bmatrix}$$

Quantum Program

A quantum gate can be fully labeled by identifying how the operation of a quantum gate transforms the computational basis, i.e. {0⟩, 1⟩ }. The NOT(X), Hadamard, CNOT(CX) are the most commonly known gates. The NOT gate can transform 0⟩ to 1⟩ and vice versa. The Hadamard gate can generate an equal superposition state, implying that the qubit's probability of collapse to either 0⟩ or 1⟩ is the same. The CNOT gate is a bit peculiar as, unlike NOT and Hadamard, it receives two qubits as input. It performs like the 'if condition' of programming (Hassija et al., 2020). This clearly explains why we need to know the quantum gates to implement the quantum algorithm.

Quantum gates can either function on a single qubit or multiple qubits. For multi-qubit gates, target qubits and control qubits have to be distinguished. The value of the target qubits is revised in the case that the control qubits are fixed to basis state |1. The Clifford+T library, which is based on the single-qubit gates H (Hadamard gate) and T (Phase shift by $\pi/4$) along with the two-qubit gate CNOT (controlled-NOT), signifies a universal set of quantum operations which are directly required to implement Quantum algorithms. To define quantum circuits, high-level quantum languages (e.g., Scaffold or Quipper, quantum assembly languages (e.g., OpenQASM 2.0 by IBM), or circuit diagrams are engaged. In a circuit diagram, qubits are denoted by horizontal lines, which are delivered through quantum gates. Contrary to conventional circuits, this does not define a joining of wires with a physical gate but states in which order the quantum gates are applied to the qubits from left to right (Liu et al., 2020).

Quantum Algorithms

In quantum computing, a quantum algorithm is an algorithm that runs on a realistic model of quantum computation, the most commonly used model being the quantum circuit model of computation. A classical (or non-quantum) algorithm is a finite sequence of instructions or a step-by-step procedure for solving a problem. Each step or instruction can be performed on a classical computer. Similarly, a quantum algorithm is a step-by-step procedure where each step can be performed on a quantum computer. Although all classical algorithms can also be performed on a quantum computer, the term quantum algorithm is usually used for those algorithms which seem inherently quantum or use some essential feature of quantum computation such as quantum superposition or quantum entanglement. The design of quantum algorithms follows completely different principles from those of classical algorithms.

To begin with, even classical algorithms need to be cast in an exceedingly special form—as reversible algorithms—before they will be run on a quantum computer. Algorithms that achieve quantum speedups use certain quantum algorithmic paradigms or building blocks that haven't any classical counterparts The Quantum algorithms depend upon one or two of quantum building blocks described within the next section and are designed to run on an idealized quantum computer. Real quantum devices are noisy, so an elaborate theory of quantum error-correcting codes and fault-tolerant quantum computing has been developed to convert noisy quantum computers to ideal quantum computers. However, this conversion incurs an overhead both within the number of qubits. The sphere is now entering the time of noisy intermediate-scale quantum (NISQ) devices (Liu et al., 2020)—the race to create quantum computers that are sufficiently large (tens to hundreds or some thousand qubits) that they can't be efficiently simulated by a classical computer, but aren't fault-tolerant so cannot directly implement the algorithms developed for ideal quantum computers. While the big interest and funding for building NISQ computers have undoubtedly moved up the calendar for scalable, fault-tolerant quantum computers, significant work remains before each milestone is met. The largest upcoming challenges are algorithmic; this includes looking for computational tasks that such computers can speed up within the near term. Developing algorithms that run on NISQ computers is as essential as creating physical devices since the machine isn't helpful without both. Within the long term, much work remains to be worn out the sector of quantum algorithms for ideal (scalable, fault-tolerant) quantum computers. The subsequent section describes significant building blocks for quantum algorithms and known algorithms for idealized quantum computers that provide speedups over the most effective classical algorithms for the identical computational tasks. The following section describes quantum error-correction and fault-tolerance techniques for converting a loud quantum computer to an idealized quantum computer. The chapter concludes by discussing the numerous algorithmic challenge presented by NISQ computers and also the most promising leads in the seek for such algorithms. The facility of quantum algorithms ultimately derives from the exponential complexity of quantum systems—the state of a system of n entangled qubits is described by (and can thus encode) $N = 2^n$ complex coefficients. Moreover, the applying of every elementary gate on, say, two qubits updates the 2^n complex numbers describing the state, thereby seeming to perform 2^n computations during a single step. On the opposite hand, when the n qubits are measured at the tip of the computation, the result's just n classical bits. The challenge of designing advantageous quantum algorithms derives from the strain between these two phenomena—one must find tasks whose operational solutions both make use of this parallelism and yield a final quantum state with a high probability of returning valuable information upon measurement. Successful approaches cash in of the phenomenon of quantum interference for generating valuable results.

Quantum Entanglement and Its Role in Quantum Computing

What is quantum entanglement?

Quantum entanglement is that the state where two systems are so strongly correlated that gaining information about one system will give immediate information about the opposite irrespective of how far apart these systems are. This phenomenon baffled scientists like Einstein, who called it "a spooky action at a distance" because it violates the rule that no information may be transmitted faster than the speed of light wave. However, further research-validated entanglement using photons and electrons.

How is entanglement employed in quantum computing

In quantum computers, changing the state of an entangled qubit will immediately change the paired qubit's state. Therefore, entanglement improves the processing speed of quantum computers. Doubling the quantity of qubits will not necessarily double the amount of processes since processing one qubit will reveal information about multiple qubits (i.e., the entangled qubits). Consistent with research, quantum entanglement is critical for a quantum algorithm to supply an exponential speed-up over classical computations.

QUANTUM CIRCUIT MODEL

The quantum circuit model has become famous because it is a convenient model used in complex environments. The Deutsch model is the fundamental theorem that helps in quantum computation to justify the model. The input, given in the form of a tensor product state, and the output will be a unitary evolution result. The input and output of the quantum circuit model are represented below:

Input -$|0\rangle \otimes \cdots \otimes |0\rangle$

Output - $|\psi\rangle = \sum_{c_1 \ldots c_n \in \{0,1\}^n} A(c_1 \ldots c_n)|c_1, c_2, \ldots, c_n\rangle$

Changing the Qbits to Qtrits does not make any changes in fundamentals. The complexity will change if initialization has done with different states. The measurements include analyzing, collecting, reporting in the intermediate stages will not make any changes in the process. The complete sets of universal gates are not acknowledged, and many other universal gates currently exist.

Qiskit as an Implementation Platform

In addition to performing complex computational tasks, quantum researchers also focus on logical systems to build fast decision-making capabilities. Quantum system designers are trying to explore the field of implementations of logic circuits. Numerous quantum programming software is available to design and implement the quantum circuits (Wille et al. 2019). Qiskit, an open-source software development kit (SDK) by IBM, is one of them (Wille et al. 2019). It was released at the end of 2017 (Black et al., 2002). Qiskit follows the circuit model for universal quantum computation, applicable for any quantum hardware (Wille et al. 2019). For such programming, the length of the quantum circuit was no longer restricted to 80, and the design of quantum functional circuits could be packaged as functions for reusing and expansion (Black et al., 2002).

Qiskit can be used alongside IBM's quantum experience. IBM quantum experience is a cloud-based platform that allows the general scientific community to access real quantum computers with 5 or 16 qubits. It also contains a cloud-based simulator of 32 qubits(Hassija et al., 2020).

Qiskit is constructed on top of python and can run locally or in online Jupyter notebooks (Hassija et al., 2020). It has four fundamental elements: Terra, Aer, Ignis, and Aqua, which deliver different stages of abstraction. Terra offers the lowest level of abstraction and can be employed to generate quantum circuits. In contrast, Aqua gives the highest level of abstraction and can help create quantum algorithms(Hassija et al., 2020).

For the people involved in design automation, Qiskit is a perfect playground. Several problems in the field of quantum computing can be solved flawlessly by automatic approaches and experiences. But, due to the lack of coordination between the design automation people and the quantum community, numerous automatic methods planned earlier either resolved the wrong problems or could not reach the end-users. In such a situation, Qiskit offers the best platform to bring together these communities (Liu et al., 2020).

Using the QISKIT in quantum computing aims to allow the researchers to develop the experiment and deal with the problems with small quantum processors. Figure 4 represents the Open Quantum Assembly Language.

Figure 4. Open Quantum Assembly Language Diagram

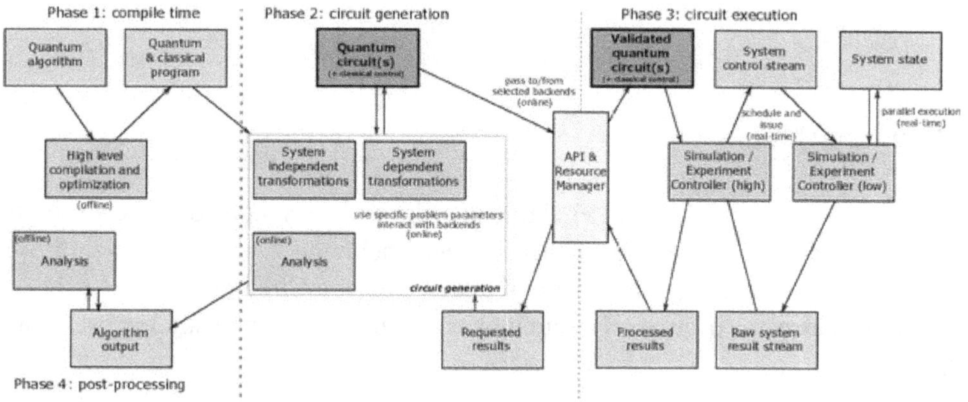

Technical Objective of Using Qiskit in Quantum Computing

The valid points are listed below about the technical objective of using the QISKIT platform in quantum computing.

- To enhance the quantum devices to increase the capability, size, and fidelity.
- The short-term circuits and pre-fault tolerance enable the investigation of error mitigation methods and algorithms regarding quantum benefits.
- To conduct practical experiments, evaluation, and simulations on the data, different interfaces, multiple simulators, analyzation tools, and circuit rewriting framework for scheduling, optimizing, and hardware mapping have been performed.
- It helps in expanding the mapping framework, increasing capabilities & characteristics over time. Moreover, allowing to maximize the abstraction level and better access and lower-level control interfaces helps on pulse shapes, timing, mapping process, etc.

Qiskit and Quantum Circuits

To execute the experimental results on quantum computing, it is necessary to install python because Qiskit is one of the packages in python. Import the files after installation.

Follow the two steps to install the Qiskit,

Install the ANACONDA - python package.
Run the command in the command prompt below after installation: pip installs qiskit.

Quantum Circuits

The quantum circuit plays a vital role in qiskit and the syntax to create an empty circuit with no outputs, and no qubits are stated below,

From qiskit import QuantumCircuit q2c = QuantumCircuit ()

For instance, the 5-qubit GHZ circuit has constructed, from qiskit import QuantumCircuit

```
q2c = QuantumCircuit(5) q2c.h (0)
q2c.cx(0, range(1, 5)) q2c,measure_all ()
```

Quantum Register

The qubits should be registered to make the circuit less trivial, and implementing a quantum register is important. The following example will show you how to register the qubits using quantum registers,

From qiskit import QuantumRegister q2c = QuantumRegister (2, ‘ b ’)

The next step is to add the quantum register to the circuit by using the following syntax,

```
qc.add_register (q2c) qc.qregs
```

Then, add another attribute using the syntax: qc.draw ()

Applying Gates and AER Simulators

Applying Gates: The gates should be applied to achieve the required output with the qubits. The following syntax will explain how to insert the values. The qr[0] and qr(Wille et al. 2019) are the two individual qubits using the controlled-NOT as cx. The two arguments, qr[0] is the control qubit, and qr(Wille et al. 2019) is the target qubit.

```
qc.cx (qr [0], qr (Wille et al. 2019))
```

Aer simulators: This states the activities of the two qubits or evaluates the output of the circuit. Using the following syntax to run the simulator, from qiskit import Aer

sv_sim = Aer.get_backend('aer_simulator')

Simplified Notations: It will make the possibility to add classical registers & multiple quantum to add in a circuit. A wide array of QuatunCircuits is rendered, and as many qubits can be added for the experiment. If the circuit has a single quantum register, the possibility with no classical register and 4 qubits is referred to below.

qc = QuantumCircuit (4)

The current devices used in the IBM Q Experience are the single-junction transmon qubits with certain gate fidelities stated for 1Q and 2Q. The 1Q gate fidelities are greater than 99%, and the 2Q gate fidelities are greater than 95%. The measurements are calculated fidelities and resulted in greater than 93%. The devices, exploited on a 5-qubit (ibmqx4) device that is convenient to access through QISKit API and Graphical User Interface (GUI). On the other hand, 16-qubit (ibmqx5) devices are convenient to access through only QISKit API. Figure 5 and Figure 6 depict the 5-qubit device and 16-qubit device.

Figure 5. 5-qubit Device

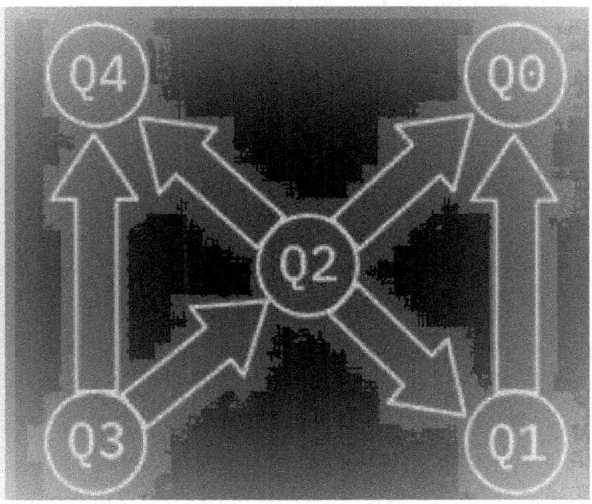

Figure 6. 16-qubit Device

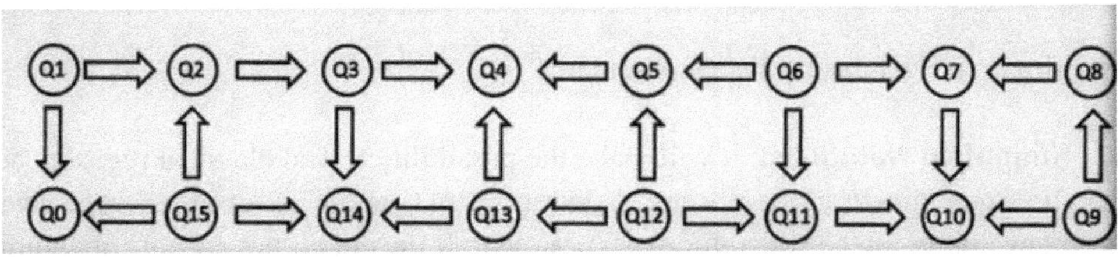

OpenQASM - Quantum Assembly Language

OpenQASM is the machine-independent language that helps to deal with the source code to compile for quantum instructions. The representation of quantum computation in OpenQASM with the quantum circuits apart from gates and qubits. The real-time classical and near-time computations are the time frames of quantum-classical interactions used for different timing purposes.

The essential features of OpenQASM are as follows:

- Allows measuring qubits.
- The barriers are used to limit optimizations in the program.
- Able to work with classically conditioned procedures.
- Able to execute register-level operations.
- Able to access other files regarding the subroutines defined.
- Possible to determine additional gates using the subroutines option.
- Classical and quantum registers are specified.

QISKit API (Application Programming Interface)

The request should be submitted in the RestAPI with the backend details, circuit set, and trials. Moreover, the results will be retrieved through various terms, such as calculating the probability of the outcome by execution time and recently collected calibration data during the execution time. Moreover, collecting backend details with static and dynamic properties will help for accurate results. Finally, to track the credits, token-based authentication procedures are used.

The examples codes on QISKit API will help you to understand the topic even better.

Runtime_program = “example_code” program_input = {”iterat”:2}

```
def run_runtime_program(program_name, program_input):
# configuing your data
params = json.dumps({
"programId": Runtime_program,
"hub": "data-q-internal",
"group": "near-time",
"project": "qiskit-runtime",
"backend": "data_montreal",
"params": [
json.dumps(program_input)
]
})
search_ID = '' response = requests.post(Runtime_API_URL + 'search',
data=params, headers=headers)
if response.status_code == 200:
search_ID = response.json()['id']
return 200, search_ID
else:
return response.status_code, None
status, search_ID = run_runtime_program(Runtime_program, program_input)
print(f' status: {status}, search: {search_ID}')
```

This source code of Qiskit can be run on the cloud. In this process, if an error persists, check the backend dictionary to update the appropriate backend data in the code. Moreover, all the information needed for the experiment will be available on the backend data. Moreover, directly use the API to run the circuit chasm code.

The implementation can be done with these three languages include Python, Swift, and Javascript. Use the following code to access in different languages.

PYTHON - qiskit-sd-py SWIFT - qiskit-sdk-swift JAVASCRIPT - qiskit-sdk-js

The central theme of the QISKit is to develop and enable research of different applications for quantum back-ends. The three central components of QISKit SKD are stated as below:

Quantum Program: Have a set of classical and quantum registers.
Transcompiler: Write the circuit transformation (new) quickly and integrate it with the existing passes. In other words - they were used for circuit optimization.
Simulators: To display results from back-ends.

How to Initialize a Quantum Program

The quantum program class is the primary interference to QISKit. It allows gathering quantum circuits and procedures to interact. Able to set up and act as a storage for quantum circuits. Moreover, it allows to export or import of text circuits in OpenQASM. Finally, able to run experiments on simulators and hardware with the backend interface.

Initialize the new program in QISKit:
CREATE -> Quantum Program
ADD -> One or More Quantum Registers
ADD -> One or More Classical Registers

Standard Extensions

The operation techniques used in circuits include - Single qubit gates, Two qubit gates, Three qubit gates, Measure, Barrier, and Reset. The additional circuit used for construction techniques - Invert gates and classical control to the gates.

Qiskit Circuit Representation

OpenQASM - The representation of circuit in texts is as follows:
Qreg q(Black et al., 2002); Cx q[0], q(Wille et al. 2019);

Quantum Circuit - The instructions in the hierarchical tree are represented in Fig 7.

Figure 7. QuantumCircuits & OpenQASM

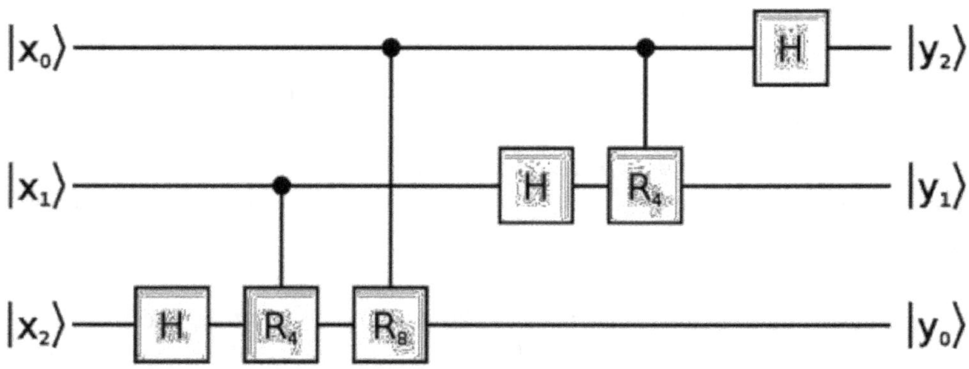

DAGCircuit - the directed graph encoded dataflow with Input nodes, Output nodes, and Instruction nodes with each gate.

Circuit Rewriting

The circuit rewriting methods uses "unroll", "swap_mapper", "cx_cancellation", "optimize_1q_gates" for different benefits.

unroll - expands loop and broaden gate definitions.
swap_mapper - insert swap gates by selecting a layout - layer by layer algorithm.
cx_cancellation - deletes the runs of CNOT gates - reduces redundancy
Optimize_1g_gates - single-qubit gates that runoff will be simplified.

The quantum algorithms, developed in quantum computers, and the applications, used in diverse fields for various reasons, and improved technologies, developed all the time. The rapid evolution of quantum theory and technology has improved a lot in diverse fields. Quantum computing develops quantum-mechanical effects to execute a computation efficiently, and its benefits reduce both the execution duration & energy consumption compared to conventional computing. The traditional network still exists, and massive information, passed in the internet world. The quantum internet techniques, implemented to save the information passed through the channels and protect online transactions. Recently, the technologies have developed and reached a maximum level, and the quantum computer can effectuate an intractable calculation on a supercomputer. The quantum algorithms proposed to search optimization, solve complex linear equations, and cryptography.

Moreover, the different quantum algorithms implemented in quantum computers enhance efficiency and speed up the process with classical algorithms. The quantum software Qiskit used to write quantum computing codes with different stages include building and execution stages. The single Qubit gates, controlled two-bit gates, and multi-controlled gates help identify the rotations of different dimensions of the plans. The three phenomena of quantum computing will be explained in detail on Superposition, quantum measurement, and entanglement to evaluate its functioning.

The two-state processes enhance faster and reliable computation, and various applications are used in different scenarios to get a prominent output. The superconducting qubits, multiple qubits, single qubits, and entangled states help to make the process quicker with high-speed static memory. The usual computer we use in our daily lives will provide only 0s and 1s as output and input. However, the qubits allow using 0s, 1s, or a combination of both during the process. The advanced systems in the storage boost the development and consistently store the data. In addition, the quantum computing features enhance the researcher's methodology and improve the process at a faster pace.

Quantum mechanics is a complicated field, and day-to-day researchers are getting more unique experiences in diverse fields. The different pack of rules followed in everyday life generates powerful technology. Every day the messages passed on the internet are processed and stored in 0's and 1's. However, the qubits allow to use of the input differently, and the information kept track while applying the gates. The matrices and vectors are the two pillars of mathematical calculation.

Moreover, different kinds of frameworks and platforms are used to generate an appropriate output. Splitting the information into bits and again processing the bits into text messages protect the data from intruders in one way or another. In this case, the quantum circuits play an important role, and the measurement happens to each qubit. The different API references include Qiskit Terra, Aer, Ignis, Aqua, and IBM quantum provider to make the coding part easy and flexible for the various processes. The API error codes help convey appropriate messages to the researchers or scientists with optimum and accurate solutions.

GROVER'S ALGORITHM

Grover's algorithms are used to sort out the significant issues in quantum techniques. Grover's Algorithm solves the unstructured search issues. The algorithm can speed up the process and acquire quadratic run time enhancements for other algorithms. This process is also known as the amplitude amplification trick. The different implementation techniques are used in a 2-qubit implementation, 3-qubit implementation and still, researchers are exhibiting Grover's algorithm for 4-qubit implementation. Moreover, it is also used in resolving the collision issues with the unordered states and binary function.

The unordered states are represented as N = 2n;X={x1, x2, ..., xN}

Binary Function: f: {0,1}. The iterations are calculated in Grover's Algorithm according to the performance of amplitude and oracle. Every iteration is denoted by O($\sqrt{}$ 1 N). Grover's techniques experiment with the capability and help accelerate the unstructured search issues; however, it is used for several purposes and processes as a subroutine to acquire quadratic improvements in run-time for various kinds of other algorithms. This trick is known as amplitude_amplification_trick.

Grover's Algorithm, segregated as follows:

- Use Hadamard gate to use the qubits in the place of superposition.
- To find the state, execute an oracle.
- The amplification should be implemented to increase the state.
- Finally, measure the qubits.

The methods and steps of Grover's Algorithm, represented in 4.1.4

Initialization

The qubits are determined in superposition, and the states are calculated with their amplitude.

Oracle Creation

The issues are solved by adding the negative phase to Grover's algorithm's result or solution states. The states will be added as follows:

Uω|x⟩ ={|x⟩ ifx≠ω−|x⟩ ifx=ω

The quantum oracle formula:

$$U_{\omega}|s\rangle = U_{\omega}\frac{1}{2}(|00\rangle + |01\rangle + |10\rangle + |11\rangle) = \frac{1}{2}(|00\rangle + |01\rangle + |10\rangle - |11\rangle)$$

or

$$U_{\omega} = \begin{bmatrix} 1 & 0 & 0 & 0 \\ 0 & 1 & 0 & 0 \\ 0 & 0 & 1 & 0 \\ 0 & 0 & 0 & -1 \end{bmatrix}$$

Amplification

The amplitudes target state will be inverted when the original states retain their actual amplitudes. Therefore the flipping of the target state increases and the other actual amplitude will decrease.

Measurement

The qubits are measured by the zigzag line, which helps prevent the backend optimization, reorganizing the gates.

Grover's Algorithm Steps:

Step - 1: Initialize qubits 2^n on the |0> state.
Step - 2: Apply Hadamard gates to individual qubit by creating superposition equally on all possible states.
Step - 3: Pass the states via the quantum oracle.
Step - 4: Then, Pass the states via Grover diffuser.

Step - 5: Repeat 2 & 3 steps.

Sample Code

Grover's algorithm sample code on Qiskit implementation is mentioned below:

```
Import matplotlib.pyplot as plt Import numpy as np
from qiskit import IBMQ, Aer, assemble, transpile
from qiskit import QuantumCircuit, ClassicalRegister, QuantumRegister from
qiskit.providers.ibmq import least_busy
from qiskit.visualization import plot_histogram
```

The following source code is used to define the Oracle: grover_circuit.cz(0,1)

```
grover_circuit.draw()
```

The diffusion operator is denoted as below, grover_circuit.h ([0,1]) grover_circuit.z([0,1]) grover_circuit.cz([0,1]) grover_circuit.h([0,1])

```
grover_circuit.draw()
```

The different experiments are performed via simulators Real Devices to measure the majority of the cases. Figure 8 represents the 3 qubits with marked states: |101⟩ & |110⟩. The phase-Oracle is used to solve the problem of the quantum circuit.

Figure 8. Phase-Oracle Init Oracle, Amplification

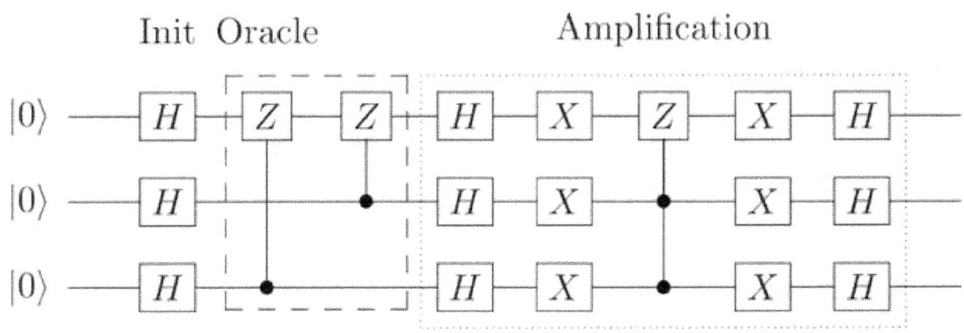

Purpose of Grover's Algorithm

The primary purpose of this algorithm is to search a database. It has a significant advantage in satisfying the logical expression using the operators to measure it. It is proven that Grover's algorithm is way better than the classical methods. Moreover, Grover's algorithm in qiskit is straightforward and practical with easy and simple steps. And also, the algorithm will speed up the computation process and in unstructured search in databases. Grover's algorithm should be modified according to the demonstration and data needed for the experiments.

COMPARISON OF QISKIT WITH OTHER QUANTUM SOFTWARE PLATFORMS

Cirq is also the same as Qiskit, but Google's software implements cirq for optimizing, coding, and manipulating quantum circuits for simulators and quantum computers. Moreover, QuTiP, Ocean (D-Wave's Tool), and ProjectQ are also the best tools that can be used in quantum computers to make your operations easy. The comparison table of QISKIT with other quantum software platforms is shown in Table 1.

Table 1. Comparison of QISKIT with other quantum software platforms

Overview	Qiskit	Forest	ProjectQ	QDK
Host Language	Python, Java-script, &Swift	Python	Python	C#
Quantum	IBMQX2,	8 qubits	IBM Back-ends	-
Hardware	IBMQX4,IBMQX5, QS1_1(qubits-5,5,16,20)			
Simulator	30qubits via cloud, 25 qubits - locally	20qubits - locally, 26 qubits - API to QVM, 30+ private access	28 qubits - locally	30 qubits - locally, 40 via Azure cloud
Features	Aqua Library, Community Slack channel, topology compiler, circuit drawer, QASM code	Quil code, community Slack channel, simulator, topology- specific compiler	Multiply library - plugins, draw circuits, IBM back-ends	Example algorithms , built-in algorithms
Institution	IBM	Rigetti	ETH Zurich	Microsoft

continued on following page

Table 1. Continued

Overview	Qiskit	Forest	ProjectQ	QDK
Requirements	Anaconda 3, Python 3.5 +, & Jupyter Notebooks	Anaconda, Python 3	Python 2 or 3	Visual Studio

CONCLUSION

A quantum circuit is a computational procedure that combines coherent quantum operations on quantum data like qubits with real-time classical computation. It's a system of quantum gates, measurements, and resets that can be conditioned on and fed from real-time classical processing. If any unitary transformation of quantum data can be efficiently approximated arbitrarily well as a sequence of gates in the set, it is said to be universal. A sequence of quantum circuits and non-concurrent classical processing can be used to describe any quantum program.

REFERENCES

Abdollahi, A., & Pedram, M. (2006, March). Analysis and synthesis of quantum circuits by using quantum decision diagrams. [). IEEE.]. *Proceedings of the Design Automation & Test in Europe Conference*, 1, 1–6.

Asfaw, A., Corcoles, A., Bello, L., Ben-Haim, Y., Bozzo-Rey, M., Bravyi, S., Bronn, N., Capelluto, L., Vazquez, A. C., Ceroni, J., Chen, R., Frisch, A., Gambetta, J., Garion, S., Gil, L., De La Puente Gonzalez, S., Harkins, F., Imamichi, T., Kang, H., Karamlou, A. H., Loredo, R., McKay, D., Mezzacapo, A., Minev, Z., Movassagh, R., Nannicini, G., Nation, P., Phan, A., Pistoia, M., Rattew, A., Schaefer, J., Shabani, J., Smolin, J., Stenger, J., Temme, K., Tod, M., Wood, S., & Wootton, J. (2020). *Learn Quantum Computation Using Qiskit.* Qiskit-community.

Aharonov, D., Kitaev, A., & Nisan, N. (1998, May). Quantum circuits with mixed states. In *Proceedings of the thirtieth annual ACM symposium on Theory of computing* (pp. 20-30).

Black, P. E., Kuhn, D. R., & Williams, C. J. (2002). Quantum computing and communication. [). Elsevier.]. *Advances in Computers*, 56, 189–244.

Burgholzer, L., Raymond, R., & Wille, R. (2020, October). Verifying results of the IBM Qiskit quantum circuit compilation flow. In *2020 IEEE International Conference on Quantum Computing and Engineering (QCE)* (pp. 356-365). IEEE.

Ferrari, D., Tavernelli, I., & Amoretti, M. (2021). Deterministic algorithms for compiling quantum circuits with recurrent patterns. *Quantum Information Processing*, 20(6), 1–21.

Han, K. H., & Kim, J. H. (2000, July). Genetic quantum algorithm and its application to combinatorial optimization problem. In Proceedings of the 2000 congress on evolutionary computation. CEC00 (Cat. No. 00TH8512) (Vol. 2, pp. 1354-1360). IEEE

Hassija, V., Chamola, V., Saxena, V., Chanana, V., Parashari, P., Mumtaz, S., & Guizani, M. (2020). Present landscape of quantum computing. *IET Quantum Communication*, 1(2), 42–48.

Iwama, K., Kambayashi, Y., & Yamashita, S. (2002, June). Transformation rules for designing CNOT-based quantum circuits. In *Proceedings of the 39th annual Design Automation Conference* (pp. 419-424).

Liu, J., Bello, L., & Zhou, H. (2021, February). Relaxed peephole optimization: A novel compiler optimization for quantum circuits. In *2021 IEEE/ACM International Symposium on Code Generation and Optimization (CGO)* (pp. 301-314). IEEE.

Liu, J., & Gregory, T. Byrd, & Zhou, H. 2020. Quantum Circuits for Dynamic Runtime Assertions in Quantum Computation. In Proceedings of the Twenty-Fifth International Conference on Architectural Support for Programming Languages and Operating Systems (ASPLOS). Association for Computing Machinery, New York, NY, USA, 1017–1030. DOI:https://doi.org/10.1145/3373376.3378488

Liu, W., Chen, J., Xu, Y., Tang, J., Tong, L., & Song, X. (2020). Quantum Algorithms and Experiment Implementations Based on IBM Q. *Computers, Materials & Continua.*

Nachman, B., Provasoli, D., de Jong, W. A., & Bauer, C. W. (2021). Quantum Algorithm for High Energy Physics Simulations. *Physical Review Letters*, 126(6), 062001.

Preskill, J. (2001). Lecture notes for ph219/cs219: Quantum information and computation.

Shao, C., Li, Y., & Li, H. (2019). Quantum algorithm design: Techniques and applications. *Journal of Systems Science and Complexity*, 32(1), 375–452.

Singh, P. N., & Aarthi, S. (2021, February). Quantum Circuits–An Application in Qiskit-Python. In *2021 Third International Conference on Intelligent Communication Technologies and Virtual Mobile Networks (*ICICV) (pp. 661-667). IEEE

Wille, R., Van Meter, R., & Naveh, Y. (2019, March). IBM's Qiskit Tool Chain: Working with and Developing for Real Quantum Computers. In *2019 Design, Automation & Test in Europe Conference & Exhibition (DATE)* (pp. 1234-1240). IEEE.

Zhang, X., Xiang, H., & Xiang, T. (2018, December). An efficient quantum circuits optimizing scheme compared with qiskit (short paper). In *International Conference on Collaborative Computing: Networking, Applications and Worksharing* (pp. 467-476). Springer, Cham.

Zulehner, A., & Wille, R. (2019, January). Compiling SU (4) quantum circuits to IBM QX architectures. In *Proceedings of the 24th Asia and South Pacific Design Automation Conference* (pp. 185-190).

Chapter 7
QKD Protocol for Securing the Communication With Real-Life Application Scenarios

Hasan Abbas Al-Mohammed
Qatar University, Qatar

Elias Yaacoub
Qatar University, Qatar

Khalid Abualsaud
Qatar University, Qatar

ABSTRACT

QKD is a technique for sharing a secret key between two parties by utilizing quantum mechanics. Two well-known protocols that contributed to securing the communication are BB84 and E91. This chapter discussed the principle of quantum security and cryptography and emphasizes the recent developments and potential applications of new and emerging applications of these techniques in security in current applications or the future; moreover, different scenarios for using QKD lengths such as seeds for generating keys to encryptions messages, using QKD as key for DES or AES algorithms, also, using QKD in real-life scenarios such as healthcare in personal area networks for protecting the privacy of the patients' data, or railway monitoring scenarios to encrypt the collected data generated by the sensors are discussed.

DOI: 10.4018/978-1-7998-9522-0.ch007

INTRODUCTION

Due to increased attack number and complexity, uniformly linked computers, attack speed, and attack tool availability and simplicity, all these facts make hacking the number one crime to worry about, as security incidents grew from 3.4 million in 2009 to reach 42.4 in 2014. Cryptanalysis is the science of stopping unofficial access to private information, as well as protecting the secrecy and security of files and other data. The difficulties of certain numerical procedures, including integer factorization or indeed the discrete logarithms problem, are the foundation of today's encryption technology. Nevertheless, because these challenges are not typically recognized to be difficult for a malevolent person with quantum computation abilities, the resulting cryptographic protocols are supposedly weak (Babber & Singh, 2021).

Code-based cryptosystems, such as the Diffie-Hellman key exchange and the Rivest-Shamir-Adleman (RSA) and ElGamal cryptosystems, are among the most promising encryptions that still rely on the hardness of the integer factorization or discrete logarithm problems (Fernandez-Carames, 2019).

Quantum computers are now the digital world's reality. It is a fact that as a new invention arrives, it acts as a solution to the current challenges, but also carries fresh security concerns as are the case for Quantum Computing. By quickly solving complex mathematical problems, these machines are able to crack the existing public key infrastructure, such that can be broken by Shor's algorithm. In addition, post-quantum cybersecurity is now one of the most widely studied areas of cryptology to model the age of the post quantum computer such as multivariate public key cryptosystem(Broadbent & Schaffner, 2016).

Hence, new attack surfaces are now being presented in the IoT environment. Such attack areas are triggered by interconnected and interdependent IoT systems. As a result, the protection against the threats is at greater danger in IoT applications than in other applications, and conventional cryptographic solutions may be inefficient for these kinds of technologies.

The formation of symmetric keys between remote parties over an insecure network is one of the most fundamental cryptographic primitives, and it underlies many modern cryptographic techniques. To do this, public-key encryption is frequently utilized.

Quantum key distribution (QKD), like classical public-key cryptography, permits key establishment over an untrusted network. This technique is identified as the distribution of the keys (quantum), therefore the abbreviation QKD. The safety of the QKD is based on quantum mechanics phenomena (natural) rather than the sophistication of numerical issues, and it can be demonstrated also in contradiction of an eavesdropper, Eve, who possesses infinite computing capacity.

If indeed the shared key is only used once and is produced for the duration of the message (one-time pad), the encrypted message cannot be cracked using any computing power, counting the greatest effective equipment. This type of security refers to the theoretical security of information. Meanwhile the introduction of the novel protocol, BB84, various quantum security protocols, such as the E91 protocol, have been created and introduced for the dissemination of the key that was established to provide confidentiality between transmitter and receiver (Pirandola et al., 2017).

In an end-to-end connection, QKD produces a private key made up of randomized photons. A sender, a receiver, and two communicating networks make up QKD communication. The first is a quantum channel, which links the generator and receiver ends for the transmission of quantum random-bit signals, and the second is a traditional channel that links the major counterparties (Gisin, 2018).

The encoder (polarization filter), decoder at the receiver side (polarization filter), and inner bits of the transmitter and receiver are all inaccessible to an attacker with limitless processing capacity; thus, the attacker (Eve) is not capable to find (know) the key.

It is critical to secure the traditional channel, often known as the public channel. The sender and receiver will detect Eve eavesdropping on the quantum channel in any form permitted by the laws of physics by monitoring the change in the threshold calculated before transmission. In comparison to the BB84 and B92 protocols, the EPR Protocol additionally uses Bell's inequality to define the presence or absence of Eve as a secret feature if the mistake exceeds a particular threshold and results in the discarding of random bits along with those limitations. The EPR quantum protocol is a three-stage protocol (Pfister et al., 2016).

On the other hand, the public map of a multivariate public key cryptosystem (MPKCs) contains a set of quadratic polynomials over a finite field. The NP-hardness of the problem of solving nonlinear equations over a finite field is provided by its primary safety assumption. This area is called to be one of the big PKC groups that even the strong quantum computers of the future could theoretically avoid. In the last two decades, there has been rapid and intense growth in Multivariate Public Key Cryptography (Malina et al., 2018). Some constructions are not as stable as was originally believed, but others are still viable. A Key Encapsulation Mechanisms (KEM) enables a symmetric key to be encapsulated under any public key inside a ciphertext, such that the symmetric key can be decapsulated again (only when) the corresponding hidden key is identified. Security-wise, indistinguishably from a random string, the ciphertext conceals the encapsulated symmetric key (Hashimoto, 2021).

In addition, QKDDifferential Phase Shift is a protocol suitable for fiber transmission systems and provides higher efficiency of key output than traditional BB84-based fiber. A photon divided into three pulses is sent from Alice to Bob in this scheme, where the phase difference between two consecutive pulses carries bit data. Via passive differential phase detection, Bob tests the phase difference.

This book chapter will discuss:

- The principle of quantum security and cryptography emphasizing the recent developments and potential applications of new and emerging applications of these techniques in security in current applications or in the future.
- Different scenarios for using QKD lengths such as: seeds for generating keys to encryptions messages, using QKD as key for DES or AES algorithms.
- Using QKD in real life scenarios such as healthcare in personal area networks for protecting the privacy of the patients' data, or railway monitoring scenarios to encrypt the collected the data generated by the sensors.

QUANTUM CRYPTOGRAPHY

The use of conventional cryptographic methods and techniques is not always secure. It provides a certain level of security that is dependent on several characteristics. Consider an encryption method: if the system or method guaranteed that the attacker or hacker could know for sure nothing about the ciphertext, it could be deemed secure; nevertheless, this is hard to do in practice.

On the other hand, quantum cryptography is the art of utilizing the quantum mechanics features on the data to do a certain act to secure it, also, it is the science of denying an attacker or unauthorized person to access a precious location (website, system, etc.). Furthermore, if the malicious nodes have a powerful computer or quantum system, theoretically, it is impossible to hack quantum cryptography.

Quantum computers potentially support the development of new, more secure, and efficient encryption methods that are impossible to create with current computing and communication structures. Though many aspects of quantum security are still theoretical rather than practical, some key scenarios where encryption systems and quantum computing collide are critical to the near-term future of cybersecurity (Staff, 2020).

Quantum mechanics has different characteristics and formalisms that give quantum cryptography's potentiality to be unbreakable. Such circumstances are like the uncertainty principle, the no-cloning theory, and quantum entanglement.

Uncertainty Principle

Uncertainty principle's notion has got prodigious importance in the development of quantum theory that describes the physical system. The uncertainty principle, first proposed in 1927 by German physicist Werner Heisenberg, holds that the more precisely a particle's position is established, the less exactly its momentum can be predicted given initial conditions, and vice versa. Earle Hesse Kennard developed the formal inequality linking the standard deviation of location σ_x and the standard deviation of momentum σ_p. After conquering early misunderstandings and ambiguity, the notion grew steadily but is still an extremely productive topic for researchers today. With the discovery of modern measurements of uncertainty or indeterminacy and the development of quantum measurement theory, fascinating new insights and correlations are discovered and created.

Heisenberg claimed how this relationship is a simple and direct mathematical repercussion of the quantum states commutation principle for the position and relating momentum operators $pq\text{-}qp=\text{-}i\hbar$, where $\hbar$ is the modified Planck constant $h/(2\pi)$; however, he typically extracted the relationship relating a semi-quantitative definition of imprecision and indeterminacy in the position coordinate q and the relating momentum p in terms of 'spreading ' of the Gaussian 'probability–amplitude packet' of a particle (like such an electron) (Sen, 2014). This phenomenon leads to the use of the photon to be utilized for security purposes such as quantum key distribution.

No-Cloning Theory

A hidden variable conjecture tries to clarify the unpredictability of testing outcomes by varying the amounts of some variables that represent the object's “real” qualities. As a result, most “no-go” proofs began by connecting probability distributions S(d) with quantum states S and random variables with quantum observables X, where $\omega \in \Omega$ - a hypothetical phase space. As a result, it was commonly considered one-to-one equivalences such as S^ ↔ S and X^ ↔ X. (Holevo, 2011). The aspect of this theory has led to the more comprehensive theory, which is called no-cloning theory. It claims that there is no way to build a piece of equipment that can accept a generic qubit as input and return the prior form with a copy of some of the information in it as output. It is not possible, for instance, to make perfect duplicates of all the input states while keeping the source undamaged. The apparatus can measure a flowing set of measurements, but it must subsequently remove the information in the initial state stored in the respective phases of the associated eigenspaces, which is a closely connected basic fact (Lindblad, 1999).

Quantum Entanglement

Entanglement is a quantum mechanics (system) phenomenon, where two or more particles or even microscopic bodies are said to be entangled if they have been characterized by a combined wave function that cannot be stated as a combination of the wave functions of each subsystem (or, for mixed states, if a density matrix cannot be written as a weighted sum of product density matrices). Even though they are arbitrarily far apart, the subsystems cannot be said to have their state. Entanglement creates linkages among subsystems that go beyond what is conventionally feasible. Quantum communication techniques like teleportation and superdense coding are made possible by this property. However, preparing, exchanging, and purifying entanglement is typically a time-consuming and costly procedure that necessitates extreme caution. As a result, it's critical to comprehend all of the processes that could affect quantum entanglement (Gingrich & Adami, 2002).

Quantum characteristics are being used in two different cryptographic applications that are currently being developed.

- **Quantum-safe cryptography:** The creation of encryption algorithms, often known as post-quantum cryptography, that are secure from quantum cyber-attacks and can be adapted to provide quantum-safe certificates, is out of this chapter scope.
- **Quantum key distribution:** The method of establishing a shared key between two trusted parties utilizing quantum communication so that an untrustworthy eavesdropper cannot discover something at all from the key; the chapter will highlight this topic.

The rest of this chapter is organized as follows. the section 2 describes the quantum key distribution. The QKD applications are described in Section 3 such as Section 3.1 utilizing the QKD for healthcare, 3.2 is the railway application of QKD, 3.3 is the application layer that used QKD, QKD in described 6G era in 3.4, and the QKD used as a pseudorandom number generator in Section 3.5. Finally, Section 4 concludes the chapter.

QUANTUM KEY DISTRIBUTION

QKD is a mechanism for sharing a secret symmetric key between two communicating parties (Alice and Bob). If an unauthorized user (Eve) attempted to obtain the key from a QKD protocol, communicators could use quantum principles to detect it (e.g., the well-known Heisenberg uncertainty theory). In quantum mechanics, it is impossible to observe a system without forcing it to be disrupted, therefore, QKD was established. Eve is interested in listening on the two participants of quantum communications, so she would leave any detectable indications. As a result, the security of the QKD protocol is ensured. However, as long as Eve remains passive, the protocol generates lengthy keys at random. Furthermore, as Eve tampers with quantum connections, the protocol detects the threat and stops the creation of keys. The protocol will not be suspended as long as Eve is inactive. For any quantum network assault, the chances of the protocol not stopping and an unauthorized person copying the created keys are vanishingly small(Zhu & Liu, 2018).

QKD produces a secret key of random photons in an end-to-end interaction (Qubit). A sender, a receiver, and communication between two networks make up QKD communication. The first is a quantum link, which connects the encoder and decoder sides to transmit quantum random-bit signals, and the second is a traditional channel. The quantum and classical networks are accessible to an attacker with infinite processing power, but not the transmitter (polarization filter) or receiver (polarization filter), hence the unauthorized party (Eve) is unable to find the key. It is critical to verify the traditional channel, often known as the transmission medium. The sender and receiver can identify Eve eavesdropping on the quantum channel by analyzing the change in the threshold that is established before the communication. The sender and receiver discontinue producing the secret key if the error exceeds a particular threshold, resulting in the discarding of random bits along with those limitations (see figure.1) (Al-Mohammed & Yaacoub, 2021).

Figure 1. A quantum link (the red line) and an authorized conventional public network are usually included in a QKD protocol (black line)

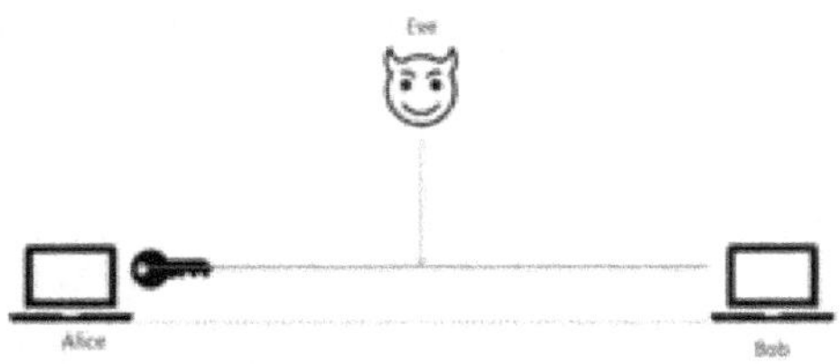

QKD APPLICATIONS

Quantum mechanics offers unconditional security by utilizing its features, authorized parties can share secret keys with complete security using QKD protocols. The security of QKD is based on physical rules, as opposed to standard cryptography approaches that rely on the computational complexity of mathematical functions. Traditional cryptography is vulnerable to computing power; however, QKD systems can withstand infinite computational power. When quantum computers become available, QKD will be able to protect our security. The No-Cloning Theorem is a result of quantum mechanics' basics, claiming that making a perfect clone of a quantum state is unachievable (Gyongyosi & Bacsardi, 2019). Moreover, significant advancement in communications technology, like the internet of things (IoT), has exceptionally transformed the sensing and prediction of the surrounding environment's parameters. IoT Innovation enables modernization that enhances the value of life and is capable of collecting, quantifying, and understanding the environmental conditions. This helps to simplify the new ways of communication between things and humans, therefore facilitating the generation of intelligent modern cities. Internet of things is also one of the fastest emerging areas from the beginning of the computing fields with approximately fifty billion devices by the start of the year 2021. IoT technology plays a critical part in improving actual life through smart technologies, such as intelligent universal health care, intelligent homes, intelligent transport, and intelligent education.

Conversely, the implementation of security in the landscape of IoT devices has faced significant difficulties. Keeping the security needs in a significant IoT threat environment is a challenge. Challenging internet of things security structures are attributable not just to limited computing, connectivity, and power consumption resources, but also to reliable interactions with the physical area, in particular the behavior of the environment in unforeseen and random modes, because the internet of things structure is correspondingly part of the cyber-physical structure. Furthermore, independent IoT systems should continue to figure out how to survive in a comprehensive and consistent manner, to safeguard operation as the main concern, especially in situations under which threatening circumstances, like the healthcare services, may arise (Di Candia et al., 2020).

QKD for Healthcare

By combining the virtual and physical worlds, the Internet of Things (IoT) creates an integrated communication ecosystem of interconnected objects and applications. The exchange of health information has become a regular routine since the introduction of remote digital healthcare-based IoT technologies. Furthermore, today's globe faces numerous concerns relating to public health issues of chronic conditions caused by dangerous viruses as COVID-19. The growing health concerns, combined with rising healthcare expenditures, pushes everyone, especially older people and disabled, to employ computer-assisted technology for remote health monitoring (Tataria et al., 2021). IoT has played an increasingly crucial role in enabling automation in various industries, including remote and smart healthcare systems, in recent years. In terms of technology, the IoT includes wireless body area networks (WBANs) and radio frequency identification (RFID) to transport data to the cloud for analysis and extraction of important data for on-time proper decision making. IoT can be explored and used as a development of new techniques in health management systems as the need to make healthcare more personalized, proactive, and cost-effective grows (Haghi Kashani et al., 2021).

As a result, an efficient model for ensuring the security and integrity of the hospital's medical data transferred and collected from the IoT environment must be developed. To do this, asymmetric encryption approaches and system encryption algorithms are combined to conceal digital information in an image (Elhoseny et al., 2018).

In fact, in a mobile healthcare (mHealth) environment, IoT sensors can communicate with an IoT controller over a body area network (BAN). A fiber optic connection can connect the controller to a remote server. QKD can be performed using this connection. The keys gained can then be transmitted to IoT sensors, which can subsequently use them to encrypt communications across wireless networks between sensors and controllers. As a result, the BAN measures can be securely provided without violating the patient's privacy (Al-Mohammed et al., 2021). Fig. 2 corresponds to a mobile healthcare (mHealth) scenario.

Figure 2. System model: mHealth scenario (adapted from Al-Mohammed et al., 2021)

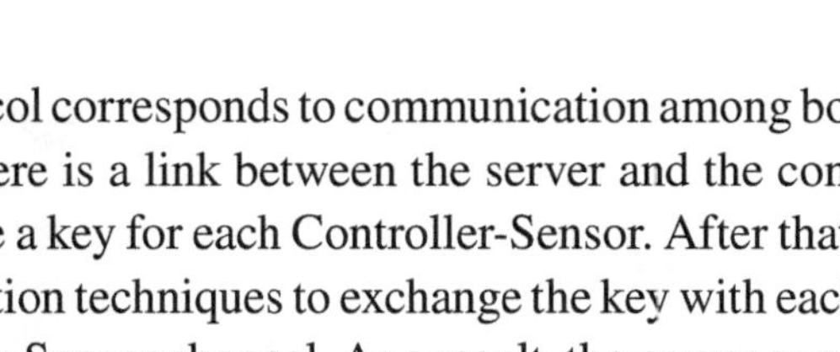

The QKD protocol corresponds to communication among both the server and the IoT controllers. There is a link between the server and the controller, and QKD is utilized to exchange a key for each Controller-Sensor. After that, the controller uses typical key distribution techniques to exchange the key with each sensor through the wire-free Controller-Sensor channel. As a result, the power consumption associated with the generation and transmission of a long secure key is the responsibility of the server and controller.

Encryption and cryptography are the technique of encrypting messages in such a way that only authorized persons may read them. AES is a symmetric encryption, which means it uses the same key on both sides. It uses keys with lengths of 128,192, or 256 bits and a fixed message block size of 128 bits of text (plain or encrypted). Larger messages are separated into 128-bit blocks and are sent. Longer keys make the encryption quite hard to crack, but they also increase the encryption and decryption process take longer, these technique can implemented between the IoT devices and the controller (Al-Mohammed et al., 2021; Daemen & Rijmen, 1999).

QKD for Railway Application

Significant advancement in communications technology, like IoT, has exceptionally transformed the perception of the surrounding environment. IoT Innovation enables modernization that enhances the value of life and is capable of collecting, quantifying, and understanding the environmental conditions (Dastjerdi, Amir Vahid and Buyya, 2016), e.g., in the railroad systems, where rail track parameters (for example, tilt, dip, temperature, and etc.,) maybe observed instantaneously with a range of sensor devices. The sensor data would be given to an IoT controller, which would then route it to a control center, where the railroad parameters would be checked in real-time to assure the track's stability. Fiber-optic connections would be used to implement QKD. In this case, QKD would be utilized to execute via fiber-optic links. Besides railway tracks, optical fibers are routinely deployed. As a result, the remote server and the local controller may conduct QKD many times over the fiber optic line (Al-Mohammed et al., 2021). Furthermore, the created keys can be used to encrypt data transferred via the "IoT Sensors – Controller" network where RF wireless communications are utilized to create linkages. Different technologies, such as millimeter-wave (mmWave) or 5G mMTC communications, use more "conventional" methods and RF frequencies for communication (Diaz, 2015).

Figure 3. railway's scenario (adapted from Al-Mohammed et al., 2021)

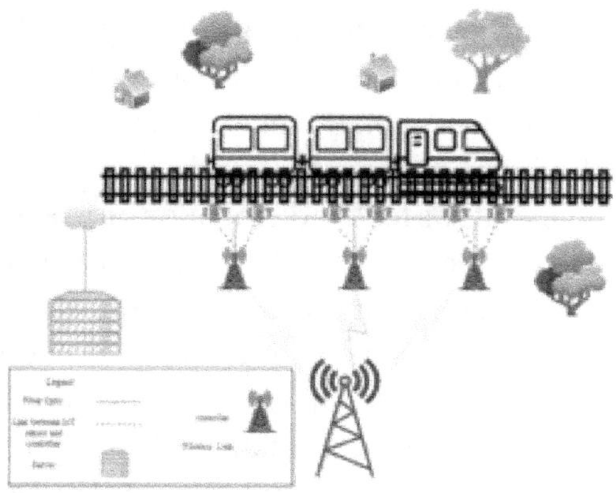

The implementation of the scenario outlined before is shown in Figure 3. Fiber can be placed at a lower cost by using micro-trenching on the road's border. Parallel to the rail track, several fibers optic cables, each with a significant number of fibers, can be run. A pair of fibers might be separated from the network and assigned to each IoT controller to perform QKD over the optical link. The RF link between the controllers and perhaps WiFi access points, Zigbee controllers, cellular base stations (BSs), and remote radio heads (RRHs) installed parallel to the longitudinal rail track corresponds to the RF link where data is encrypted using symmetric keys transferred by QKD (Yaacoub, 2021). A wireless link to the BS is shown in Figure 3; this is also connected through fiber to the 5G network infrastructure.

QKD for Application Layer

The QKD Platform is planned to be used in a variety of applications, including point-to-point and multi-party to multi-party networks, like those that proposed system smartphones communication between various endpoints. An application-independent crypto-key supply that matches multi-party communication is a crucial requirement for the QKD platform. Because there are numerous types of QKD protocols, such as BB84, the second assumption is crypto-key management that correlates to various forms of QKD systems. The QKD platform should also be able to accommodate a wide range of network topologies (Liu et al., 2018).

Therefore, to address the above objectives, a three-layer QKD platform design was presented in (Sasaki, 2018), which includes a quantum layer, a key management layer, and a security layer. The key management server is a centralized management server, and the key supply agents supply secure keys to applications. The key management actors are agent networks that get quantum keys through QKD connections, the key management server is indeed a centrally controlled software, and the key supply agents ensure that the keys to programs are secure. Each quantum layer QKD connection generates quantum keys in its way. The keys are subsequently forwarded to the key management layer's key management providers (servers). The key management layer's responsibilities include storing and relaying the quantum key as well as network administration.

The key relay technique allows for key distribution on a range of network designs including long-distance connections. The key supply layer serves two purposes: software key supply and secure key transmission out from the QKD system to target communication methods (key consumers). A centrally managed strategy is used by the QKD platform. Moreover, the key distribution server also keeps a list of the QKD system's status, including the errors of all QKD links and the total number of produced keys that have been collected and then used. The key distribution server in the QKD system also selects the key relay route and controls re-routing. The Tokyo QKD Network's network design has been updated and is presently in operation (*Tokyo QKD Network*, 2021).

QKD With 6G Era

Currently, network capacity is being legislated to be 1000x expanded on a constant basis, which is essentially driving the evolution of wireless networks. While growth of wireless capacity will keep rising, the increase of the Internet of Everything (IoE), which connects large numbers of people and huge amount of devices, has resulted in a revolutionary fundamental change away from rate-centric 5G enhanced mobile broadband (eMBB) facilities moving to 6G mobile broadband reliable low latency communications (MBRLLC) (Saad et al., 2019). Despite the fact that now the 5G cellular system was promoted as a crucial IoE enabler, the essential premise of 5G — by means of a real carrier of IoE services — has yet to be fulfilled, despite concerted 5G standardization efforts that resulted to the first 5G new radio (5G NR) milestone and subsequent 3GPP releases. The emergence of mobile IoE services is at an all high. Extended reality (XR) services (augmented reality/virtual reality AR/VR) with applications to telemedicine, haptics, flying vehicles, brain-computer interfaces, and networked independent systems are just a few examples. The initial 5G goal of delivering short-packet, sensing-based URLLC services will be disrupted by these applications, in order to run IoE services like XR and connected autonomous vehicles successfully (Viswanathan & Mogensen, 2020).

A revolutionary sixth generation (6G) wireless system, on the other hand, is required, with a design that is inherently adapted to the performance requirements of IoE applications and their associated technical trends. A convergence of past trends (e.g., densification, greater rates, and enormous antennas) with current trends (e.g., new services and the recent revolution in wireless devices artificial intelligence (AI) (Tomkos et al., 2020), computing, and sensing will drive 6G. In (Yaacoub & Alouini, 2020), the authors identify some applications, trends, and disruptive technologies that will propel the 6G revolution ahead. The security difficulties associated with IoT devices have been increased by the breakthroughs in computations that have accompanied the sixth generation, particularly in the domain of quantum communications. These devices necessitate to be protected from security breaches in the front of adversaries that possibly have quantum capacities (Lohachab et al., 2020).

As a result, the authors in (Al-Mohammed & Yaacoub, 2021) investigated the use of quantum cryptography to improve the security of these technologies in the sixth generation. Due to their limited power and processing capabilities, it would be unreasonable to expect IoT devices to manage quantum communications on their own. Many IoT applications use IoT controllers to receive, gather, and process data from several IoT sensors before transmitting it over the network to be processed and analyzed on the server-side (Dawy et al., 2017). The authors expand on this concept by examining a broader framework that can be used with free space optics (FSO) or fiber optics. They also studied over the QKD procedure in depth and determined the minimum and maximum key lengths for a specific simulated case. The controllers are frequently highly advanced that use encryption algorithms to interact safely with the server (Al-Mohammed & Yaacoub, 2021). They also used these controllers to run QKD and then distribute the produced keys to IoT devices, allowing them to encrypt data through regular radio frequency (RF) communication connections. So, controllers use QKD to protect the vulnerable IoT device-controller connection from unauthorized users. (Mavromatis et al., 2018) used software defined networking to demonstrate an IoT QKD experiment using fiber optics to connect the controller to the network. It has been proved that this leads reduced power consumption in IoT devices.

Figure 4 illustrates the main arrangement between both the IoT device and the server, which exhibits three IoT controllers, each connected to a large number of IoT sensors that communicate with the server.

Figure 4. The quantum key is shared by the controller, IoT devices, and server/base station (quantum channel is the red line)

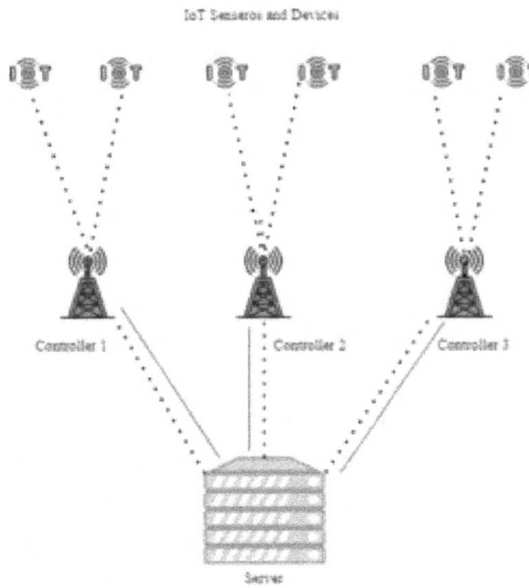

QKD as a Pseudorandom Number and Salting Generator

Every part of cryptography relies on random numbers, cryptographers design techniques like RC4 and DSA, as well as protocols like SET and SSL, assuming that random numbers are easily accessible. Even a typical application like encrypting a document on a disk with a pass requires random integers for the salt to be hashed into the pass and the initialization vector to be utilized in the encryption. Furthermore, random numbers can be used in encryption and decryption processes such as symmetric ciphers like triple-DES, seeds for routines that generate mathematical values, as well as initialization vectors for block cipher chaining modes, salts to combine with passwords to frustrate offline password guessing programs, and random values for specific instances of many digital signature schemes like DSA(Kelsey et al., 2000).

Salt is a set of random data that can be generated using many procedures available in computer programming languages (Kent & Liebrock, 2011). These techniques are pseudo-random number generators (PRNGs), which generate random data using previously used data (Dodis et al., 2013). As a result, they are not generating random data that is cryptographically secure. For that type of applications with important security needs, cryptographically secure pseudo random number generator techniques are necessary to create encrypted random data that can be more secure against attacks (McEvoy et al., 2007). Moreover, hashing techniques can employ these random bits as a supplement to avoid collisions and provide more complicated encoded outputs by using them as salt during the hashing process. Salted hash encryption is the name for this type of hashing. This cryptography type creates a salt hash password by appending a salt to the user's password, which is produced using a trusted random process, and then storing the generated password in the database (Achuthan & Murali, 2015; Alattar et al., 2018).

On the other hand, the QKD relies on quantum mechanics and it is generated by physical phenomena (polarization angle), the final key length that has to arrive at the destination holds the polarization angles state, then represent them as bits (zeros and ones), later, can be used as a seed (with small amount of photons such as 100 initial photons) to PRNG or it can be used as PRNGs directly with the longer initial photons, so, if the QKD used for seed or PRNGs either one will satisfy the requirements of the randomness.

The scenario to use QKD for generating a random number is illustrated in fig. 4. It can be generated at the sender (server) that has the powerful capability to run a quantum algorithm and handle the requirements for generating and producing QKD. Moreover, it can be used in a pool to generate more than one random number at a time like the one used in Tokyo Japan (*Tokyo QKD Network*, 2021).

In (Al-Mohammed & Yaacoub, 2022) the authors proposed a new method for simulating QKD and the result was promising. The new method is simple to use and can accommodate any number of photons (Qubit). In addition, the study compared the novel technique to prior studies, and the results indicate that it is interesting. This approach, as well as the algorithm described in the work (Al-Mohammed & Yaacoub, 2021), for any initial photons that want to utilize to share a secret key, the sender produces a determined number of polarization photons (generated randomly), at the receiver, a polarization filter is used to get the photons that are collected to make the final key length. The most important thing the final key length doesn't exceed the threshold that has been calculated before, and each initial photon has a different final key length even for the same initial photon for other iteration to generate a shared key, so, algorithms in (Al-Mohammed & Yaacoub, 2021) can be used to obtain thresholds for 100 initial photons.

The minimum final key length that will reach the destination is 7 photons, and the maximum key length is 48 photons, as shown in Fig. 5.

Figure 5. The thresholds for 100 initial photons

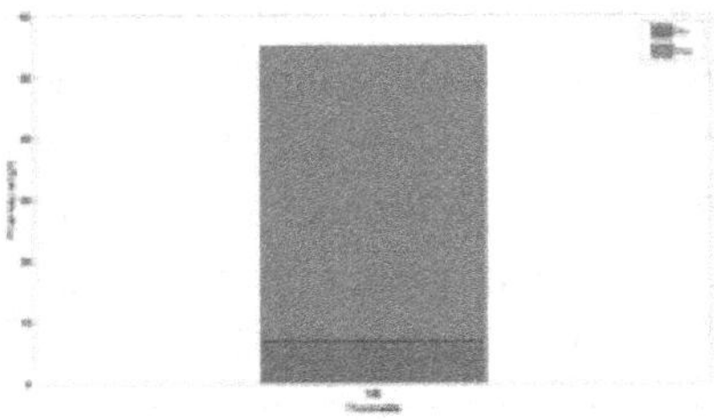

Furthermore, fig. 6 shows the distribution of the final key length for 100 initial photons, which was generated several times to analyze the sequence of the final key lengths, and the results show that the majority of them are more than 12 bits(Anderson, 2020), actually it shows the 99% of them more than 12 bits, indicating that such initial photons are suitable for salting and seed for the pseudorandom number

Figure 6. The distribution of the final key lengths of 100 initial sending photons for QKD

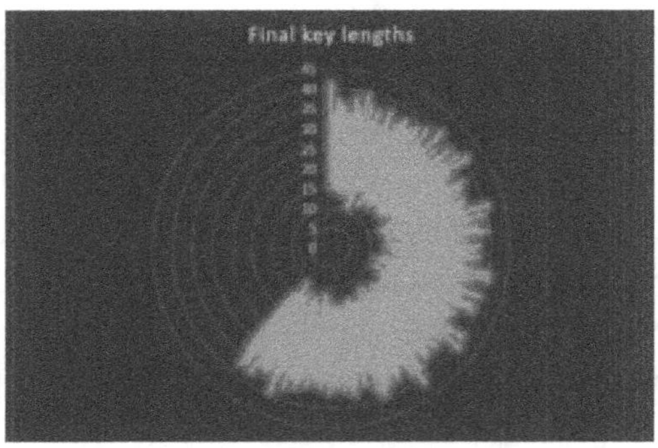

CONCLUSION

QKD is a technique for sharing a secret key between two parties by utilizing quantum mechanics, two well-known protocols that contributed to securing the communication are BB84 and E91. This book chapter discussed: The principle of quantum security and cryptography emphasizes the recent developments and potential applications of new and emerging applications of these techniques in security in current applications or the future; different scenarios for using QKD lengths such as seeds for generating keys to encryptions messages, using QKD as key for DES or AES algorithms, also, using QKD in real-life scenarios such as healthcare in personal area networks for protecting the privacy of the patients' data, or railway monitoring scenarios to encrypt the collected data generated by the sensors.

ACKNOWLEDGMENT

This publication was jointly supported by Qatar University and IS-Wireless - International Research Collaboration Co-Fund Grant no. IRCC-2021-003.

REFERENCES

Achuthan, K., & Murali, S. S. (2015). Software Engineering in Intelligent Systems. In *Advances in Intelligent Systems and Computing* (Vol. 349). https://www.scopus.com/inward/record.url?eid=2-s2.0-84942759426&partnerID=tZOtx3y1

Al-Mohammed, H. A., Al-Ali, A., Yaacoub, E., Qidwai, U., Abualsaud, K., Rzewuski, S., & Flizikowski, A. (2021). Machine Learning Techniques for Detecting Attackers During Quantum Key Distribution in IoT Networks With Application to Railway Scenarios. *IEEE Access : Practical Innovations, Open Solutions*, 9, 136994–137004. 10.1109/ACCESS.2021.3117405

Al-Mohammed, H. A., & Yaacoub, E. (2021). On the Use of Quantum Communications for Securing IoT Devices in the 6G Era. *2021 IEEE International Conference on Communications Workshops, ICC Workshops 2021 - Proceedings*, 1–6. 10.1109/ICCWorkshops50388.2021.9473793

Al-Mohammed, H. A., & Yaacoub, E. (2022). New Way to Generating and Simulation QKD. In *Proceedings of Sixth International Congress on Information and Communication Technology* (pp. 801--809). springer. 10.1007/978-981-16-1781-2_69

Alattar, M. H., Al Farawn, A., & Ali, N. S. (2018). Anti-continuous collisions user-based unpredictable iterative password salted hash encryption. *International Journal of Internet Technology and Secured Transactions*, 8(4), 619–634. 10.1504/IJITST.2018.095944

Anderson, R. (2020). Security engineering: a guide to building dependable distributed systems. *John Wiley & Sons*. 10.1002/9781119644682

Babber, K., & Singh, J. P. (2021). Quantum cryptography and security analysis. *Journal of Discrete Mathematical Sciences and Cryptography*. 10.1080/09720529.2019.1692452

Broadbent, A., & Schaffner, C. (2016). Quantum cryptography beyond quantum key distribution. In *Designs, Codes, and Cryptography* (Vol. 78, Issue 1). Springer US. 10.1007/s10623-015-0157-4

Daemen, J., & Rijmen, V. (1999). *AES proposal: Rijndael*. Academic Press.

Dastjerdi, Amir Vahid and Buyya, R. (2016). Fog Computing: Helping the Internet of Things Realize Its Potential. *Computer, 49*(8), 112–116. 10.1109/MC.2016.245

Dawy, Saad, Ghosh, Andrews, & Yaacoub. (2017). Toward massive machine type cellular communications. *IEEE Wireless Communications*, 120–128.

Di Candia, R., Yi itler, H., Paraoanu, G. S., & Jäntti, R. (2020). Two-way covert microwave quantum communication. In *arXiv*. 10.21203/rs.3.rs-93750/v1

Diaz, V. (2015). Backhauling with fibre? *Fibre Systems, 6*.

Dodis, Y., Pointcheval, D., Ruhault, S., Vergniaud, D., & Wichs, D. (2013). Security analysis of pseudo-random number generators with input: /dev/random is not robust. *Proceedings of the ACM Conference on Computer and Communications Security*, 647–658. 10.1145/2508859.2516653

Elhoseny, M., Ramírez-González, G., Abu-Elnasr, O. M., Shawkat, S. A., Arunkumar, N., & Farouk, A. (2018). Secure Medical Data Transmission Model for IoT-Based Healthcare Systems. *IEEE Access : Practical Innovations, Open Solutions*, 6, 20596–20608. 10.1109/ACCESS.2018.2817615

Fernandez-Carames, T. M. (2019). From Pre-Quantum to Post-Quantum IoT Security: A Survey on Quantum-Resistant Cryptosystems for the Internet of Things. *IEEE Internet of Things Journal*, 7(7), 1–1. 10.1109/JIOT.2019.2958788

Gingrich, R. M., & Adami, C. (2002). Quantum entanglement of moving bodies. *Physical Review Letters*, 89(27), 2704021–2704024. 10.1103/PhysRevLett.89.27040212513186

Gisin, N. (2018). Quantum communications. *Optics InfoBase Conference Papers: Vol. Part F112*. 10.1364/SPPCOM.2018.SpM3G.5

Gyongyosi, L., Bacsardi, L., & Imre, S. (2019). A survey on quantum key distribution. *Infocommunications Journal*, 11(2), 14–21. 10.36244/ICJ.2019.2.2

Haghi Kashani, M., Madanipour, M., Nikravan, M., Asghari, P., & Mahdipour, E. (2021). A systematic review of IoT in healthcare: Applications, techniques, and trends. *Journal of Network and Computer Applications*, 192(July), 103164. 10.1016/j.jnca.2021.103164

Hashimoto, Y. (2021). Recent Developments in Multivariate Public Key Cryptosystems. In *International Symposium on Mathematics, Quantum Theory, and Cryptography*. Springer. 10.1007/978-981-15-5191-8_16

Holevo, A. S. (2011). *Probablilistic and Statistical Aspects of Quantum Theory*. Springer Science & Business Media. 10.1007/978-88-7642-378-9

Kelsey, J., Schneier, B., & Ferguson, N. (2000). Yarrow-160: Notes on the design and analysis of the yarrow cryptographic pseudorandom number generator. *Lecture Notes in Computer Science (Including Subseries Lecture Notes in Artificial Intelligence and Lecture Notes in Bioinformatics), 1758*, 13–33. 10.1007/3-540-46513-8_2

Kent, A. D., & Liebrock, L. M. (2011). Secure communication via shared knowledge and a salted hash in Ad-hoc environments. *Proceedings - International Computer Software and Applications Conference*, 122–127. 10.1109/COMPSACW.2011.30

Lindblad, G. (1999). A general no-cloning theorem. *Letters in Mathematical Physics*, 47(2), 189–196. 10.1023/A:1007581027660

Liu, W., Huang, P., Peng, J., Fan, J., & Zeng, G. (2018). Integrating machine learning to achieve an automatic parameter prediction for practical continuous-variable quantum key distribution. *Physical Review. A*, 97(2), 1–9. 10.1103/PhysRevA.97.022316

Lohachab, A., Lohachab, A., & Jangra, A. (2020). A comprehensive survey of prominent cryptographic aspects for securing communication in post-quantum IoT networks. *Internet of Things : Engineering Cyber Physical Human Systems*, 9(Mar), 100174. 10.1016/j.iot.2020.100174

Malina, L., Popelova, L., Dzurenda, P., Hajny, J., & Martinasek, Z. (2018). On Feasibility of Post-Quantum Cryptography on Small Devices. *IFAC-PapersOnLine*, 51(6), 462–467. 10.1016/j.ifacol.2018.07.104

Mavromatis, A., Ntavou, F., Salas, E. H., Kanellos, G. T., Nejabati, R., & Simeonidou, D. (2018). Experimental Demonstration of Quantum Key Distribution (QKD) for Energy-Efficient Software-Defined Internet of Things. *European Conference on Optical Communication, ECOC,* (1), 1–3. 10.1109/ECOC.2018.8535267

McEvoy, R., Curran, J., Cotter, P., & Murphy, C. (2007). *Fortuna: cryptographically secure pseudo-random number generation in software and hardware*. 10.1049/cp:20060479

Pfister, A. D., Salz, M., Hettrich, M., Poschinger, U. G., & Schmidt-Kaler, F. (2016). A quantum repeater node with trapped ions: A realistic case example. *Applied Physics. B, Lasers and Optics*, 122(4), 1–19. 10.1007/s00340-016-6362-7

Pirandola, S., Laurenza, R., Ottaviani, C., & Banchi, L. (2017). Fundamental limits of repeaterless quantum communications. *Nature Communications*, 8(1), 15043. Advance online publication. 10.1038/ncomms1504328443624

Saad, W., Bennis, M., & Chen, M. (2019). A vision of 6G wireless systems: Applications, trends, technologies, and open research problems. *ArXiv*, (June), 134–142.

Sasaki, M. (2018). Quantum key distribution and its applications. *IEEE Security and Privacy*, 16(5), 42–48. 10.1109/MSP.2018.3761713

Sen, D. (2014). The uncertainty relations in quantum mechanics. *Current Science*, 107(2), 203–218. 10.18520/cs/v107/i2/203-218

Staff, S. (2020). *Quantum Cryptography & Encryption: What It Is & How It Works*. Sectigo. https://sectigo.com/resource-library/quantum-cryptography

Tokyo QKD Network. (2021). Tokyo QKD. http://www.tokyoqkd.jp/

Tomkos, I., Klonidis, D., Pikasis, E., & Theodoridis, S. (2020). Toward the 6G Network Era: Opportunities and Challenges. *IT Professional*, 22(1), 34–38. 10.1109/MITP.2019.2963491

Viswanathan, H., & Mogensen, P. E. (2020). Communications in the 6G Era. *IEEE Access : Practical Innovations, Open Solutions*, 8, 57063–57074. 10.1109/ACCESS.2020.2981745

Yaacoub, E. (2021). Ad Hoc Networks Travel Hopping Enabled Resource Allocation (THEResA) and delay tolerant networking through the use of UAVs in railroad networks. *Ad Hoc Networks*, 122(June), 102628. 10.1016/j.adhoc.2021.102628

Yaacoub, E., & Alouini, M. S. (2020). A Key 6G Challenge and Opportunity—Connecting the Base of the Pyramid: A Survey on Rural Connectivity. *Proceedings of the IEEE, PP*, 1–50. 10.1109/JPROC.2020.2976703

Chapter 8
The Impact of Key Lengths on QKD Security:
An ML Study

Hasan Abbas Al-Mohammed
Qatar University, Qatar

Afnan S. Al-Ali
Qatar University, Qatar

Elias Yaacoub
Qatar University, Qatar

Khalid Abualsaud
Qatar University, Qatar

ABSTRACT

Charles Bennett and Gilles Brassard, in 1984, proposed the first QKD protocol and called BB84. It is assumed the protocol shares a quantum key safely (between two parties). In 2000 it was implemented easily and showed a significant method for detecting an attacker that trying to get the shared key by utilizing the final key length. When the length exceeds a certain value, that is calculated before the transmitting the key, the majority of prior works agree with that, but this chapter showed a significant threat that affects the final key to same. There is no attacker at the middle. Moreover, this chapter analyzed the final key lengths and showed the harmed value of the final key length with the attacker effect. It also showed how often these values could be within the threshold. Furthermore, a solution was found to detect the attacker by using a machine learning technique. The results showed a promising accuracy to detect the attacker relying on the final key length.

DOI: 10.4018/978-1-7998-9522-0.ch008

1. INTRODUCTION

Quantum Key Distribution (QKD) is a technique involving two members (Alice and Bob) to exchange a private symmetric key. If an attacker (Eve) decided to steal the private key in a QKD protocol, communicators can see it using appropriate quantum rules (e.g., the well-known Heisenberg uncertainty theory) (Busch et al., 2007).

The first QKD protocol BB84 was proposed. BB84 is currently the most general and powerful quantum cryptography protocol for transmitting data utilizing photon polarization states. Moreover, as Eve tries to interfere with quantum networks, the protocol detects the attack and stops key generation. Furthermore, the protocol will not be suspended as long as Eve is passive. For any quantum network assault, the likelihood that the protocol does not stop and an attacker duplicates the generated keys is extremely low (Shor & Preskill, 2000).

Alice and Bob run a single search before the shared key stream may be utilized confidently. If they find a large number of mismatches, they publicly select and check few bits at random from their key streams, and if the error rate exceeds a predetermined threshold, they trash the entire shared key and generate a new one. The thresholds rely on the initial photons sent from the transmitter (Alice) in order to obtain the final key lengths used for generating the secret key (Jeong et al., 2020). Moreover, the ratio between the maximum and minimum range of the final key length plays a role in terms of utilizing the key in a given function or encryption algorithm. The effect of raising the initial photons sent by the server on the final key length at the destination needs to be quantified, for example in the case of IoT devices. Therefore, it is important to use several values for the number of initial photons and compare them with the corresponding final key length, in order to demonstrate the effect of increasing the number of initial photons on the final key length for utilizing it for securing the communication or making it hard to detect by an attacker (S. K. Singh et al., 2020).

However, sometimes the attacked final key length remains within the threshold. In this case, the receiver does not know that there is an attacker in the middle trying to detect the key that is shared between the sender and receiver, because the final key length does not exceed the thresholds range that has been agreed between them. In this case, the vulnerabilities of QKD lead to losing the possibility of detecting the attacker and changing the keys. In addition, the percentage for the attacked key within threshold depends on the initial photons. Therefore, the best way to discover an attacker even when the final length is within threshold is to use machine learning techniques (Zbinden et al., 1998).

To achieve this aim, a support vector machine (SVM) classifier is used. For our given problem and the type of our data which consists of one feature represented by the key to each sample, SVM is the most suitable algorithm because it maps the input data into a high dimensional space. The fundamental principle of SVM is to find the decision surface that partitions the best vectors extracted from the data in vector space into two groups. Furthermore, the accuracy of the machine learning depends on the number of initial photons, as each one has different percent of QKD that are attacked within the threshold (Biamonte et al., 2017).

The contribution of this chapter will be:

- o Calculating the thresholds of many initial photons (minimum and maximum range).
- o Calculating and analyzing the effect of the increasing the initial photons on the final key lengths, the minimum and maximum range of the threshold.
- o Calculating the effects of the attacker on the final key length, how many photons can the attacker change; moreover, study the effect of increasing initial photons on the exposed part by the attacker.
- o Calculating and studying the likelihood of attacked final key within threshold, using machine learning and other approach to detect the attacks, calculating and studying the accuracy of detecting an attacker even within thresholds, and analyzing the machine learning accuracy while changing the initial photons.

2. QUANTUM KEY DISTRIBUTION

QKD in quantum mechanics is typically based on the difficulty to examine a system without upsetting it. Because Eve needs to eavesdrop on Alice and Bob's quantum correspondence, an eavesdropper (Eve) would undoubtedly leave any detectable evidence. As a result, the QKD protocol achieves security. If Eve is passive, on the other hand, the protocol generates long keys at random. As a general convention in quantum cryptography, Alice communicates quantum states to Bob over a quantum channel. Eve, an eavesdropper, could be listening in on the quantum communication. Alice and Bob can generate a shared key, which they can then employ using a symmetrical encryption mechanism like a one-time pad.

This chapter analyzes the final key length in the presence of an attacker. The rest of this chapter is structured as follows. The BB84 protocols for distributing quantum keys are described in Section 2. Section 3 presents the QKD under attack, such as Section 3.1, which uses the classic approach for detecting an attacker. Section 3.2, which uses machine learning algorithms to detect the attacker, and Section 3.3, which utilizes a support vector machine to detect the attacked final key length. Finally, Section 4 draws the chapter to a conclusion.

3. BB84 PROTOCOL

Following the pioneering argument put forward by Charles Bennett and Gilles Brassard, in 1984, the first QKD protocol BB84 was proposed (C.H. Bennett and G. Brassard, 1984). At the moment, BB84 is the most general and powerful quantum cryptography protocol that transmits data using photon polarization states (Zhang et al., 2019). In addition, the protocol identifies the attack and terminates the generation of keys as Eve tampers with quantum networks. Also, as long as Eve is passive, the protocol will not be suspended. The likelihood that the protocol does not stop and an attacker copies the generated keys is so tiny for any attack on quantum networks (Mavromatis et al., 2018).

In quantum cryptography, as a universal convention, Alice transfers quantum states through a quantum channel to Bob. An eavesdropper (Eve) might be listening to the quantum channel. Alice and Bob can come up with a shared key that they can then use with a symmetrical encryption method, such as a one-time pad. Photons are used as qubits in the initial formulation of the protocol, the information being stored in their polarization (Mina & Simion, 2020), (H. Singh et al., 2014). In the basic design of the protocol, photons are utilized as qubits, with information stored in their polarization, the remaining bits are then used to extract a shared key stream. Eve takes the same strategy as Bob, sometimes utilizing the correct detector and sometimes the incorrect detector. If she is correct in her decision, the photon will retain its previous polarization on its path. If her choice is incorrect, she will change the polarity of the photons she sends to Bob. Due to Eve's interference at the conclusion of Bob's detection, the resulting bit stream would be incorrect in any way, and after checking with Alice to delete incorrectly seen photons (Al-Mohammed & Yaacoub, 2022).

To explain the BB84 protocol, there are four polarization states (quantum bits) to represent the two classical bits and two types of filters. Alice has four polarized filters where each stream has one of four distinct polarizations, rectilinear polarization of 0 °, 90 ° and diagonal polarizations of 45 ° and 135 °. The output of the photon (Qubit) will be one of these four represented equations (C.H. Bennett and G. Brassard, 1984; Haitjema, 2007):

$$|\psi_{00} = |0\rangle \tag{1}$$

$$|\psi_{10} = |1\rangle \tag{2}$$

$$|\psi_{01} = \frac{1}{\sqrt{2}}|0\rangle + \frac{1}{\sqrt{2}}|1\rangle \tag{3}$$

$$|\psi_{11} = \frac{1}{\sqrt{2}}|0\rangle - \frac{1}{\sqrt{2}}|1\rangle \tag{4}$$

For the binary state 0", Alice and Bob settle on |→ > or | ↑> to represent it, whereas |↗>or |↖> will represent the binary state 1", there are stages to complete the BB84 protocol between these two parties and detect the attacker:

The first stage: Alice is the main source, and generates a polarized photon stream, choosing their polarization at random.

The second stage: Alice transmits these photons one polarization at a time through the quantum channel; Alice makes a note of the unpredictability polarization chain that Bob has no way of guessing what polarization any of the photons has.

The third stage: at Bob (the receiver side): Bob has two detectors:

- The one with rectilinear Photons with a rectilinear position move unchanged into this filter; the rectilinear filter switches to a random rectilinear state (|→ > or | ↑>) the state of a diagonally polarized photon.
- The second with a diagonal filter to detect Photons with a diagonal state pass unchanged across a diagonal filter; the diagonal filter alters the state to a random diagonal state (|↗>or |↖>) of a rectilinear polarized photon, as described in figure 1 (Al-Mohammed & Yaacoub, 2021).

Figure 1. Quantum Key Distribution Process: (a) Without Attacker; (b) With Effect of the Attacker

(a)

Bit Number	1	2	3	4	5	6	7	8
Alice's choice of scheme	+	x	x	x	+	+	x	+
Photons spin from Alice	↕	⤢	⤡	⤢	↔	↔	⤢	↕
Alice's random key stream	1	1	0	1	0	0	1	1
Bob's choice of scheme	x	x	+	+	+	+	x	x
Orientation of photons measured by Bob	⤢	⤢	↕	↔	↔	↔	⤢	⤡
Bob's received key stream	1	1	1	0	0	0	1	0

(b)

Bit Number	1	2	3	4	5	6	7	8
Alice's choice of scheme	+	x	x	x	+	+	x	+
Photons spin from Alice	↕	⤢	⤡	⤢	↔	↔	⤢	↕
Alice's random key stream	1	1	0	1	0	0	1	1
Eve's choice of detector	x	x	+	x	+	x	+	+
Orientation of photons from Eve	⤢	⤢	↔	⤢	↔	⤡	↔	↕
Bob's choice of scheme	x	x	+	+	+	+	x	x
Orientation of photons measured by Bob	⤢	⤢	↔	↔	↔	↕	⤡	⤡
Bob's received key stream	1	1	0	0	0	1	0	0

The fourth stage: When each photon arrives, Bob arbitrarily guides it to one of his two detectors. Then, Alice and Bob interact over the normal contact channel to address Bob's detector preference, discarding all the wrong bits that Bob acquired from an incorrect filter range, to extract a shared keystream, the remaining bits are then used.

The fifth stage: Eve follows the same approach used by Bob, sometimes using the right detector and sometimes the wrong detector. If her decision is correct, the photon will continue with its prior polarization on its path. If her decision is false, she is going to change the polarization of the photon that she passes on to Bob (Al-Mohammed, Al-Ali, Yaacoub, Abualsaud, et al., 2021).

The resulting bitstream would not be right in any way due to Eve's interference at the end of Bob's detection, and after consulting with Alice to discard wrongly observed photons, Alice and Bob execute a single search before the mutual keystream is used comfortably. If they discover a bunch of mismatches, they publicly pick and compare several bits picked at random from their key streams, and if the error rate crosses a negotiated threshold, they delete the whole keystream and produce a new one.

4. QKD (BB84) UNDER ATTACK

If an attacker has the quantum power to detect photon polarization using a polarization filter to determine the entire key length, the key can be used to decrypt the message that will be delivered. It should be noted that the attacker should detect the photon polarization of the sender and the one that will be used in the receiver for correct detection; therefore, if Alice (the sender) and Bob (the receiver) have used a different polarization, as a consequence that bit will be discarded from the final key length (Al-Mohammed & Yaacoub, 2021). The rest of the section describes the two methods for detecting an attacker (conventional, and the new one that relies on machine learning).

4.1. The Traditional Way for Detecting the Attacker

To detect an attacker, the sender and receiver can use a mechanism that calculates the minimum and maximum values of the final key length by repeatedly generating the QKD between the sender and receiver, and then getting the smallest amount of the final key length that reached the destination, as well as getting the largest value of the key length. Furthermore, it will be simple to detect the attacker by comparing the final key length with the threshold; if the length exceeds the boundaries, the receiver will discard the entire key (Al-Mohammed, et al., 2021). In (Al-Mohammed & Yaacoub, 2021) the algorithms proposed are used to simulate the attacker trying to detect the QKD, where the attacker is trying to capture the polarization of the photons. By implementing the algorithm by using MATLAB and by considering 500 initial photons, the results in Figure 2 are obtained.

Figure 2. Threshold of the Initial Photons 500, With Attacked and Non-Attacked Final Key Length

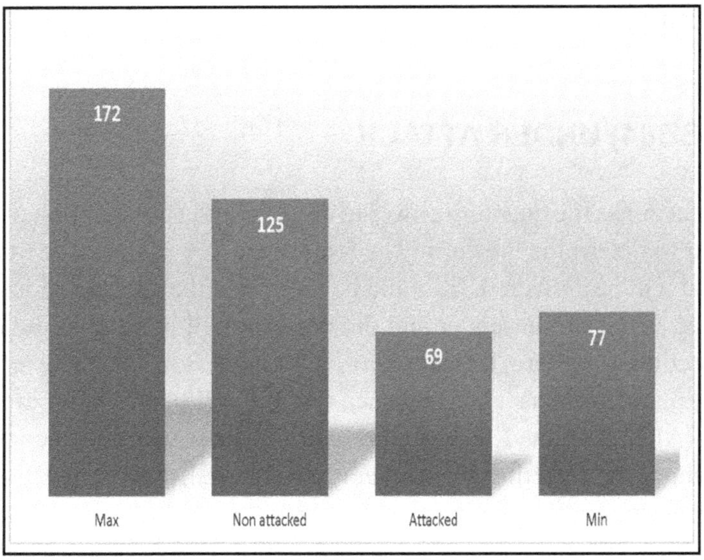

Figure 2 shows the initial photons 500 thresholds; the greatest value of the final key length that may be reached is 172, while the minimum value is 77. The final key length for this simulation shared between the sender and receiver is 125, which is within the threshold and will be used to encrypt the message by the receiver. The final key length that was received with the effect of the attacker is 69, therefore it exceeds the criteria (below the minimum threshold of 77). In this situation, the receiver will discard the key and the sender will generate a new one to distribute.

4.2. The New Way for Detecting the Attacker by Machine Learning

The majority of prior works agree with the fact of detecting the attacker by exceeding the threshold, although this might not always be the case. According to the authors' investigation and analysis of the data, they have discovered a contradiction of previous works, such that when resharing a quantum key between two parties for the same example as before (the initial photons 500), the final key length with the effect of the attacker is within the threshold (see figure 3). This is a serious issue because even if the attacker is attempting to identify (listen for) the sharing key, the sender and receiver are oblivious of that as well.

Figure 3. Threshold of the initial photons 500, with attacked (the final key exceeds the threshold) and non-attacked final key length

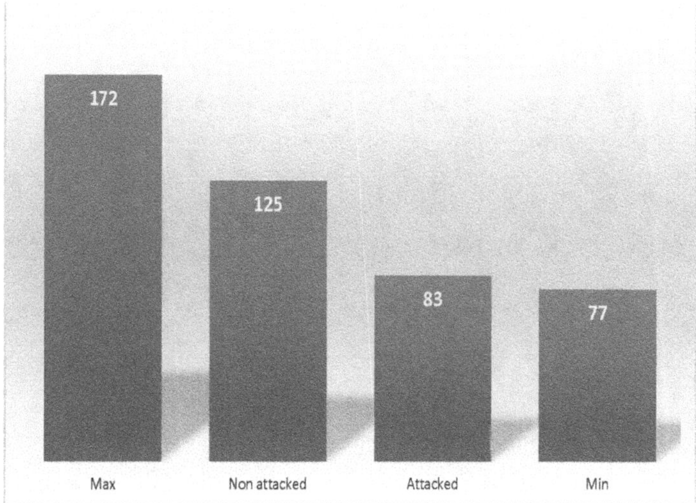

The thresholds corresponding to 500 initial photons are shown in Figure 3; the maximum final key length that may be reached is 172, while the lowest is 77. The final key length exchanged between the sender and receiver for this simulation is 125, which is within the threshold and will be utilized by the receiver to encrypt the message. The final key length received under the presence of the attacker 83, which does not exceed the thresholds, but there is an attacker in the middle listening for communication between the sender and the receiver. The receiver will not discard the key, and the sender will not produce a new one to distribute in this circumstance.

Figure 4 shows another example similar to the one that has been used in (Al-Mohammed, et al., 2021) and in (Al-Mohammed & Yaacoub, 2021), the initial photons are 1000, the maximum final key length that can be attained using the similar procedures is 321, while the minimum is 188. For this scenario, the final key length transferred between the sender and receiver is 256, which is within the threshold and will be used to encrypt the message by the receiver. The attacker obtains a final key length of 197, which is within the thresholds. Hence, although there is an attacker in the middle listening for communication between the sender and the recipient, the receiver will not discard the key, and the sender will not produce a new one to distribute in this circumstance.

Figure 4. Threshold of the initial photons 1000, with attacked (the final key exceeds the threshold) and non-attacked final key length

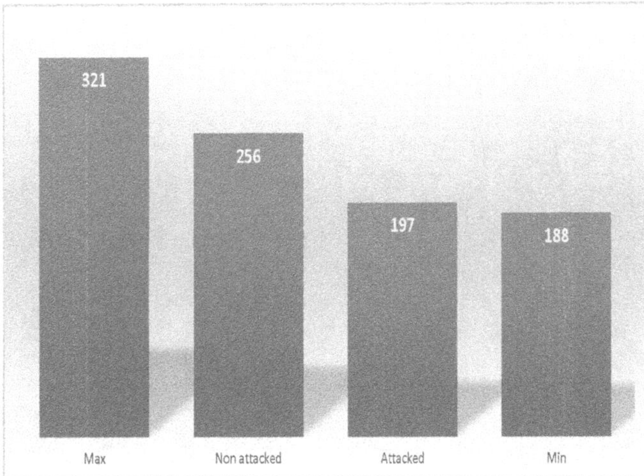

Figure 5. Variation of the probability

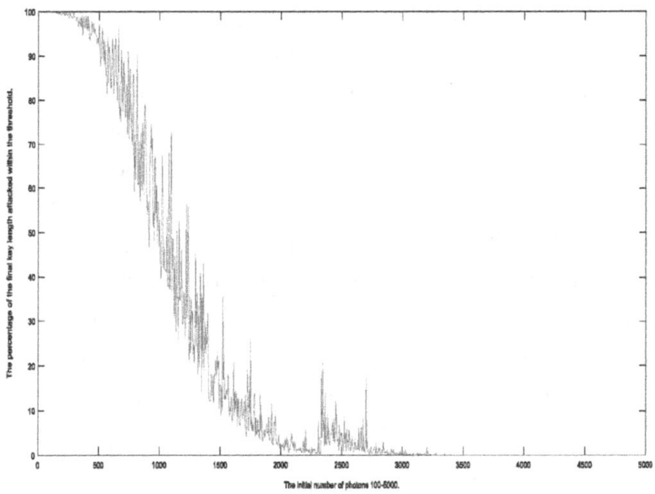

These are the abnormal behaviors observed during the distribution of the Keys when the initial photons are less than 5000 photons by analyzing the QKD proposed in (Al-Mohammed & Yaacoub, 2022). Additionally, the probability of finding the key with the effect of the attacker within the threshold varies between a small ratio and a high ratio of the likelihood of getting a key length with the influence of the attacker, which will change the usual method for identifying the harmed key. The overall trend is in decreasing probability from 99.9% at 100 to 0% at 3425. Figure 5 illustrates the variation of the probability.

In figure 6, the probability of finding an undetected attacker by using the old mechanism (passing threshold) is very low when the initial photons are 100. It is approximately 99.9 percent. For instance, if the sender generates 1000 times (each time a new key) to share with the receiver and an attacker is listening at that time, the final key length will be within acceptable length 999 times, and only one time there will be a chance to knowing there is an attacker. Furthermore, the graph shows the line of the probability decreases when the initial photons will be 510, to reach 87.60% of the attacked key will not be caught.

Figure 6. Initial photons to the corresponding probability (more than 50%) attacked final key length within the threshold

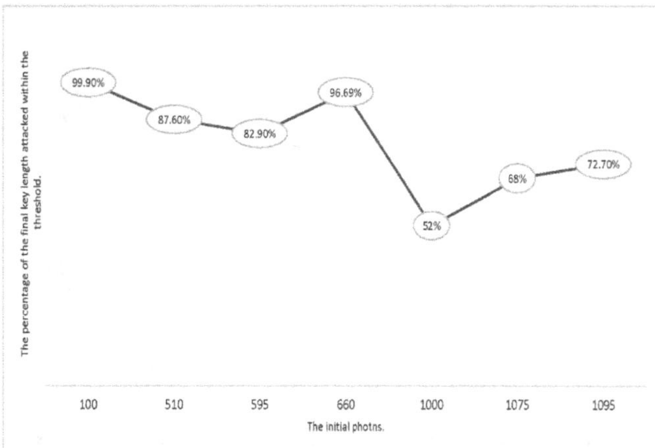

The line keeps dipping to reach 82.9% at 595 initial photons until the probability rises again at 660 to be 96.69% ; also, the figure shows a decline at 1000 photons where the probability will be 52% at this point. Moreover, the fluctuation of the probability of passing the attacked final key length within the threshold fluctuates between increasing and decreasing at 68% and 67.72% for 1075 and 1095 respectively. These percentages of the probability that find an attacked key within threshold were chosen based on the significant values that were reached after analyzing data and the percentages are more than the half.

Figure 7. shows the probabilities that less than 50% of the attacked key lengths are within threshold. The figure shows the line of the probability decreased when the initial photons are 1155 to reach 25.03% of the attacked key that will not be caught by the conventional one, but at 1220 the probability increased to pass the half, that is 56.32%. The likelihood of the attacked key to be within threshold drops at 1830 initial photons to be 13.66%, and it keeps fluctuating between 11.60%, 3%, 21%, 14.08%, 0.85%, and to be 0.00%, at the values for initial photons 1925, 2185, 2345, 2365, 2535, and 3425, respectively. Moreover, the initial photons from 3425 up to less than 5000 will keep a variant but small amount of probability near zero to keep still at 0.00%.

Figure 7. Initial photons to the corresponding probability (less than 50%) attacked final key length within the threshold

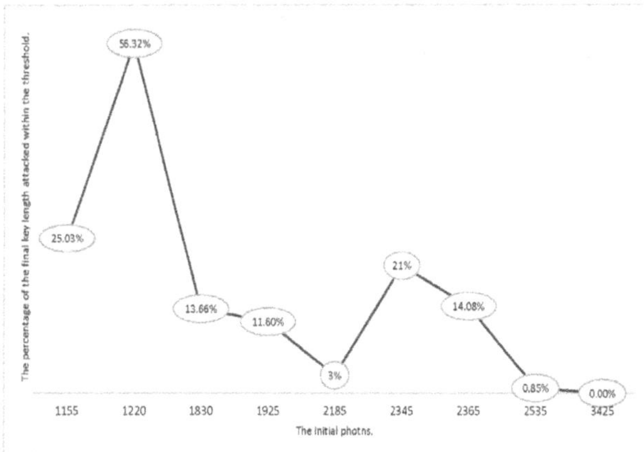

The conclusion from figures 5 to 7 is that it is important to find a better solution to detect an attacker. Therefore, a machine learning technique is used to detect an attacker leading to a key within the thresholds and thus the receiver does not know it.

4.3. Support Vector Machine (SVM)

For the given problem and the type of the data which consists of one feature represented by the key to each sample, SVM is the most suitable algorithm because it maps the input data into a high dimensional space. The fundamental principle of SVM is to find the decision surface that partitions the best vectors extracted from the data in vector space into two groups. This decision surface is known as the hyperplane that distinguishes between the two classes as shown in figure 8, where the two classes are separated by the maximum margin (Aqsazafar, 2020), (Kotpalliwar & Wajgi, 2015). As the dataset is a one-dimensional feature space, SVM was found to be the best algorithm because it mapped the input data into a high dimensional space.

Figure 8. Support vector machine with maximum margin (adapted from Aqsazafar, 2020)

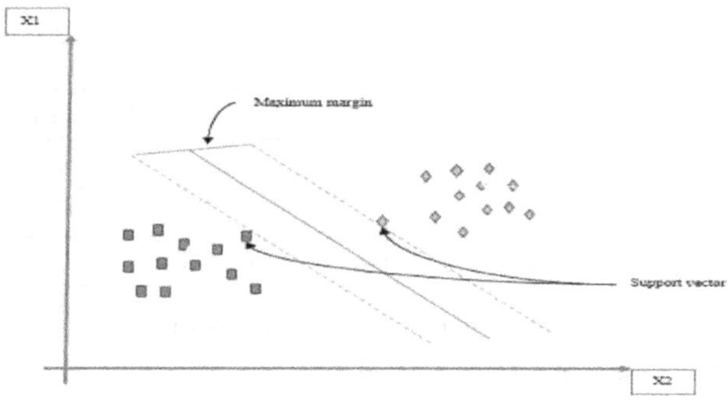

A decision surface, or hyperplane, is designed in this feature space. In the case of linearly separable data, this decision surface maximizes the "margin" between two groups (which means it has max distance from both classes. The sum of these two classes must be maximized to make this line as maximum margin).

The unique attributes of this decision surface (hyperplane) guarantee the learning machine's high generalization performance. The mathematical representation of this hyperplane is illustrated in equation (5) (Aqsazafar, 2020), (Noble, 2006).

$$\vec{w}\vec{x} + b = 0 \tag{5}$$

Where: Vector W and constant b are the learned parameters from the training dataset. Vector $\vec{x}$ is the datapoint.

The SVM transfers the dataset to a high dimensional space based on the concept of Kernel trick to compute the inner products between all pairs of data in the feature space. There is no need to explicitly increase the dimensions by explicitly transforming into feature vector representations in higher space.

The data set contains a total of 20,000 final key lengths which are classified into two categories, attacker and no attacker. We have classified the various types of generated keys to their belonging class. Classification is done using SVM classifier. Because our dataset is only one dimension, the scatter plot will not work. In order to make it work, we need to add dummy variable such that we can plot using scatter plot and train using SVM.

Four traditional assessment metrics, namely: accuracy, F1 score, precision and recall, were chosen. Any of these metrics is determined using the True Positive (TP), True Negative (TN), False Positive (FP) and False Negative (FN) values calculated and represented by the uncertainty matrix throughout the test process. For a binary classification query, table 1 shows the general confusion matrix.

Table 1. Binary Classification Confusion Matrix

		Predicted	
		Positive	**Negative**
Actual	Positive	TP	FN
	Negative	FP	TN

Any of the chosen metrics will offer some insights into the model's results, which will strengthen the assessment process. A brief overview of each is shown below:

1) Accuracy: The ratio of accurate predictions to the total number of predictions is calculated. This can be measured in a binary hierarchy as:

$$Accuracy = \frac{TP + TN}{TP + TN + FN + FP} \tag{6}$$

2) Precision: The ratio between the correctly expected data and the overall optimistic predicted details. This ensures that a high-precision model is capable of accurately defining much of the expected results (see equation 7).

$$Precision = \frac{TP}{TP + FP} \tag{7}$$

3) Recall: This metric gives an analysis of the model's sensitivity. That is, the percentage of the positive data that was accurately defined as positive and the positive total data (see equation (8).

$$Recall = \frac{TP}{TP + FN} \tag{8}$$

4) F1 score: Using precision and recall, the fourth evaluation metric is calculated using Equation (9).

$$F1score = 2 * \frac{Recall * Precision}{Recall + Precision} \tag{9}$$

The F1 score is used to demonstrate the model's overall success in relation to both accuracy and recall. The benefit of using the F1 score for assessing a model's overall success is that the F1 score takes into account the distribution of data and the unequal class situation where false positive and false negative are at stake, which is typically the case with all the algorithms.

Figure (9) illustrates the data distribution, where Var1 is the generated keys and Var2 is the class. For training, the result is as shown in table 2, and figures 10 and 11.

Figure 9. The Data Distribution using Scatter Plot

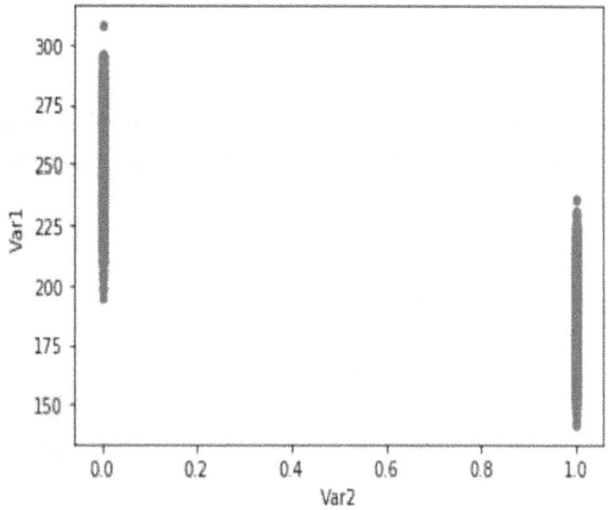

It is clear from the evaluation metrics results that we obtained high accuracy referring to the high integrity and confidentiality while transferring the photons, and the small error that appeared is negligible (table 2).

Table 2. The accuracy results for testing model in SVM

Dataset- split	Accuracy	precision	Recall	F1 score	
70%-30%	99.08%	99.1%	99.0%	99.04%	

There is no over fitting because we already did different experiments by using other machine learning algorithms as well as using another type of SVM kernels, see figures (10 and 11).

Figure 10. The Correlation Between the Actual Result and the Predicted One

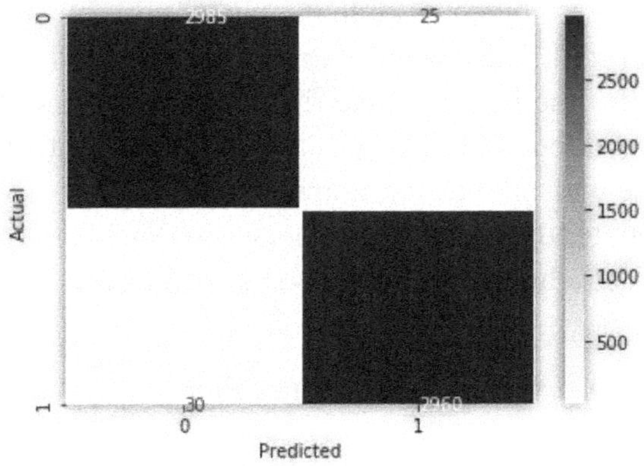

where “0” corresponds to the final quantum keys length without attacker, and “1” corresponds to the scenario with the attacker effect, also, the testing and training accuracy are close.

Figure 11. The Results after Classification by SVM

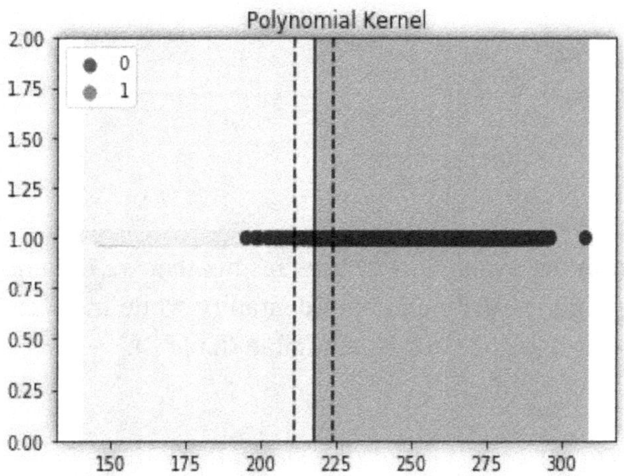

This result is promising and proved our goal from this test. All the results showed high accuracy for the other initial photons that have shown significant changes or fluctuations of the probability to be attacked while having the final key within the threshold, such that 100, 660, 1220, 2345, etc.

After these capable results showing that the proposed method can detect the attacker, this led to the next section where we aim to prevent the attacker from getting the shared key by making it hard to get. It needs to quantify the effect of raising the number of initial photons that are sent by the server on the final key length at the destination. In this book chapter, several initial photons are used and compared with the corresponding final key length. Figure 12 illustrates the number of initial sent photons versus the final key length. It is apparent that increasing the number of the sent photons will raise the final key length, but the figure shows this increase in a way not to be expected.

Figure 12. The Effect of Raising the Number of Initial Photons

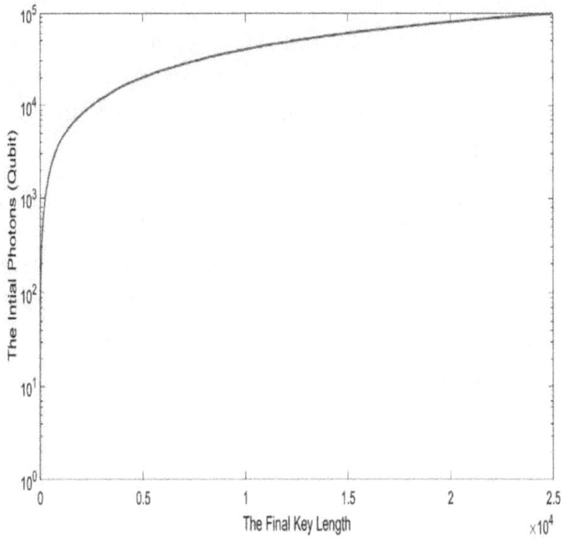

For example, for the initial number of 1000 photons, the final key length will be up to 322 in the best case. In this way we can estimate that when sending 2000 photons the final key length will be 644 in the best case, but this is not the calculation: We can see from the figure the final key length will be more than 700 in case of 2000 initial photons, and so on and so forth. In this way, after detecting an attacker by the proposed method we can be raising the number of the initial photons.

5. CONCLUSION

The advent of Quantum Key Distribution (QKD), initiated by Charles Bennett and Gilles Brassard through the BB84 protocol in 1984, heralded an unprecedented shift towards secure communication, promising an unbreakable method of exchanging quantum keys. As we approached the era, the practical deployment of this protocol highlighted its critical role in the detection of security breaches by monitoring the length of the generated key. Traditional beliefs and research have long held that significant alterations in key length are indicative of tampering attempts. Yet, this chapter exposes a more complex threat landscape where deviations in key length might not always indicator an attacker, with some variations remaining undetectably within acceptable margins.

This chapter not only uncovers this subtle vulnerability but also delves into the complexities surrounding final key lengths and their impact on the assumed security of QKD systems. A detailed examination reveals how certain compromised key lengths could appear indistinguishable from a secure exchange, thus questioning the reliability of traditional detection methods. In an innovative turn, the chapter proposes a machine learning-based approach that promises to redefine the detection of such sophisticated threats. The presented empirical evidence attests to the remarkable precision of this method in identifying intrusions through subtle differences in key length.

Furthermore, the text explores the strategic increase of initial photon counts, offering a quantitative analysis of its effect on bolstering QKD protocols against hidden dangers. This thorough investigation not only contests established notions in the field but also sets the stage for a more secure quantum communication framework, strengthened by the combined forces of quantum cryptography and cutting-edge machine learning technologies.

ACKNOWLEDGMENT

This publication was jointly supported by Qatar University and IS-Wireless - International Research Collaboration Co-Fund Grant no. IRCC-2021-003. The findings achieved herein are solely the responsibility of the authors.

REFERENCES

Al-Mohammed, H. A., Al-Ali, A., Yaacoub, E., Abualsaud, K., & Khattab, T. (2021). Detecting Attackers during Quantum Key Distribution in IoT Networks using Neural Networks. *GLOBECOM 2021-2021 IEEE Global Communications Conference*.

Al-Mohammed, H. A., Al-Ali, A., Yaacoub, E., Qidwai, U., Abualsaud, K., Rzewuski, S., & Flizikowski, A. (2021). Machine Learning Techniques for Detecting Attackers During Quantum Key Distribution in IoT Networks With Application to Railway Scenarios. *IEEE Access : Practical Innovations, Open Solutions*, 9, 136994–137004. 10.1109/ACCESS.2021.3117405

Al-Mohammed, H. A., & Yaacoub, E. (2021). On the Use of Quantum Communications for Securing IoT Devices in the 6G Era. *2021 IEEE International Conference on Communications Workshops, ICC Workshops 2021 - Proceedings*, 1–6. 10.1109/ICCWorkshops50388.2021.9473793

Al-Mohammed, H. A., & Yaacoub, E. (2022). New Way to Generating and Simulation QKD. In *Proceedings of Sixth International Congress on Information and Communication Technology* (pp. 801-809). Springer. 10.1007/978-981-16-1781-2_69

Aqsazafar. (2020). *SVM machine learning*. https://www.mltut.com/svm-machine-learning/

Bennett, C. H., & Brassard, G. (1984). Quantum cryptography: public key distribution and coin tossing. *Proceedings of the International Conference on Computers, Systems & Signal Processing*, 175–179.

Biamonte, J., Wittek, P., Pancotti, N., Rebentrost, P., Wiebe, N., & Lloyd, S. (2017). Quantum machine learning. *Nature*, 549(7671), 195–202. 10.1038/nature2347428905917

Busch, P., Heinonen, T., & Lahti, P. (2007). Heisenberg's uncertainty principle. *Physics Reports*, 452(6), 155–176. 10.1016/j.physrep.2007.05.006

Haitjema, M. (2007). *A Survey of the Prominent Quantum Key Distribution Protocols*. https://www.cse.wustl.edu/~jain/cse571-07/quantum.htm

Jeong, Y. C., Ji, S. W., Hong, C., Park, H. S., & Jang, J. (2020). Deterministic secure quantum communication on the BB84 system. *Entropy (Basel, Switzerland)*, 22(11), 1–13. 10.3390/e2211126833287036

Kotpalliwar, M. V., & Wajgi, R. (2015). Classification of attacks using support vector machine (SVM) on KDDCUP'99 IDS database. *Proceedings - 2015 5th International Conference on Communication Systems and Network Technologies, CSNT 2015*, 987–990. 10.1109/CSNT.2015.185

Mavromatis, A., Ntavou, F., Salas, E. H., Kanellos, G. T., Nejabati, R., & Simeonidou, D. (2018). Experimental Demonstration of Quantum Key Distribution (QKD) for Energy-Efficient Software-Defined Internet of Things. *European Conference on Optical Communication, ECOC,* (1), 1–3. 10.1109/ECOC.2018.8535267

Mina, M., & Simion, E. (2020). A Scalable Simulation of the BB84 Protocol Involving Eavesdropping. *Cryptology EPrint Archive, 1*, 1–19. https://eprint.iacr.org/2020/1074

Noble, W. S. (2006). What is a support vector machine? *Nature Biotechnology*, 24(12), 1565–1567. 10.1038/nbt1206-156517160063

Shor, P. W., & Preskill, J. (2000). Simple proof of security of the BB84 quantum key distribution protocol. *Physical Review Letters*, 85(2), 441–444. 10.1103/PhysRevLett.85.44110991303

Singh, H., Gupta, D. L., & Singh, A. (2014). Quantum Key Distribution Protocols: A Review. *IOSR Journal of Computer Engineering, 16*(2), 1–9. 10.9790/0661-162110109

Singh, S. K., El Azzaoui, A., Salim, M. M., & Park, J. H. (2020). Quantum Communication Technology for Future ICT - Review. *Journal of Information Processing Systems*, 16(6), 1459–1478. 10.3745/JIPS.03.0154

Zbinden, H., Bechmann-Pasquinucci, H., Gisin, N., & Ribordy, G. (1998). Applied Physics B Lasers and Optics Quantum cryptography. *Applied Physics. B, Lasers and Optics*, 67(6), 743–748. 10.1007/s003400050574

Zhang, H., Ji, Z., Wang, H., & Wu, W. (2019). Survey on quantum information security. *China Communications*, 16(10), 1–36. 10.23919/JCC.2019.10.001

Chapter 9
Embarking on Quantum Horizons

Shyam Sihare
https://orcid.org/0000-0003-2096-8273
Dr. A.P.J. Abdul Kalam Government College, India

ABSTRACT

The multifaceted landscape of quantum technologies, including their advances, challenges, and ethical implications, is explored in the chapter. It covers the diverse applications of quantum computing, collaborating across scientific domains, impacting society and the critical role of quantum literacy. The chapter explores different aspects of quantum technology, including interdisciplinary research, quantum communication networks, the relationship between quantum computing and artificial intelligence, and ethical considerations. It also highlights the societal responsibilities of the quantum community. Topics covered include ethical supply chains, media representation, disaster response, and international relations. The chapter advocates ethical governance, responsible innovation, and inclusive access to ensure harmonious integration into the fabric of our society as quantum technologies continue to develop.

NAVIGATING THE QUANTUM LANDSCAPE

Quantum mechanics, a revolutionary branch of physics born in the early 20th century, introduces new principles governing the behavior of particles at the atomic and subatomic level, where classical mechanics falter (Griffiths & Schroeter, 2018).

DOI: 10.4018/978-1-7998-9522-0.ch009

This realm challenges our everyday intuition and redefines our classical understanding of the physical world (Weinberg, 1992; Sihare, S. R (2023 (b))).

One of its fundamental pillars is the concept of superposition, allowing quantum particles like electrons and photons to exist in multiple states simultaneously (Griffiths & Schroeter, 2018). The iconic Schrödinger's cat experiment illustrates this, where the cat's state (alive or dead) remains undetermined until observed, highlighting the probabilistic nature of quantum reality (Greenberger et al., 1983; Sihare, S. R. (2022 (a))). This very feature underpins the remarkable potential of quantum computing, enabling parallel processing of information at scales far exceeding classical computers (Mabesoone, 2019).

Beyond superposition, quantum entanglement further blurs the boundaries of our classical intuition. This phenomenon, where two particles become intrinsically linked, dictates that measuring the state of one particle instantly determines the state of its entangled partner, regardless of their spatial separation (Horodecki et al., 2009; Sihare, S. R. (2022 (b))). This non-local correlation defies classical notions of locality and holds immense promise for revolutionary applications like quantum teleportation and secure communication channels (Bennett & Brassard, 2014).

Adding another layer of complexity is the uncertainty principle, formulated by Werner Heisenberg. It states that certain pairs of properties, like a particle's position and momentum, cannot be simultaneously known with perfect precision (Heisenberg, 1927; Sihare, S. R. (2022 (c))). This inherent uncertainty at the quantum level has profound implications for measurement accuracy and has paved the way for innovative technologies like quantum sensors and imaging devices (Giovannetti et al., 2011).

Navigating the quantum landscape necessitates a deep dive into the bedrock principles that sustain its enigmatic nature. As we unravel these mysteries, we unlock the potential for groundbreaking quantum technologies poised to revolutionize our technological landscape and redefine the bounds of possibility (Deutsch & Ekert, 2000).

One pivotal development in quantum information science is the shift from classical bits to their quantum counterparts, qubits. Unlike their classical counterparts confined to the binary realms of 0 and 1, qubits possess the remarkable ability to exist in a superposition of both states simultaneously (Nielsen & Chuang, 2000). This inherent duality empowers quantum computers to process information in parallel, simultaneously exploring all potential solutions, leading to potential exponential speedup compared to classical algorithms (Aaronson & Gottesman, 2002). Prime examples include Shor's algorithm for factoring integers with revolutionary efficiency (Shor, 1997; Sihare, S. R. (2022 (d))) and Grover's algorithm for searching immense databases exponentially faster (Grover, 1996).

Further blurring the lines of classical intuition is quantum entanglement, an intrinsic correlation between entangled qubits that transcends physical locality (Horodecki et al., 2009). This eerie phenomenon allows the creation of entangled states that defies classical description, paving the way for groundbreaking applications like quantum teleportation, where the quantum state of one qubit is instantaneously transmitted to another, even at vast distances (Bennett & Brassard, 1984; Sihare, S. R. (2023 (a))). Such feats hold immense promise for secure and unbreakable quantum communication networks.

Manipulating these fragile quantum states is fundamental to quantum computation, with techniques like quantum gates enabling qubit operations (Nielsen & Chuang, 2000). However, maintaining their delicate status poses a significant challenge due to environmental factors like temperature fluctuations and electromagnetic noise (Preskill, 2018). Researchers are actively developing error correction techniques and quantum error correction codes to combat these disturbances, paving the way for robust and scalable quantum computation (Fowler et al., 2012; Sihare, S., & Khang, A. (2023)).

The transition from classical bits to quantum qubits extends to the realm of communication as well. Quantum key distribution (QKD) utilizes the principles of quantum mechanics to establish unbreakable communication channels. Unlike classical cryptography, where the security relies on keeping the encryption key secret, QKD relies on the inherent randomness and fragility of quantum states to guarantee security. Even attempting to eavesdrop on a QKD transmission will inevitably disturb the quantum states, alerting the sender and receiver to the intrusion (Ekert, 1991). This makes QKD a highly secure solution for sensitive data transmission in fields like finance, healthcare, and national security (Scarani et al., 2009).

Beyond communication, quantum states offer exceptional sensitivity, paving the way for sensors that surpass the limitations of classical counterparts. By harnessing entangled states or exploiting the superposition principle, quantum sensors can achieve unprecedented levels of precision and accuracy. For example, quantum magnetometers based on nitrogen-vacancy centers in diamond have demonstrated superior sensitivity for detecting magnetic fields, with applications in medical imaging and mineral exploration (Walsworth et al., 2010). Similarly, quantum clocks based on trapped ions have the potential to redefine timekeeping with unmatched accuracy, impacting navigation, telecommunications, and scientific research (Hänsch & Wallman, 2006).

Quantum algorithms, leveraging the unique properties of qubits, hold the potential to revolutionize fields like cryptography and optimization. Shor's algorithm, a groundbreaking quantum algorithm developed in 1994, can factor large numbers exponentially faster than classical algorithms, posing a significant threat to existing cryptographic systems that rely on the difficulty of factoring (Shor, 1994). Grover's algorithm, another influential quantum algorithm, offers a quadratic speedup for searching unsorted databases, making it a powerful tool for data retrieval and optimization problems (Grover, 1997).

One of the most compelling advantages of quantum computers lies in their ability to efficiently simulate the behavior of complex quantum systems. This opens up new avenues for studying phenomena that are currently intractable for classical computers, such as the behavior of materials, chemical reactions, and biological processes (Aspuru-Huerva et al., 2013). Quantum simulations have the potential to revolutionize fields like drug discovery, materials science, and fundamental physics by providing insights into complex systems at an unprecedented level (McClean et al., 2016).

To address the challenges of noisy intermediate-scale quantum (NISQ) devices, variational quantum algorithms (VQAs) offer a promising approach. These hybrid algorithms combine classical optimization techniques with quantum circuits to tackle complex optimization problems. One prominent example is the quantum approximate optimization algorithm (QAOA), which has shown significant promise for solving combinatorial optimization problems in areas like logistics and scheduling (Farhi et al., 2014). VQAs offer a versatile solution for real-world optimization challenges, paving the way for practical applications of quantum computing in diverse fields.

As the field of quantum algorithms matures, researchers are actively exploring their potential to enhance machine learning (ML) and artificial intelligence (AI). The promise lies in leveraging quantum computing's unique features, namely parallelism and superposition, to significantly improve the efficiency of specific ML tasks. This could offer breakthroughs in pattern recognition, data analysis, and natural language processing (Biamonte et al., 2017; Schuld & Killback, 2019). For instance, quantum support vector machines have been shown to achieve exponential speedup in certain training scenarios compared to classical counterparts (Babbush et al., 2013).

However, harnessing the full potential of quantum ML requires overcoming the current limitations of quantum hardware. One key challenge is qubit noise, arising from environmental factors and imperfections in hardware, which can introduce errors during computation. These errors can significantly degrade the performance of quantum algorithms, making robust error correction techniques crucial for reliable operation (Fowler et al., 2015; Preskill, 2018). Recent developments in fault-tolerant quantum computing offer promising strategies for mitigating noise and preserving quantum information, paving the way for practical large-scale quantum ML applications (Harty et al., 2017; Bravyi & Kitaev, 2005).

REFLECTING ON COMPUTATIONAL PARADIGMS

Quantum supremacy is the ability of quantum computers to perform calculations that are practically impossible for classical computers. The term gained prominence in 2012 when physicist John Preskill coined it, emphasizing the potential of quantum computers to surpass classical computers in solving certain problems. In October 2019, Google claimed to have achieved quantum supremacy, demonstrating that quantum processors can perform computations at a scale and speed beyond classical computers. This milestone is significant in the development of quantum computers.

$$|Q = \sum_i \alpha_i |i \tag{1}$$

Where, $|Q\rangle$ is the quantum state, $|i\rangle$ are the basis states, and α_i are the complex amplitudes. The pursuit of quantum supremacy extends beyond academia and research institutions, with major technology companies investing heavily in quantum hardware and algorithms. National governments and international organizations acknowledge the strategic significance of quantum technologies and are dedicating substantial resources to quantum research. The quantum ecosystem is rapidly expanding, promoting collaboration and competition that stimulate innovation in the pursuit of computational power beyond classical limits.

$$V_q(n) = 2^{n_q} V_c(n_c) \tag{2}$$

Where V_q and V_c are the computational volumes of quantum and classical computers, respectively, n_q is the number of qubits, and n_c is the number of classical bits. The achievement of quantum supremacy raises significant questions about the societal impact of quantum technologies. Quantum computers have the potential to revolutionise fields such as cryptography, optimisation, and drug discovery. However, they also pose challenges, particularly in the realm of cybersecurity, as they could break widely used encryption schemes. Addressing these challenges requires a multidisciplinary approach involving not only physicists and computer scientists but also experts in ethics, policy, and cybersecurity.

$$P(breaking\_encryption) = 1 - e^{-\frac{T}{T_c}} \tag{3}$$

Where P is the probability of breaking encryption, T is the computation time, and T_c is the classical computation time for the same task. As we consider the journey towards quantum supremacy, it is clear that the goal is not only to achieve computational superiority, but also represents a paradigm shift in our approach to information processing. Quantum computers operate on principles that are fundamentally different from classical computers, utilizing the quantum nature of particles to perform computations in ways that were previously thought to be impossible. Quantum supremacy poses a challenge to our understanding of what is computationally achievable and pushes us into a new era where the limits of classical computational paradigms are surpassed.

$$|\psi = \alpha|0...0 + \beta|1...1 + \sum_i \gamma_i|\Phi_i \quad (4)$$

where $|\psi\rangle$ is the quantum state, $|0...0\rangle$ and $|1...1\rangle$ represent classical basis states, $|\Phi_i\rangle$ represent superpositions of basis states, and α, β, and γ_i are complex amplitudes.

Hybrid computing approaches are a promising way to enhance computational capabilities by combining the strengths of classical and quantum computing systems. The aim is to overcome the limitations inherent in each individual paradigm, such as the delicate nature of qubits and the need for long-term stability and error correction. The concept of quantum co-processors involves a quantum processor collaborating with a classical processor to perform specific tasks more efficiently than either could achieve alone. This collaboration is orchestrated through hybrid algorithms, which partition a computational problem into segments best suited for quantum processing and classical processing, respectively.

$$C_q(n) + C_c(n) - A\left(C_q, C_c\right) \approx T_h(n) \quad (5)$$

Where C_q and C_c represent the classical and quantum complexities of subproblems, $A\left(C_q, C_c\right)$ is the communication overhead and synchronization cost, and $T_h(n)$ is the overall hybrid complexity.

Hybrid computing has found a prominent application in optimization problems. Quantum algorithms, such as the QAOA, are used to optimize overall computational efficiency. The hybrid quantum-classical optimization (HQCO) framework is an excellent example of the synergy between classical and quantum computing in tackling optimization challenges. This collaborative effort aims to utilise quantum advantage for optimisation tasks, providing faster and potentially more optimal solutions than classical approaches.

$$H(Q,C) = \max(H(Q) + H(C) - K(Q,C)) \quad (6)$$

where $H(Q)$ and $H(C)$ represent the individual entropies of the quantum and classical subsystems, $K(Q, C)$ is the mutual information shared between them, and $H(Q, C)$ is the joint entropy of the combined system.

Hybrid computing also demonstrates its prowess in quantum chemistry simulations, where it combines classical methods for certain calculations with quantum processors for the most intricate quantum aspects. IBM and Rigetti Computing are developing quantum processors that can be integrated with classical computing systems, creating a hybrid infrastructure that facilitates the transition from classical to quantum processing.

$$E_{mol}(q, c) = E_{class}(c) + \Delta E_q(q, c) + \Omega_{qc} \tag{7}$$

where $E_{mol}(q, c)$ is the total molecule energy, $E_{class}(c)$ is the classical energy contribution, $\Delta E_q(q, c)$ is the quantum correction from the hybrid simulation, and Ω_{qc} is the error term due to qubit coherence losses and classical-quantum communication.

Challenges persist in achieving seamless integration and effective collaboration between quantum and classical processors. To mitigate these challenges, researchers are actively developing techniques for error correction and optimizing the orchestration of quantum and classical processes. Advancements in both quantum hardware and classical computing infrastructure are necessary to ensure the scalability of hybrid computing architectures.

$$C_{total} = C_q + C_c + \alpha C_q c \tag{8}$$

where C_{total} is the overall hybrid computation cost, C_q and C_c are the individual quantum and classical costs, and $\alpha C_q c$ represents the communication and synchronization overhead, with α being a factor dependent on hardware architecture and software optimization.

Quantum cloud computing is a significant advancement in computational paradigms, providing access to quantum processing power to a wider audience. This approach democratizes access to quantum processors, enabling researchers, developers, and businesses to leverage their potential without significant upfront investments in hardware infrastructure. IBM, Microsoft, and Rigetti Computing are among the companies that offer users remote access to quantum processors. This allows researchers and developers to run quantum algorithms, simulate quantum circuits, and experiment with quantum applications through cloud-based interfaces.

Quantum cloud computing accelerates the development and adoption of quantum algorithms by promoting a more inclusive and diverse community of users. It enables access to quantum processors without physical proximity to the hardware. Platforms usually offer programming interfaces and software development kits (SDKs) that simplify the complexities of quantum mechanics. This enables users without specialized knowledge to experiment with quantum algorithms and explore their capabilities.

$$N_{users} \approx \frac{A}{t_{access}} + \beta q_{access} \tag{9}$$

where N_{users} is the estimated number of users, A is the total availability of quantum processing power, t_{access} is the average access time per user, q_{access} is the minimum quantum resources required per user application, and β is a factor accounting for user diversity and application complexity

Quantum cloud computing also has implications for businesses that want to explore potential applications of quantum computing. Integrating quantum processing capabilities into cloud-based infrastructures allows companies to experiment with quantum algorithms to address industry-specific challenges in a scalable and cost-effective manner, without significant upfront investments.

$$B_Q(n) = aB_c\left(\frac{n}{q}\right) - C_q c - dQ_n \tag{10}$$

where $B_Q(n)$ is the potential business value from using a quantum algorithm for a problem of size n, $B_c\left(\frac{n}{q}\right)$ is the classical solution value if quantum resources are limited to q qubits, $C_q c$ is the overhead from cloud access and communication, and dQ_n is the cost of utilizing n qubits in the cloud.

The feasibility of quantum cloud computing is closely tied to advancements in quantum hardware and error correction techniques. As quantum processors become more robust and error rates decrease, the reliability and stability of quantum computations in the cloud improve. Ongoing research and development efforts are focused on enhancing the performance of quantum processors, extending coherence times, and implementing error correction protocols to ensure the accuracy of results obtained through cloud-based quantum computing.

$$R_{QC}(t) = e^{-\frac{t}{t_c}} \cdot F_{EC}(t) \tag{11}$$

where $R_{QC}(t)$ is the reliability of a quantum computation at time t, T_c is the qubit coherence time, and $F_{EC}(t)$ is the effectiveness of the error correction protocol used.

Regarding environmental impact, quantum cloud computing introduces complexity due to the need to optimize the energy consumption of supporting classical infrastructure. The integration of quantum computing resources into cloud-based infrastructures aligns with the trend of hybrid cloud computing, enabling users to utilize both classical and quantum processing capabilities based on specific task requirements.

$$E_{total} = E_{cl}(u_c) + E_q(u_q) + \gamma(u_c, u_q) \tag{12}$$

where E_{total} is the total energy consumption, $E_{cl}(u_c)$ is the energy consumed by classical components used by u_c users, $E_q(u_q)$ is the energy consumed by quantum components used by u_q users, and $\gamma(u_c, u_q)$ represents the additional energy overhead due to resource allocation and scheduling for hybrid cloud computing (see Figure 1 and Figure 2).

Figure 1. Key concepts in quantum computing, including comparing quantum and classical computational volumes, encryption breaking probabilities over computation time, and hybrid complexity related to classical and quantum complexity. Diagrams provide insight into the relationships between various parameters, helping to illuminate the complex nature of quantum computational models.

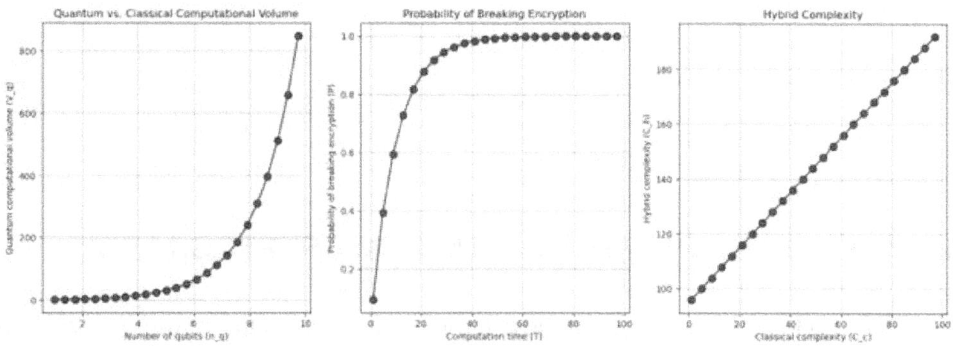

Figure 2. Exploring the reliability of quantum cloud computing: A grid of subplots illustrating the reliability of quantum computation over time for different qubit coherence times (T_c) and error correction effectiveness (F_{EC}) values. Each curve, represented by dashed lines with markers, shows the dynamic behaviour of the reliability function ($R_{QC}(t)$) under varying quantum parameters, highlighting the intricate relationship between qubit coherence, error correction and time in the evolving landscape of quantum cloud computing.

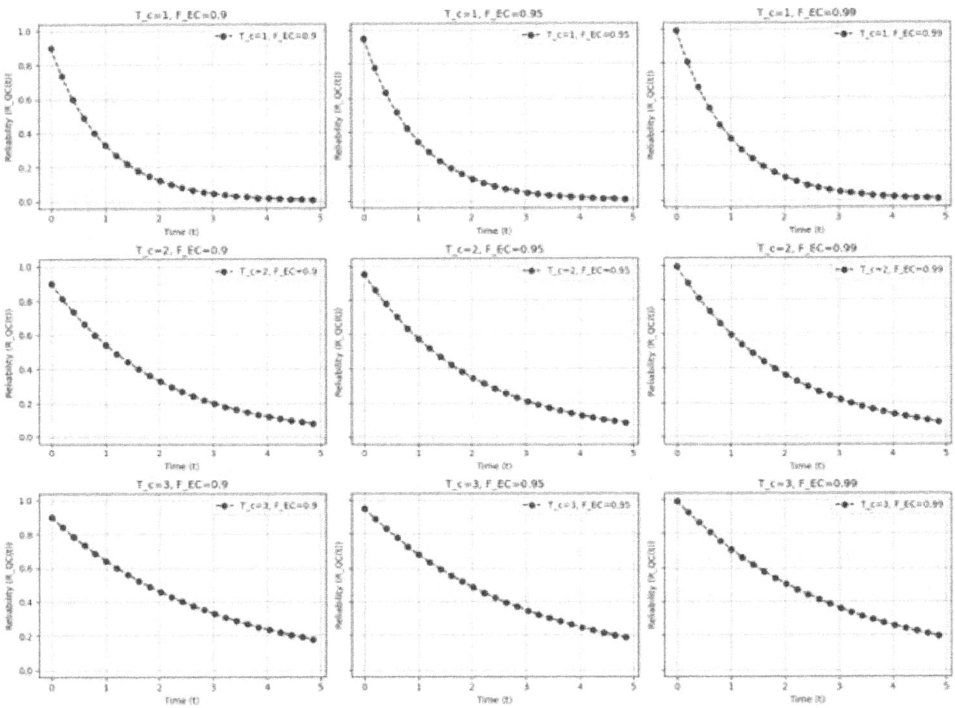

UNVEILING THE POWER OF QUANTUM CRYPTOGRAPHY

QKD is a technology that uses the principles of quantum mechanics to establish cryptographic keys and ensure the confidentiality of communication channels. QKD's foundation can be traced back to the work of physicists Stephen Wiesner and Artur Ekert in the 1980s, who introduced the concept of quantum entanglement as a means to secure communication channels.

$$\psi_A = \alpha|H + \beta|V \tag{13}$$

Eq. (13) represents the Alice's quantum state in superposition.

$$|K_{AB} = \sum_i \alpha_i \beta_i |i_A| i_B \quad (14)$$

Eq. (14) entangled state shared by Alice and Bob.

$$I(A; Eve) = H(A) - H(A|Eve) \quad (15)$$

Eq. (15) mutual information between Alice and Eve, quantifying Eve's knowledge about the key.

The photon state in superposition for Alice is represented by ψ_A, where α and β denote amplitudes for horizontal and vertical polarizations. The entangled state shared between Alice and Bob is captured by $|K_{AB}$, where i_A and i_B denote basis states. The information gain of Eve about the key through eavesdropping attempts is measured by $I(A; Eve)$. The security of QKD relies on the no-cloning theorem. This theorem states that $|K_{AB}$ cannot be perfectly copied. Any attempt by Eve to measure the photons destroys the entanglement, revealing her presence through increased $I(A; Eve)$.

QKD operates on the principle of quantum superposition and the no-cloning theorem. In a QKD protocol, Alice prepares a quantum state in superposition and encodes information in the polarization of the photons. The recipient, Bob, measures the polarization of the received photons, with the measurement basis chosen randomly. The uncertainty principle in quantum mechanics ensures that any attempt by an eavesdropper (Eve) to intercept the quantum states will disturb the delicate superposition, thus revealing her presence.

$$P_e = 1 - QBER - f(n) \quad (16)$$

Eq. (16) represents the probability of key compromise, considering quantum QBER and other factors.

$$\chi^2(n) = \frac{(N_{AB} - N_{EE})}{(N_{AB} + N_{EE})} \quad (17)$$

Eq. (17) represents the chi-squared test statistic for comparing Alice and Bob's key bits.

$$D_{BW}(\rho_A, \rho_B) = 1 - F(tr[\rho_A \rho_B]) \quad (18)$$

Eq. (18) Bures distance between Alice and Bob's density matrices, quantifying their state similarity.

P_e captures the probability of compromise due to QBER and other factors such as channel imperfection. The statistical test $\chi^2(n)$ is used to measure the agreement between the publicly revealed bases of Alice and Bob, which helps in detecting eavesdropping. $D_{BW}(\rho_A, \rho_B)$ measures the similarity between the final states of Alice and Bob (ρ_A and ρ_B). A low D_{BW} indicates successful entanglement distribution and minimal eavesdropping interference.

One of the main QKD protocols is the BB92 (Bennett-Brassard 1992) protocol, also known as the BB84 protocol. In this protocol, Alice sends a sequence of photons to Bob, randomly choosing one of two complementary bases for each photon. After transmission, Bob randomly selects a measurement basis for each received photon. Alice and Bob then publicly disclose their chosen bases for a subset of the transmitted photons. This is used to estimate the portion of the key that is known to both parties and could potentially be accessed by an eavesdropper.

The security of QKD against eavesdropping attempts has been rigorously analysed through information-theoretic principles. This includes the concept of 'QKD security proofs', which provide a robust foundation for the deployment of QKD systems in real-world scenarios.

$$\varepsilon_{sec}(X, Y, Z) = min_T\left[max_Q P_T(Z|XY) - P_{ideal}(Z|XY)\right] \tag{19}$$

Eq. (19) represents the min-max entropic uncertainty measure of security.

$$\Theta_{secrecy}(X, Y, Z) = max_T\{I(X;Z) - I(Y;Z)\} \tag{20}$$

Eq. (20) represents the secrecy rate, quantifying the extractable secret key rate.

$$\varepsilon_{privacy}(X, Y, Z) = max_T min_Q I(X;Z|Y) \tag{21}$$

Eq. (21) represents the min-max entropic privacy measure.

The function $\varepsilon_{sec}(X, Y, Z$ quantifies the uncertainty that an eavesdropper has about the key based on her observations (X, Y) and the final outcome (Z). The function $\Theta_{secrecy}(X, Y, Z)$ calculates the key generation rate at which Eve gains negligible information. The function $\varepsilon_{privacy}(X, Y, Z)$ quantifies the privacy of Alice's information from Eve's perspective.

QKD holds immense potential for revolutionizing secure communication, particularly in the era of quantum computers that pose threats to classical cryptographic systems. The distribution of cryptographic keys through QKD provides a secure mechanism for protecting sensitive information against the computational power of quantum adversaries. As the field of quantum computing advances, the importance of QKD in securing communication channels becomes increasingly critical.

$$P_{break}(C) = 1 - e^{\left(-\frac{t_B}{t_Q}\right)} \quad (22)$$

Eq. (22) represents the probability of breaking classical cryptosystem C by a quantum computer, with t_B as classical break time and t_Q as quantum break time.

$$QBER_{tol}(n) = \alpha P_{dark} + \beta P_{mis} + \gamma P_{error} \quad (23)$$

Eq. (23) represents the tolerable quantum bit error rate for secure QKD, accounting for dark counts, misalignment, and other errors.

$$K_{ent}(n) = H(\rho_{AB}) - I(A;E|\rho_{AB}) \quad (24)$$

Eq. (24) represents the extractable secret key rate from entangled state, considering Eve's information gain through channel errors.

$P_{break}(C)$ demonstrates the exponential advantage of quantum computers over classical ones in cracking cryptography. $QBER_{tol}(n)$ stresses the significance of low error rates in QKD for secure key generation. $K_{ent}(n)$ measures the extractable secret key rate ensured by entanglement, even with channel imperfections and eavesdropping attempts.

The emergence of quantum computers presents a substantial challenge to classical cryptographic systems, highlighting the importance of post-quantum cryptography as a crucial area of research and development. The objective of this field is to develop cryptographic methods that can withstand the computational capabilities of quantum adversaries. The vulnerability of classical cryptographic systems to quantum attacks is due to the unique capabilities of quantum computers. These computers can solve certain mathematical problems exponentially faster than classical computers. As quantum technologies continue to advance, the need for cryptographic approaches that are resilient to quantum attacks becomes increasingly urgent.

$$f(N) = \frac{2^{t_Q}}{2^{t_C}} \quad (25)$$

Eq. (25) represents the quantum speedup for factoring integers, with t_Q and t_C as break times for quantum and classical algorithms.

$$QBER_{crit}(n) = \frac{1}{\sqrt{N}} \quad (26)$$

Eq. (26) represents the critical quantum bit error rate for Grover's search algorithm, where N is the search space size.

$$t_Q(n) = O(log^2 n) \tag{27}$$

Eq. (27) represents the quantum speedup for solving linear systems.

The susceptibility of classical cryptography to quantum attacks, including the exponential speed-up of cracking RSA encryption using Shor's algorithm and the significant acceleration of brute-force searches using Grover's algorithm. Additionally, the HHL algorithm demonstrates the quantum advantage for solving certain mathematical problems that are crucial for cryptographic primitives.

Post-quantum cryptography is a diverse field that involves designing and analysing cryptographic algorithms that are resilient to quantum attacks. Researchers in this field explore various approaches to secure communication and data integrity in the post-quantum era. Key categories of post-quantum cryptographic primitives and protocols have emerged, each addressing specific challenges posed by quantum adversaries.

$$G = P \cdot T, H = Q \cdot G \tag{28}$$

Eq. (28) represents the key generation McEliece cryptosystem, where P and Q are random matrices and T is a sparse secret permutation matrix.

$$BqBv = qI + u \tag{29}$$

Eq. (29) represents the key encapsulation Lattice-based cryptography, where B is a basis matrix, q is the underlying field modulus, u is a noise term, and v is the encapsulated secret.

$$f(x) = x^4 + ax^3 + bx^2 + cx + d \tag{30}$$

Eq. (30) represents the example multivariate public key function with quadratic terms.

The McEliece cryptosystem is based on code-based problems, whereas lattice-based cryptography leverages the geometry of lattices. Multivariate cryptography uses complex polynomial equations for secure key exchange and digital signatures.

The transition to post-quantum cryptography requires the development of quantum-resistant algorithms and their integration into existing cryptographic infrastructures. This process involves several challenges and considerations, including algorithm standardization, quantum-safe cryptographic protocols, integration challenges, quantum-resistant cryptographic key management, and quantum cryptographic hygiene.

$$C_{tot}(n) = C_{alg}(n) + C_{int}(n) + C_{key}(n) + C_{hyg}(n) \tag{31}$$

Eq. (31) represents the total cost of transition to post-quantum cryptography, considering algorithm implementation cost (C_{alg}), integration cost (C_{int}), key management cost (C_{key}), and cryptographic hygiene cost (C_{hyg}).

$$\Omega_{std}(T,N) = \frac{N \cdot p_{rej}}{\sqrt{T}} \tag{32}$$

Eq. (32) represents the expected delay in standardization due to rejections (p_{rej}) and number of candidates (N), considering time for evaluation (T).

$$\sigma_{sec}(Q,C) = min\left(H(K_Q), H(K_C) - I(E;K_C|Q)\right) \tag{33}$$

Eq. (33) represents the hybrid protocol security, considering entropies of quantum $\left(H(K_Q)\right)$ and classical $\left(H(K_C)\right)$ keys and Eve's information gain ($I(E;K_C|Q)$) given quantum channel.

The cost factors involved are captured by $C_{tot}(n)$, which emphasizes the need for efficient implementations and hybrid approaches. The potential delay in standardization due to the rigorous evaluation process is highlighted by $\Omega_{std}(T,N)$. The security considerations for hybrid protocols are showcased by $\sigma_{sec}(Q,C)$, which balances quantum key entropy with Eve's knowledge gained through the classical channel.

Algorithm standardization is a crucial step in ensuring interoperability and widespread adoption of post-quantum cryptographic algorithms. The national institute of standards and technology (NIST) plays a pivotal role in soliciting, evaluating, and standardizing post-quantum cryptographic algorithms through open competitions. Quantum-safe cryptographic protocols are essential for securing communication channels against both classical and quantum adversaries. Hybrid cryptographic approaches that combine classical and post-quantum cryptographic techniques are being explored to facilitate a smooth transition.

$$k \leq \log_2\left(\frac{t_{sec}}{\varepsilon}\right) \tag{34}$$

Eq. (34) represents the minimum key size requirement based on desired security level (t_{sec}) and tolerable failure probability (ε).

$$f_Q(N) = \frac{2^{\sqrt{3}\cdot ln(2)\cdot N^{\frac{1}{3}}}}{\sqrt{\pi}} \quad (35)$$

Eq. (35) represents the Shor's algorithm quantum speedup for factoring, indicating classical attack resistance of algorithms with larger key sizes.

$$\pi_{hybrid}(n) = \pi_{cl}(n) + QBER \cdot h(\rho_{AB}) \quad (36)$$

Eq. (36) represents the Lattice-based protocol hybrid protocol performance, combining classical processing cost (π_{cl}), QBER, and entanglement entropy ($h(\rho_{AB})$) (see Figure 3).

Figure 3. Various aspects of post-quantum cryptographic algorithms, including the minimum key size requirement, Shor's quantum speedup algorithm for factoring, and the performance of a lattice-based hybrid protocol under different parameter values. Each subplot shows the impact of key parameters on cryptographic security and computational efficiency.

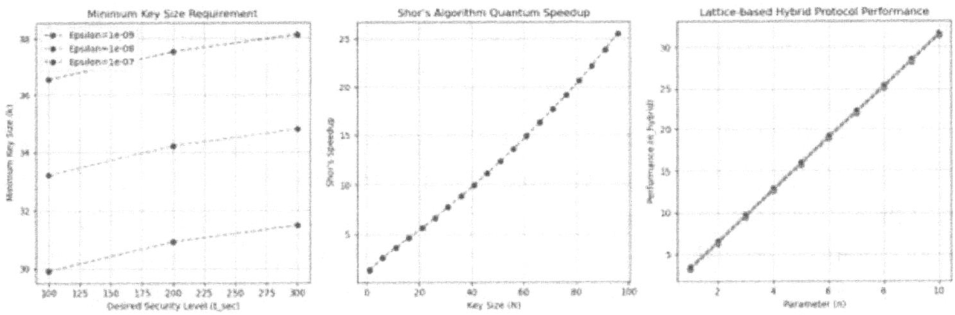

NIST's competition criteria ensure quantum resistance, as does the use of larger key sizes in algorithms such as Shor's algorithm. Hybrid protocols, such as $\pi_{hybrid}(n)$, balance classical cost and quantum resource utilization while accounting for errors.

Quantum-resistant cryptography aims to enhance the security of current cryptographic systems against classical attacks and prepare for a future where large-scale quantum computers are a reality. As quantum computing advances, research and development in post-quantum cryptography is crucial for protecting the security and privacy of digital communication.

$$P_{breach}(t) = 1 - e^{-\lambda\cdot t} \quad (37)$$

Eq. (37) represents the probability of cryptographic system breach within time t, with λ as failure rate.

$$\rho_{future}(n) = H\left(K_{fut}\right) - I\left(E; K_{fut} \mid \rho_{fut}\right) \tag{38}$$

Eq. (38) represents the future security of a new algorithm, considering key entropy $\left(H\left(K_{fut}\right)\right)$ and Eve's information gain $\left(I\left(E; K_{fut} \mid \rho_{fut}\right)\right)$ with future advancements in quantum computing.

$$\Omega_{adapt}(Q, C) = f\left(R_Q, T_{obs}\right) \tag{39}$$

Eq. (39) represents the cost of adapting cryptographic infrastructure to evolving quantum threats, considering rate of quantum progress (R_Q) and observation time (T_{obs}).

$P_{breach}(t)$ emphasises the importance of algorithms with minimal failure rates over time. $\rho_{future}(n)$ demonstrates the need to consider future quantum capabilities when designing new algorithms. $\Omega_{adapt}(Q, C)$ highlights the cost of adapting infrastructure to evolving threats, which requires continuous research and development (see Figure 4).

Figure 4. The probability of a cryptographic system breach over time for different failure rates (λ). The subplot in the bottom left examines the impact of key size on future algorithm security by simulating Eve's information gain for varying key sizes. The subplot in the bottom right depicts the cost of adapting infrastructure to quantum threats through adaptation cost curves for different rates of quantum progress (R_Q) over observation time (T_{obs}). Each subplot provides insights into the dynamic aspects of vulnerabilities in cryptographic systems, future security considerations, and the costs associated with adapting to emerging quantum threats.

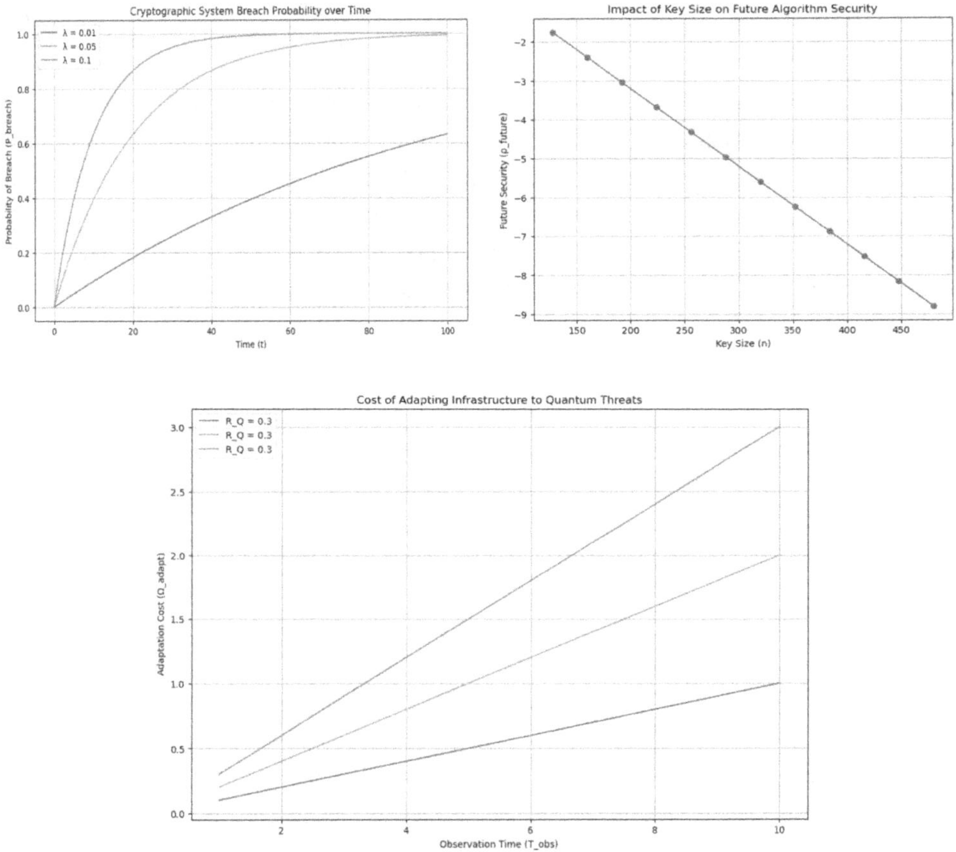

Quantum-safe protocols are created to resist the computational abilities of quantum adversaries, ensuring secure and private digital interactions in the quantum era. These protocols are based on quantum-resistant cryptographic primitives, including lattice-based, code-based, hash-based, multivariate polynomial, and isogeny-based cryptography. The authors adopt a hybrid approach that combines classical and post-quantum cryptographic techniques. This approach provides a transitional path towards quantum-safe communication while maintaining compatibility with current systems.

$$P_{break}(C_q, C_c) = min(P_{break}(C_q), P_{break}(C_c) - \alpha \cdot I(E; K_c | \rho_{AB})) \tag{40}$$

Eq. (40) represents the probability of breaking the hybrid protocol, considering classical break probability ($P_{break}(C_c)$), quantum break probability ($P_{break}(C_q)$), Eve's information gain ($I(E; K_c | \rho_{AB})$) from entanglement, and a factor (α) accounting for communication overhead.

$$QBER_{tol}(L, C) = \frac{\alpha}{\sqrt{L}} + \frac{\beta}{\sqrt{C}} \tag{41}$$

Eq. (41) represents the tolerable QBER for lattice (L) or code-based (C) cryptosystems, balancing lattice size (L) and codeword length (C) with error tolerance (α, β).

$$f(x) = \sum_i a_i x^{p_i} + \sum_j b_j y^{q_j} + c \tag{42}$$

Eq. (42) represents the Multivariate public key function polynomial with multiple variables (x, y), demonstrating resistance to brute-force and factoring attacks (see Figure 5).

The $P_{break}(C_q, C_c)$ protocol highlights the interplay between classical and quantum vulnerabilities in hybrid protocols. Meanwhile, $QBER_{tol}(L, C)$ emphasizes the trade-off between error tolerance and resource efficiency in lattice and code-based cryptography.

Figure 5. (a) Impact of Entanglement Gain on Hybrid Protocol Security. The top subplot shows the hybrid break probability concerning classical break probability and varying entanglement gains $(\alpha I(E; K_c))$. The break probability is minimized as entanglement gains increase, demonstrating the influence of entanglement on the hybrid protocol's security. (b) Balancing Cryptosystem Size and Error Tolerance. The subplot in the middle displays the tolerable Quantum Bit Error Rate ($QBER_{tol}$) for different cryptosystems, namely Lattice and Code-based, as a function of their respective sizes (L or C). The plot demonstrates the trade-off between the size of each cryptosystem and its tolerance to quantum errors. The example Multivariate Public Key Function Polynomial is shown in (c). The bottom subplots depict contour plots of the same example multivariate public key function polynomial for different degrees (x, y). The contour plots offer visual representations of the multivariate polynomial functions with different degrees, demonstrating the intricacy of public key functions in cryptography.

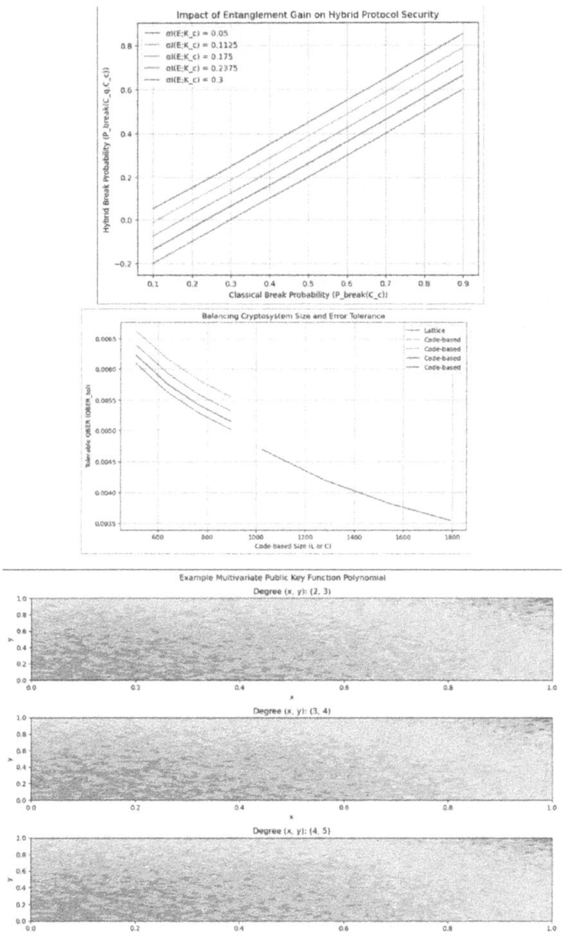

Quantum-safe protocols prioritize forward secrecy, quantum-resistant key management, efficiency, and scalability. The protocols aim to be efficient and scalable, minimizing computational overhead while providing a high level of security. Notable quantum-safe protocols include Newhope, Kyber, Dilithium, Falcon, NTRUEncrypt, and Classic McEliece. The standardization process involves multiple rounds of evaluation and public scrutiny, leading to the selection of quantum-safe algorithms that demonstrate both security and practicality.

$$P_{compromise}\left(K_{fut} \mid K_{past}\right) \leq \varepsilon \tag{43}$$

Eq. (43) represents the Forward secrecy future key compromise probability given past keys, with a low bound ε.

$$H(K) - I(E; K \mid Q) \geq \lambda \tag{44}$$

Eq. (44) represents the Quantum-resistant key management entropy ($H(K)$) minus Eve's information gain ($I(E; K \mid Q)$) through quantum channel exceeds security threshold (λ).

$$c = aK_A + H(b)G + e \tag{45}$$

Eq. (45) represents the Kyber key encapsulation ciphertext (c) generated from Alice's secret key (K_A), public parameters (a, G), Bob's randomness (b), and error term (e)

$$\sigma = H(m)S_A + e_1 + e_2 \tag{46}$$

Eq. (46) represents the Dilithium digital signature (σ) on message (m) using Alice's secret key (S_A), public parameter ($H(m)$), and error terms (e_1, e_2).

The use of forward secrecy guarantees that past key exposure risks are minimized. Additionally, quantum-resistant key management ensures sufficient information-theoretic security against eavesdropping. Kyber's key encapsulation and Dilithium's digital signature generation are also noteworthy (see Figure 6).

Figure 6. (a) Impact of past key exposure on forward secrecy probability for different security levels (ε). (b) Balancing key entropy and information gain for quantum resistance with varying Eve's information gain.

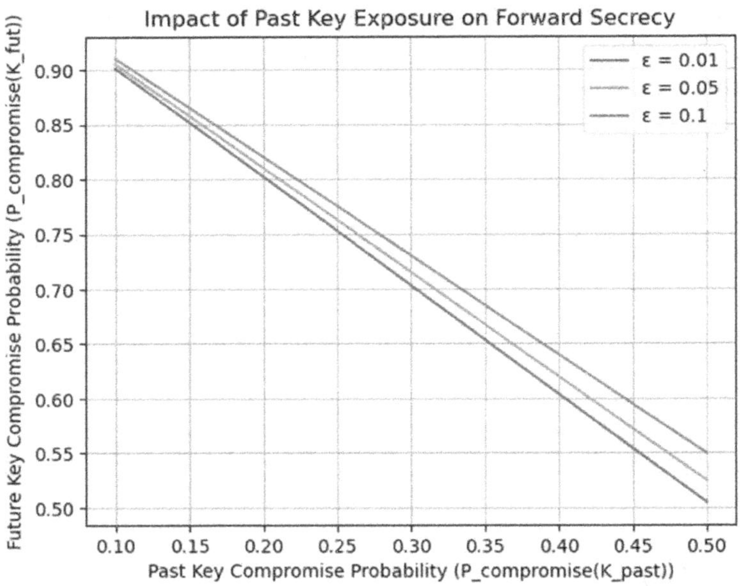

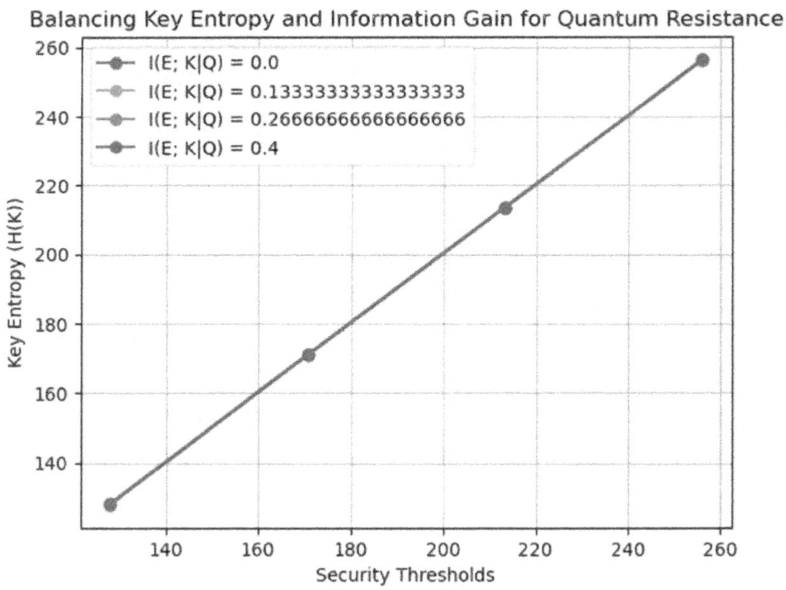

The transition to quantum-safe protocols requires careful planning and coordination among various stakeholders, including government agencies, industries, and service providers. Key aspects of this transition include risk assessment, planning, phased integration, algorithm agility, education, awareness, collaboration, and interoperability. Organizations must conduct risk assessments to evaluate their exposure to quantum threats and plan for the adoption of quantum-safe protocols. The integration of quantum-safe protocols is likely to occur in phases to minimize disruptions to existing systems.

$$C_{transition}\left(T_{phase}, A_{disruption}\right) = \alpha T_{phase} + \beta A_{disruption} + \gamma C_{training} \tag{47}$$

Eq. (47) represents the transition cost considering phasing time (T_{phase}), disruption level ($A_{disruption}$), and training costs ($\gamma C_{training}$).

$$\Omega_{collaboration}\left(N_{stakeholders}\right) = 1/N^{\gamma}_{stakeholders} \tag{48}$$

Eq. (48) represents the collaboration effectiveness, increasing with the number of stakeholders ($N_{stakeholders}$) raised to a factor (γ) accounting for communication and coordination challenges.

$$R\_risk\_assessment\left(T_{obs}, Q_{progress}\right) = f(T_{obs}, \varepsilon_{threshold}) + h\left(Q_{progress}\right) \tag{49}$$

Eq. (49) represents the risk rating based on observation time (T_{obs}), tolerable error threshold ($\varepsilon_{threshold}$), and rate of quantum progress ($Q_{progress}$).

Algorithm agility is crucial in the evolving field of post-quantum cryptography. Organizations must design cryptographic infrastructures that can easily accommodate changes in algorithms as new quantum-safe protocols are developed and standardized. Education and awareness are essential, as stakeholders must be well-informed about the implications of quantum computing on current cryptographic systems. Collaboration and interoperability are also necessary to facilitate a smooth transition.

$$C_{adapt}(n) = f\left(T_{switch}, N_{alg}, C_{training}\right) \tag{50}$$

Eq. (50) represents the cost of algorithm adaptation in round n, considering switching time (T_{switch}), number of candidate algorithms (N_{alg}), and training costs ($C_{training}$).

$$\mu_{agility}(\lambda, t) = \lambda \cdot e^{-\beta \cdot t} \tag{51}$$

Eq. (51) represents the agility metric, decreasing with time (t) and factor (β) representing resistance to change, modulated by security threshold (λ).

$$H(K) = H(K_q) \oplus H(K_c) \tag{52}$$

Eq. (52) represents the Modular PQC architecture hybrid key entropy with independent quantum $(H(K_q))$ and classical $(H(K_c))$ components, facilitating modular algorithm updates)

C_{adapt} captures the trade-offs between adaptation speed, algorithm options, and training costs. $\mu_{agility}$ measures the organization's resistance to change over time in relation to security requirements (see Figure 7).

Figure 7. The generated plots collectively portray key aspects of the transition to quantum-safe protocols, providing insights into the challenges and considerations involved. The first graph illustrates the dynamic relationship between phasing time (T_{phase}) and training cost ($C_{training}$), offering a nuanced understanding of their impact on the overall transition cost ($C_{transition}$). Subsequently, the second graph explores the crucial element of collaboration effectiveness ($\Omega_{collaboration}$), revealing its dependence on the number of stakeholders ($N_{stakeholders}$) and various factors (γ). Moving forward, the third plot delves into the realm of risk assessment (R_risk_assessment), presenting risk ratings influenced by observation time (T_{obs}) and quantum progress rates ($Q_{progress}$). The fourth graph investigates the adaptation cost (C_{adapt}), taking into account switching time (T_{switch}) and the number of candidate algorithms (N_{alg}), with distinct curves corresponding to different algorithmic scenarios. Additionally, the fifth plot introduces the concept of an agility metric ($\mu_{agility}$) over time (t), demonstrating its evolution with varying security thresholds. Lastly, a scatter plot visually represents the hybrid key entropy ($H(K)$) in a modular PQC architecture, showcasing the interplay between quantum key entropy ($H(K_q)$) and classical key entropy ($H(K_c)$). Together, these visualizations offer a comprehensive narrative on the multifaceted considerations essential for a successful and strategic transition to quantum-safe protocols.

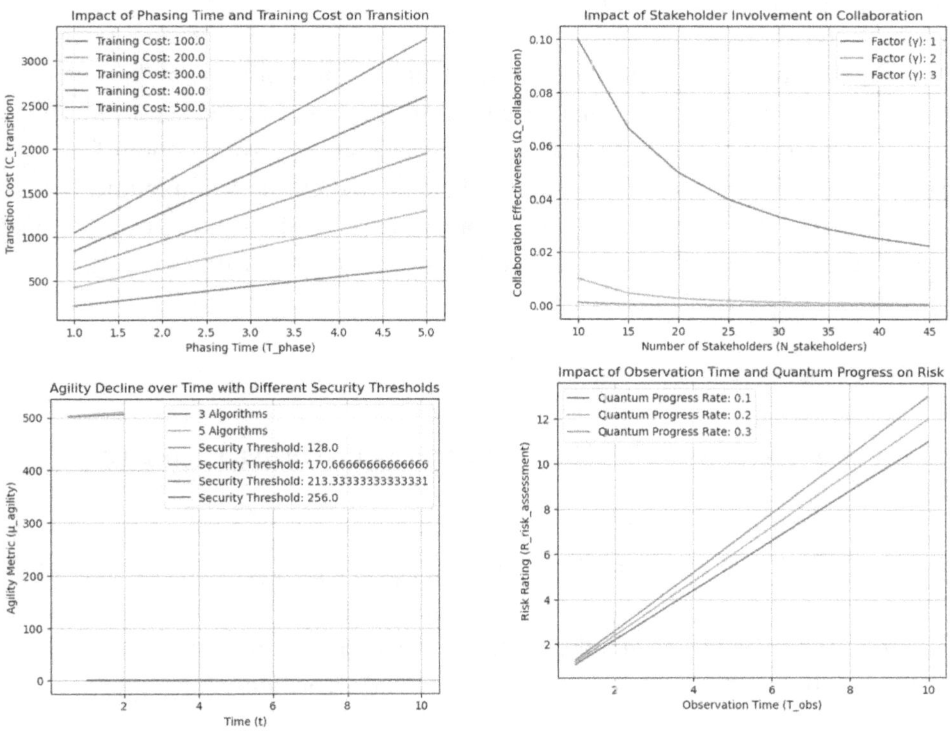

The journey ahead involves further research and development in post-quantum cryptography, ongoing standardization efforts, and strategic planning for the seamless integration of quantum-safe protocols into existing systems. Quantum-safe protocols provide a defence against quantum attacks and contribute to the resilience and longevity of secure communication in the face of evolving technological landscapes. In conclusion, the power of quantum cryptography extends beyond cryptographic primitives to encompass the protocols that define secure communication.

$$\Omega_{awareness}\left(N_{stakeholders}, P_{adoption}\right) = N_{stakeholders} \cdot \left(1 - P_{adoption}\right) \tag{53}$$

Eq. (53) represents the awareness gap considering number of stakeholders ($N_{stakeholders}$) and adoption percentage ($P_{adoption}$).

$$\sigma_{collaboration}\left(T_{planning}, R_{communication}\right) = \left(\frac{T_{planning}}{R_{communication}}\right)^2 \tag{54}$$

Eq. (54) represents the collaboration success rate based on planning time ($T_{planning}$) and communication effectiveness ($R_{communication}$).

$$I_{ij} = f\left(C_{comp}\left(A_i, B_j\right), S_{alg}\left(A_i, B_j\right)\right) \tag{55}$$

Eq. (55) represents the interoperability matrix compatibility score between systems A_i and B_j, considering computational cost (C_{comp}) and shared algorithms (S_{alg}).

Effective communication and education are necessary to drive adoption of $\Omega_{awareness}$. Successful collaborative efforts require planning and efficient communication, as emphasized by $\sigma_{collaboration}$. The interoperability matrix demonstrates the importance of considering compatibility between different systems for seamless integration.

$$R_{development}\left(F_{funding}, C_{research}\right) = \left(\frac{F_{funding}}{C_{research}}\right)^2 \tag{56}$$

Eq. (56) represents the development rate based on funding allocation ($F_{funding}$) and research cost ($C_{research}$).

$$\Omega_{standardization}\left(T_{rounds}, N_{candidates}\right) = T_{rounds} \cdot \left(1 - \varepsilon_{success}\right) \cdot N_{candidates} \tag{57}$$

Eq. (57) represents the standardization delay due to number of rounds (T_{rounds}), tolerable failure probability ($\varepsilon_{success}$), and candidate algorithms ($N_{candidates}$).

$$G(V, E) = \left(V_{cl} \cup V_q, E_{cl} \cup E_q\right) \tag{58}$$

Eq. (58) represents the quantum-resistant communication network graph representation of a hybrid network with classical nodes (V_{cl}), quantum nodes (V_q), classical channels (E_{cl}), and quantum channels (E_q).

$R_{development}$ emphasizes the role of funding and efficient research practices in accelerating progress. $\Omega_{standardization}$ captures the potential delay due to the rigorous evaluation process (see Figure 8).

Figure 8. The visualizations convey several critical aspects of strategic planning and collaboration in the transition to quantum-safe protocols. The initial plot delineates the inverse relationship between the number of stakeholders ($N_{stakeholders}$) and the awareness gap ($\Omega_{awareness}$), with distinct curves depicting varied adoption percentages. This emphasizes the pivotal role of stakeholder engagement in mitigating the awareness gap. The subsequent graph explores the correlation between planning time ($T_{planning}$) and collaboration success rate ($\sigma_{collaboration}$). Different communication rates delineate how effective communication influences collaboration success over diverse planning durations. An illustrative interoperability matrix highlights the impact of computational costs and shared algorithms on interoperability, offering insights into system compatibility. The subsequent plot delves into the interplay between funding allocation ($F_{funding}$) and development rate ($R_{development}$), showcasing nuanced relationships under different research cost scenarios. Another graph explores the influence of evaluation rounds (T_{rounds}), failure tolerance, and the number of candidate algorithms on standardization delay ($\Omega_{standardization}$), providing valuable insights into standardization considerations. Finally, a network graph visually portrays a hybrid infrastructure, seamlessly integrating classical and quantum components, offering a clear representation of the complex network architecture involved in the transition process.

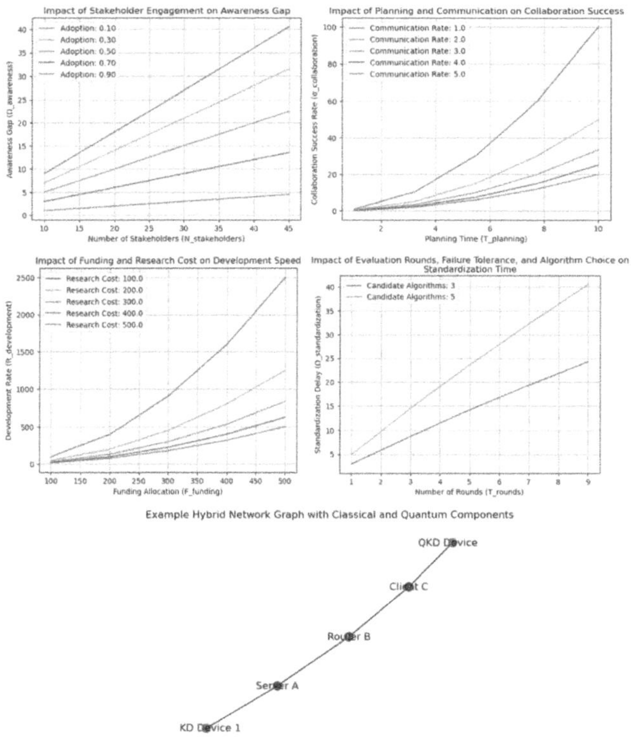

PROSPECTS FOR FUTURE QUANTUM COMPUTING

Over the past decade, quantum hardware has significantly advanced the development of quantum processors. Key advancements in quantum computing include superconducting qubits, topological qubits, and other emerging technologies. Superconducting qubits leverage superconductivity to create and manipulate quantum states with remarkable precision. Recent advancements in superconducting qubit technology have been aimed at overcoming challenges such as coherence times, gate fidelity, and qubit connectivity.

$$T_c(B) = T_0 e^{-\frac{\Delta B}{k_B}} \tag{59}$$

Eq. (59) represents the coherence time as a function of magnetic field (B), with T_0 as intrinsic coherence time, Δ as activation energy, and k_B as Boltzmann constant.

$$F_{gate}(\tau, \varepsilon) = 1 - e^{-\frac{\tau^2}{2\sigma^2}} - \varepsilon \tag{60}$$

Eq. (60) represents the gate fidelity with pulse duration (τ), standard deviation (σ), and error rate (ε).

$$Q_{connectivity}(n) = \alpha n + \beta n^{\frac{3}{2}} + \gamma \tag{61}$$

Eq. (61) represents the qubit connectivity for an n-qubit system, incorporating nearest-neighbor (α), next-nearest-neighbor (β), and further connections (γ).

$T_c(B)$ demonstrates the trade-off between magnetic field strength and coherence time, which is a crucial factor for reliable quantum operations. $F_{gate}(\tau, \varepsilon)$ highlights the significance of optimizing pulse duration and minimizing gate errors. $Q_{connectivity}$ (n) illustrates the scaling challenges of effectively connecting increasing numbers of qubits (see Figure 9).

Figure 9. Impact of pulse duration on gate fidelity for different standard deviations. Each curve represents a distinct standard deviation, showcasing how gate fidelity varies with pulse duration. The second figure explores the scaling of qubit connectivity with the number of qubits for various nearest-neighbor factors. Each curve corresponds to a different nearest-neighbor factor, illustrating the relationship between the number of qubits (n) and qubit connectivity ($Q_{connectivity}$). Both figures provide insights into the critical parameters that influence quantum gate performance and qubit connectivity in quantum systems.

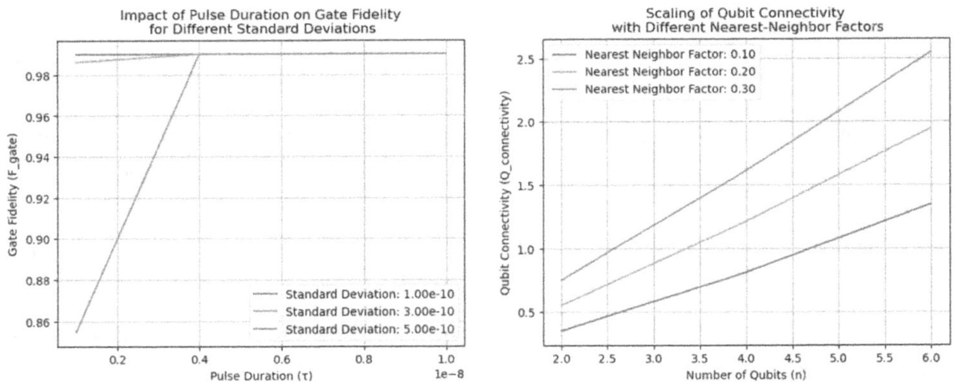

Topological qubits are an innovative approach to quantum computation, inspired by the field of topological quantum computing. Majorana fermions are exotic particles that exist as their own antiparticles. They possess unique braiding properties that make them less susceptible to certain types of errors, offering a promising avenue for fault-tolerant quantum computation. Recent breakthroughs in the pursuit of topological qubits include the experimental observation of Majorana fermions in condensed matter systems and the exploration of integrated photonic circuits for qubit manipulation.

$$E_M = \frac{\hbar \omega_F}{\sqrt{N}} \tag{62}$$

Eq. (62) represents the Majorana fermion energy gap for N degenerate modes, with $\hbar$ as reduced Planck constant and ω_F as characteristic frequency.

$$\tau_{braiding}(L, B) = \frac{L^2}{D\pi\hbar k_B T} \tag{63}$$

Eq. (63) represents the braiding error time for Majorana fermions, considering distance (*L*), magnetic field (*B*), diffusion constant (*D*), and temperature (*T*).

$$H(q) = \sum_{i,j} J_{i,j} q_i q_j + \sum_i \varepsilon_i q_i + h.c \tag{64}$$

.

Eq. (64) represents the integrated photonic circuit Hamiltonian for a photonic qubit circuit, including coupling constants (J_{ij}), on-site energies (ε_i), and Hermitian conjugate terms.

E_M emphasises the significance of a wide energy gap in safeguarding Majorana fermions. The function $\tau_{braiding}(L, B)$ demonstrates the balance between braiding distance and error rates (see Figure 10).

Figure 10. The first figure shows how the Majorana Fermion energy gap varies with the number of degenerate modes, indicating a decrease in the gap as the number of modes increases. The second figure examines the effect of braiding distance and magnetic field on braiding error time. Each curve represents different diffusion constants and magnetic fields at a fixed temperature. The third figure illustrates the eigenvalues of a sample photonic circuit Hamiltonian, displaying the energy levels of the system.

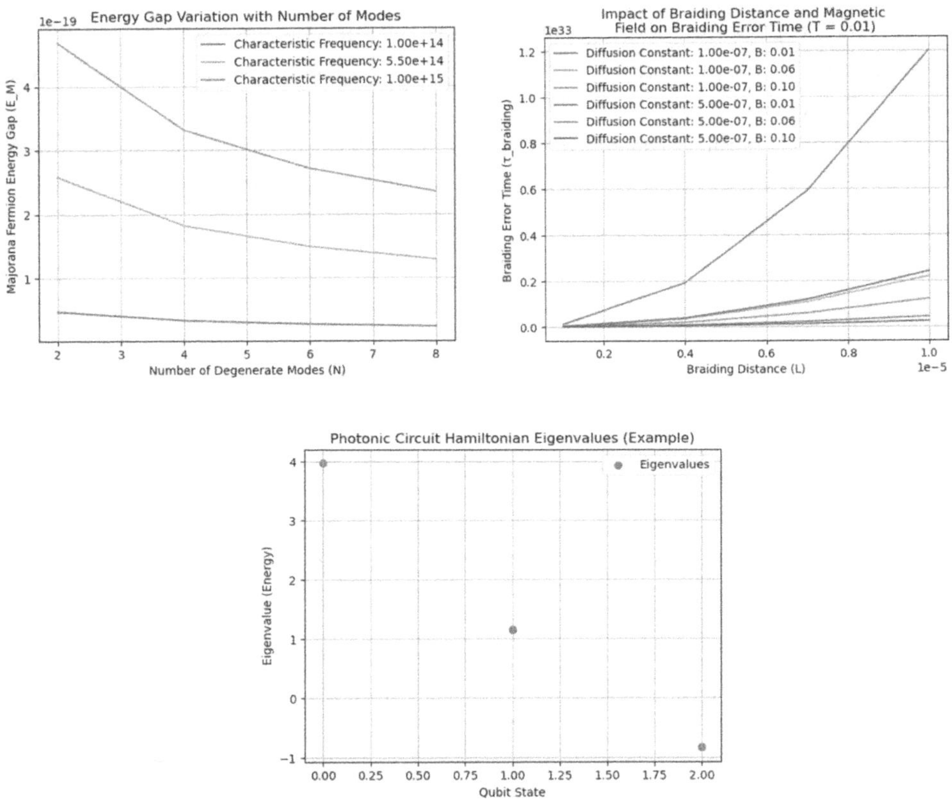

Trapped ions and photonic qubits are two distinct approaches to quantum computing, each with its own advantages and challenges. Trapped ions offer remarkable coherence times and precise qubit control, while photonic qubits use photons as quantum bits. Hybrid quantum systems combine different types of qubits or quantum processors, such as superconducting qubits with trapped ions, to achieve enhanced performance.

$$T_{motional}(\omega) = \frac{2}{\omega^2} \tag{65}$$

Eq. (65) represents the motional coherence time of trapped ion based on trap frequency (ω).

$$C_{ph} = B \cdot log_2\left(1 + \frac{S}{N}\right) \tag{66}$$

Eq. (66) represents the photonic channel capacity information capacity of a photonic channel with bandwidth (B), signal power (S), and noise power (N).

$$H_{hybrid} = H_{sc} + H_{ion} + J_{sci}\sigma_s q_i \tag{67}$$

Eq. (67) represents the superconducting-ion hybrid Hamiltonian combining superconducting (H_{sc}) and ion (H_{ion}) Hamiltonians with coupling term (J_{sci}) between superconducting state (σ_s) and ion qubit (q_i).

$T_{motional}(\omega)$ demonstrates the impressive coherence times that can be achieved with trapped ions. C_{ph} highlights the significance of maximizing bandwidth and signal-to-noise ratio for photonic qubits (see Figure 11).

Figure 11. (a) The graph shows the motional coherence time of trapped ions as a function of trap frequency, highlighting the remarkable coherence times achievable with trapped ions. (b) The graph illustrates the photonic channel capacity in relation to bandwidth, emphasizing the importance of maximizing both bandwidth and signal-to-noise ratio for photonic qubits. The bar chart shows the eigenvalues of the hybrid quantum system's superconducting-ion Hamiltonian, which displays the energy states of the system. The Hamiltonian combines superconducting and ion components, with a coupling term between the superconducting state and the ion qubit.

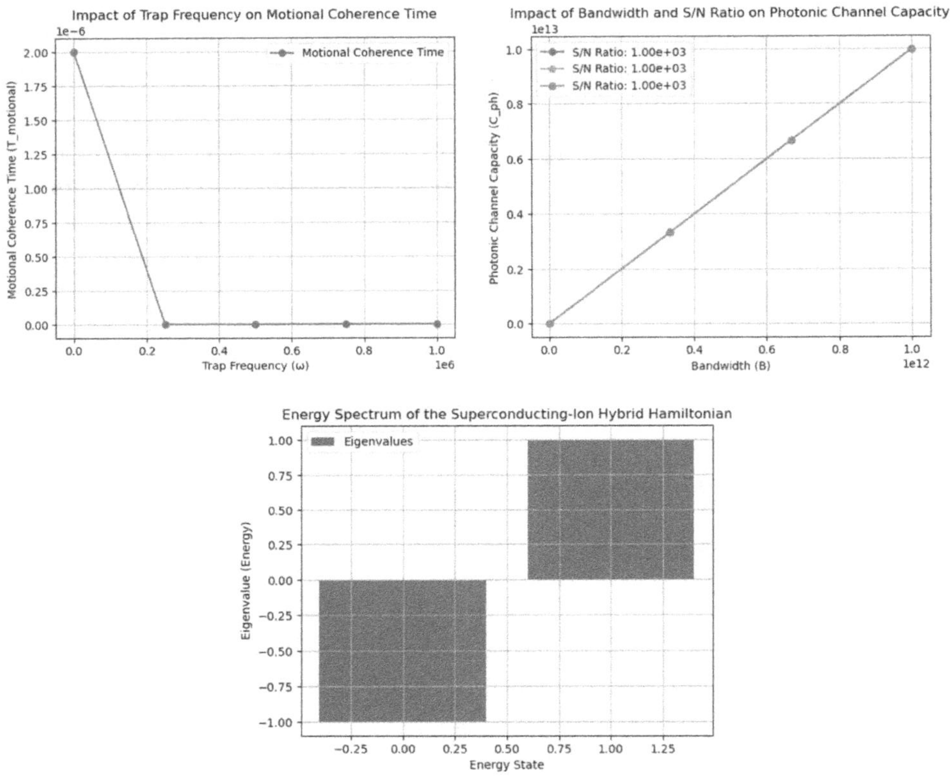

Semiconductor-based quantum technologies, such as quantum dot qubits and silicon spin qubits, have the advantage of scalability and compatibility with existing semiconductor fabrication techniques. Quantum dot qubits have made progress in achieving high-fidelity single-qubit gates and two-qubit interactions. Meanwhile, silicon spin qubits utilize the intrinsic spin properties of electrons in silicon-based materials. Recent advancements in silicon spin qubits have led to improvements in qubit coherence times and the implementation of multi-qubit gates.

$$\dot{D}(R) = \frac{k_B T}{\hbar \omega_c} \tag{68}$$

Eq. (68) represents the dot occupancy with radius R, temperature T, Boltzmann constant k_B, and confinement energy $\hbar \omega_c$.

$$F_{gate}(\tau, \varepsilon) = e^{-\pi^2(\frac{\sigma}{\tau})^2} - \varepsilon \tag{69}$$

Eq. (69) represents the single-qubit gate fidelity with pulse duration τ, standard deviation σ, and error rate ε (see Figure 12).

Figure 12. Left: The plot shows how the dot occupancy rate changes with varying dot radii at temperatures of 0.01 K, 0.055 K, and 0.1 K. The occupancy rate is influenced by the interplay of dot radius, temperature, and confinement energy. Right: The graph illustrates the single-qubit gate fidelity as a function of pulse duration for different error rates. The graph showcases the impact of pulse duration on single-qubit gate fidelity, considering a fixed standard deviation. Three error rates, 0.01, and 0.05, are represented to demonstrate the trade-off between pulse duration and fidelity in quantum gate operations.

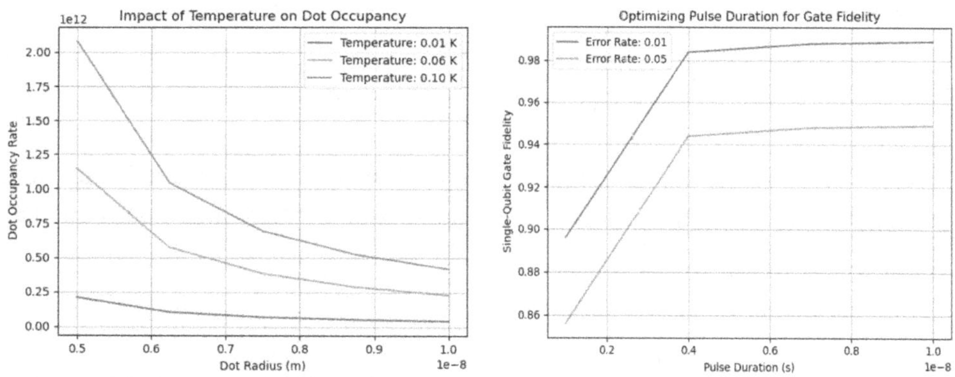

$$U_{CNOT} = |0000 + |0101 + |1010 + i|1100 \tag{70}$$

Eq. (70) represents the CNOT gate matrix representation of controlled-NOT gate for two-qubit interaction.

$\dot{D}(R)$ demonstrates the impact of dot size on occupancy and control, which is crucial for gate operations. $F_{gate}(\tau, \varepsilon)$ highlights the significance of optimizing pulse duration and minimizing gate errors. The CNOT gate matrix illustrates the fundamental building block for two-qubit interactions.

However, despite the progress made in quantum hardware, there are still several challenges and opportunities for improvement on the horizon. These include error correction and fault tolerance, quantum interconnects, noise mitigation, scalability, and quantum software development. The potential of quantum computing is vast, with possibilities for quantum advantage and beyond, quantum cloud computing, and quantum ML.

$$P_{threshold}(QBER) = 1 - 2^{n-1} \cdot QBER \cdot (1 - QBER)^n \tag{71}$$

Eq. (71) represents the threshold error probability for fault-tolerant QEC with QBER and n qubits.

$$L(d) = \alpha d^2 + \beta d + \gamma \tag{72}$$

Eq. (72) represents the loss function for quantum interconnect modeling interconnect loss as a function of distance d, with α, β, and γ representing different contributors.

$$F_{opt}(\omega_0, \omega_c) = e^{-\gamma t} sin^2\left(\frac{\pi t}{\omega_c - \omega_0}\right) \tag{73}$$

Eq. (73) represents the quantum noise mitigation optimal filter function for mitigating noise at frequency ω_0 with cutoff frequency ω_c and decay rate γ (see Figure 13).

Figure 13. The graph on the left shows the fault-tolerant QEC threshold probability decreasing as the number of qubits increases, with QBER values of 0.01, 0.03, and 0.05. The graph on the right models quantum interconnect loss as a function of distance using sample coefficients. The scatter plot illustrates the impact of distance on quantum communication channel quality by depicting the interconnect loss at varying distances. The optimal filter function for quantum noise mitigation is shown at the bottom of the graph. It considers different noise frequencies and cutoff frequencies with respect to time. The filter aims to mitigate the impact of noise in quantum systems and shows variations based on the chosen frequencies.

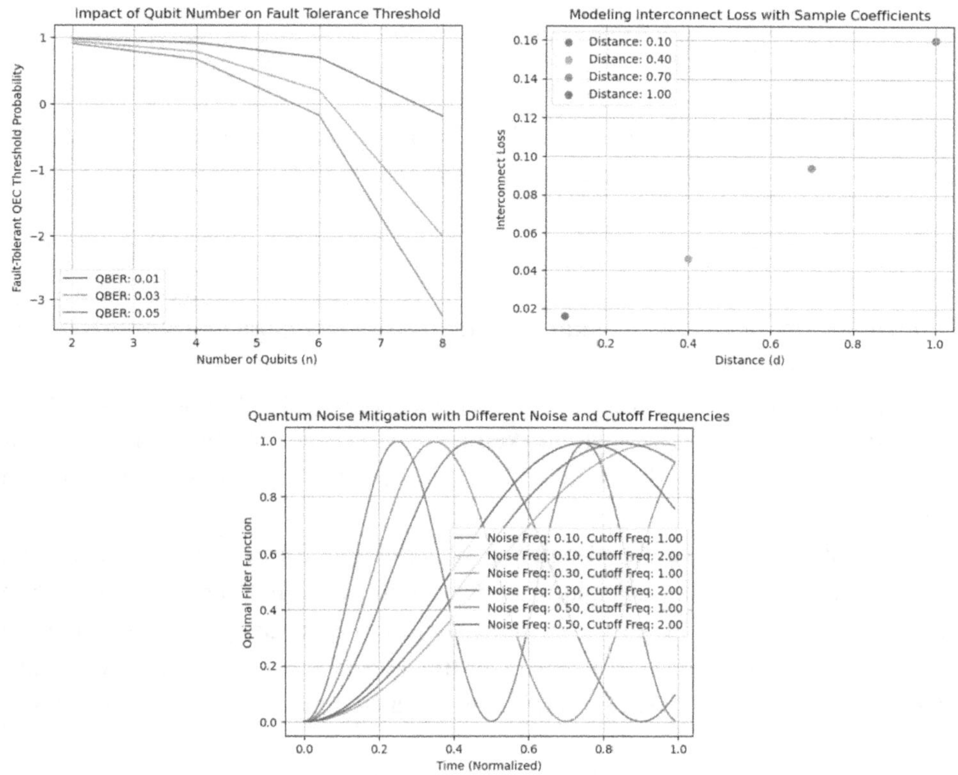

$$G(n) = \frac{n}{T_C(n)} \tag{74}$$

Eq. (74) represents the scalability metric growth rate of circuit size n relative to coherence time $T_C(n)$). The importance of error correction in achieving fault tolerance is demonstrated by $P_{threshold}(QBER)$. Efficient signal transmission across large-scale systems is emphasized by the interconnect loss function. The theoretical framework for suppressing unwanted noise is illustrated by the quantum noise mitigation. The scalability metric highlights the trade-off between circuit size and coherence time.

Quantum error correction (QEC) is a crucial field of research aimed at mitigating errors and improving the reliability of quantum computations. Quantum systems are susceptible to errors due to environmental factors, imperfections in hardware components, and the delicate nature of quantum states. The fundamental challenge in quantum computation is the preservation of delicate quantum information against these error sources. Classical error correction techniques, such as redundancy and parity checks, are insufficient in the quantum realm due to the principles of superposition and entanglement.

$$C_{qcloud} = T_q \cdot C_q + T_c \cdot C_c \quad (75)$$

Eq. (75) represents the cost model for quantum cloud computing considering quantum time T_q and cost C_q, classical time T_c and cost C_c.

$$\Delta(f_q, f_c) = f_q(x) - f_c(x) \quad (76)$$

Eq. (76) represents the quantum advantage metric comparing quantum function f_q and classical function f_c.

The cost model for quantum cloud computing demonstrates the trade-offs between on-premise and cloud-based quantum resources. By comparing the performance of quantum and classical models, quantum ML illustrates how to measure the quantum advantage.

QEC relies on the principles of superposition, entanglement, and quantum encoding to detect and correct errors in quantum computations. Key principles include quantum codes, entanglement and parity checks, specialized quantum gates, and quantum error detection and correction. Innovative strategies in QEC include surface codes, topological quantum codes, cat codes, concatenated codes, and repetition codes.

$$d_{code} = min\{ w_X + 1 | X \subseteq G, X \text{is a generator set of the stabilizer group} \} \quad (77)$$

Eq. (77) represents the code distance measure for an error-correcting code, indicating its ability to correct errors affecting up to d_{code} qubits.

$$H_E = \sum_{i,j} h_{ij} \otimes Z_i \otimes Z_j \quad (78)$$

Eq. (78) represents the entanglement-parity check matrix representation of entanglement-based parity checks, where h_{ij} are parity check coefficients and Z_i, Z_j are Pauli operators on different qubits.

$$U_{teleport} = (|00 + |11) \otimes CNOT_{2,3} \otimes CNOT_{1,2} \otimes H_1 \tag{79}$$

Eq. (79) represents the teleportation circuit diagram for teleporting an unknown qubit state, involving controlled-NOT (CNOT) gates, Hadamard (H) gate, and single-qubit measurement.

The code distance is used to evaluate the error correction capability of a code. The entanglement-parity check matrix demonstrates how entanglement can be utilized for error detection. The teleportation circuit diagram illustrates a crucial tool for error correction and communication.

Despite progress in QEC, several challenges and avenues for improvement persist. Decoherence mitigation, hardware limitations, scalability, quantum interconnects, and software integration remain significant challenges. Mitigating decoherence through strategies like dynamical decoupling techniques and quantum error avoidance are actively explored. Overcoming hardware limitations, such as gate imperfections and qubit coupling, is crucial for achieving fault-tolerant quantum computation. Scaling QEC to large quantum processors presents challenges related to qubit connectivity, gate fidelities, and resource requirements.

$$\gamma = a + b e^{-\frac{\Delta E}{k_B} T} \tag{80}$$

Eq. (80) represents the modeling decoherence rate with constant (a), exponential term (b), energy gap (ΔE), Boltzmann constant (k_B), and temperature (T).

$$P_{success} = F_{gate}^{2t+n} \tag{81}$$

Eq. (81) represents the quantum circuit accuracy, success probability of a quantum circuit with t controlled operations and n single-qubit gates, factoring in gate fidelities.

$$L(d) = \alpha d^2 + \beta d + \gamma e^{-\delta d} \tag{82}$$

Eq. (82) represents the Interconnect loss function, modeling interconnect loss with distance dependence $(\alpha d^2, \beta d)$, and exponential decay $(\gamma e^{-\delta d})$.

$$C_{software} = T_c + \alpha Q + \beta C_{mem} + \gamma C_{comp} \tag{83}$$

Eq. (83) represents the Quantum software integration metric, cost of integrating quantum software, considering classical time T_c, qubit resources (αQ), memory usage (βC_{mem}), and computational cost (γC_{comp}) (see Figure 14) .

Figure 14. The graph on the left shows the success probability of a quantum circuit plotted against the number of controlled operations, taking into account an example gate fidelity of 0.95 and the impact of gate fidelity and the number of single-qubit gates. On the right, interconnect loss is presented as a function of distance with different exponential decay factors. The plot illustrates the loss of interconnect, which is modelled with distance dependence and varying decay factors. It provides insights into how the loss varies across different distances. The bottom graph shows the integration cost of quantum software plotted against qubit resources. The graph showcases the integration cost with varying coefficients (α) for classical time, memory cost per qubit, and computational cost per gate. It reveals the influence of qubit resources on the overall integration cost in quantum software.

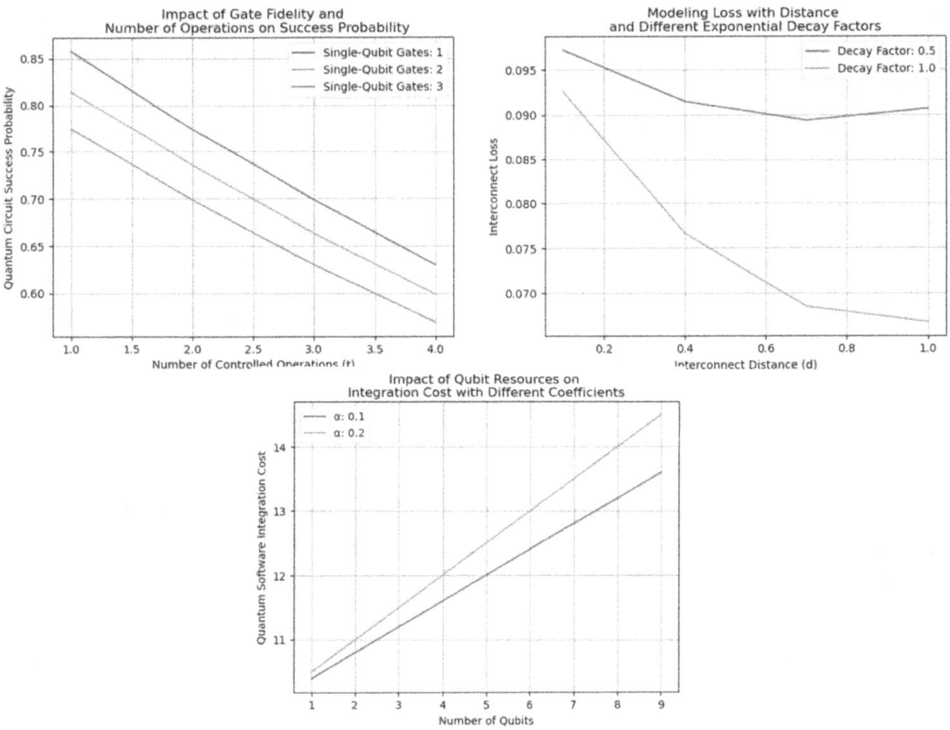

The decoherence rate shows the dependence of noise on temperature and energy gap. The quantum circuit accuracy emphasises the importance of high gate fidelity. The interconnect loss function models signal attenuation in large systems. The quantum software integration metric highlights the need for efficient software optimisation.

Establishing efficient quantum interconnects between qubits is essential for the creation of large-scale quantum processors. Quantum software integration is a key consideration for realizing the full potential of quantum computers.

Future prospects for QEC include fault-tolerant quantum processors, adaptive error correction strategies that dynamically respond to changing error environments, quantum ML for error correction, and quantum communication and error correction.

$$p_{th} = \frac{\alpha}{1 + \beta t} \tag{84}$$

Eq. (84) represents the threshold error probability for fault-tolerant QEC with constant (α) and decay term (βt).

$$A_{EC}(e_t, p_e) = \{C_{code_1}, C_{code_2}\}\psi(e_t, p_e) \tag{85}$$

Eq. (85) represents the dynamically choosing an error-correcting code (C_{code_1}, C_{code_2}) based on error type (e_t) and probability (p_e).

$$L(\theta) = \sum_i (y_i - f_\theta(x_i))^2 \tag{86}$$

Eq. (86) represents the training a quantum ML model with loss function (L) based on predicted ($f_\theta(x_i)$) and actual (y_i) outputs for data points (x_i).

$$I_{quantum}(\rho_{AB}) = H(A) + H(B) - H(AB) \tag{87}$$

Eq. (87) represents the quantum mutual information measuring correlation between entangled qubits (A, B), relevant for secure communication and error correction (see Figure 15).

Figure 15. (a) The dynamically chosen error-correcting codes, Code 1 - Type 1 and Code 2 - Type 2, which are based on different error probabilities (p_e). The error correction matrix values for each code are visualized to illustrate the adaptive nature of the error correction strategy. (b) The training process of a quantum ML model, where actual output values (scatter points) are compared with predicted output values (line plot) for various theta values, demonstrating the learning process of the quantum model. (c) The calculated quantum mutual information ($I_{quantum}$) values are derived from a function that combines a sinusoidal signal with a random component. These values represent the correlation between entangled qubits (A and B) that are relevant for secure communication and error correction. (d) Impact of evaluation rounds, failure tolerance, and algorithm choice on standardization time. (e) Adaptive error correction

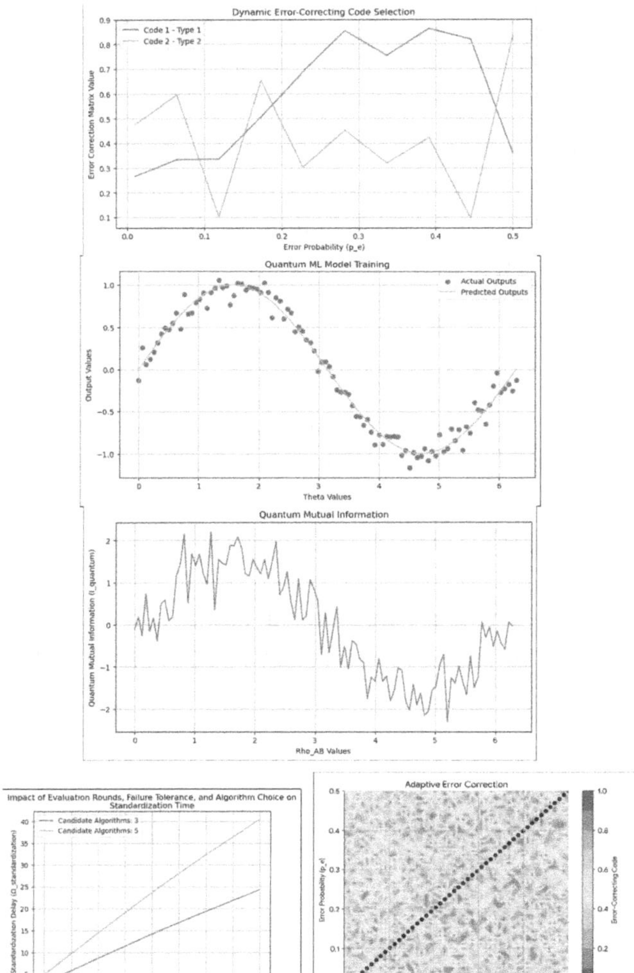

The fault-tolerant threshold is the maximum error rate for reliable quantum computation. The adaptive error correction algorithm adjusts codes dynamically based on real-time error information. Quantum ML demonstrates the use of ML for improved error correction. Quantum communication explores the role of entanglement in both communication and error correction.

Quantum computing has the potential to revolutionise various industries, including pharmaceuticals, finance, logistics, energy, aerospace, AI, telecommunications, environmental modelling, automotive, healthcare, and security.\

$$E_{discovery}\left(T_q, T_c\right) = \alpha e^{-\beta T_q} + \gamma T_c^{-\delta} \tag{88}$$

Eq. (88) represents the model for drug discovery efficiency with quantum time (T_q), classical time (T_c), decay rate (β), and classical scaling factor (γ, δ).

$$VaR(q) = f\left(P_f, \sigma_q, \rho_q\right) \tag{89}$$

Eq. (89) represents the quantum Value-at-Risk for portfolio (VaR), considering loss probability (P_f), quantum standard deviation (σ_q), and quantum correlations (ρ_q) (see Figure 16).

$$L_{route}(G, Q) = min \sum_i w_i d_i Q_i \tag{90}$$

Eq. (90) represents the optimize logistics routes on a graph (G) with weights (w_i), distances (d_i), and quantum resource requirements (Q_i).

Figure 16. (a) The efficiency model for drug discovery illustrates the trade-off between quantum time (T_q) and classical time (T_c). The blue markers indicate selected data points. (b) The quantum VaR for a portfolio demonstrates its dependence on loss probability (P_f), quantum standard deviation (σ_q), and quantum correlations (ρ_q). The blue markers highlight specific data points.

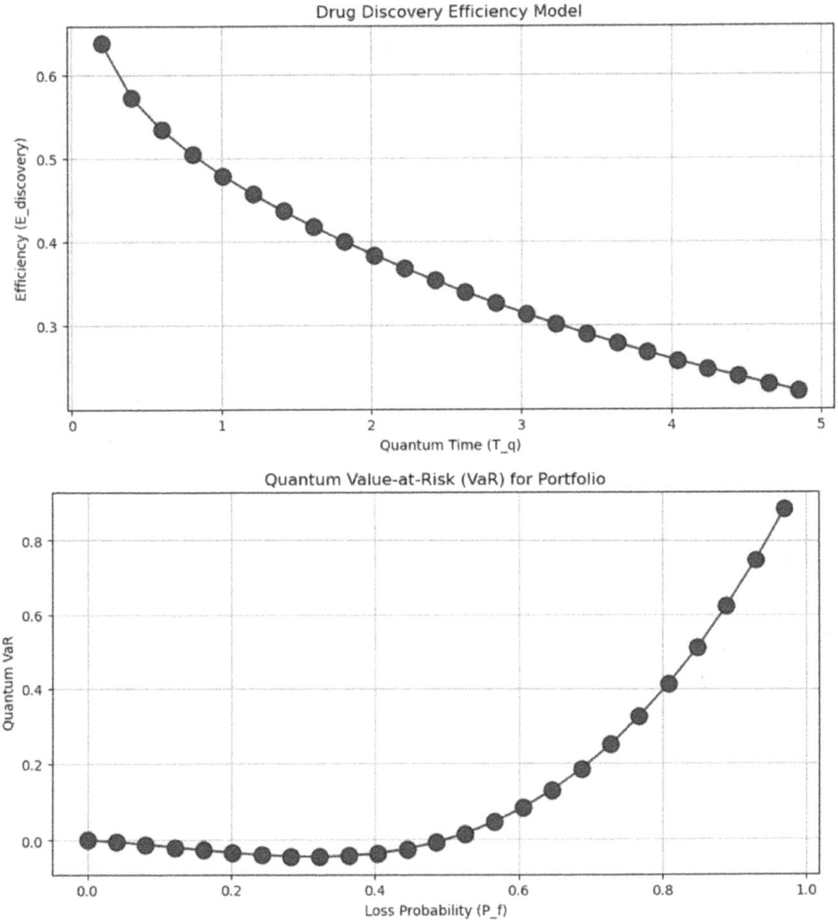

The model for drug discovery efficiency highlights the trade-off between accelerated quantum simulations and classical costs. The use of quantum correlations to calculate VaR demonstrates its potential for risk assessment in financial analysis.

Quantum computing has the potential to accelerate the simulation of molecular structures and interactions, leading to more efficient drug discovery processes. Additionally, it can aid in the discovery of new materials for energy storage, optimize energy production processes, and model complex quantum states that are relevant to materials science. Quantum simulations have the potential to contribute to the development of advanced materials and more efficient energy systems, promoting sustainability and innovation in the energy sector.

$$H_{mol} = \sum_i T_i + \sum_{i,j} V_{ij} + \sum_n T_n V_n + \ldots \tag{91}$$

Eq. (91) represents the quantum Hamiltonian modeling molecular interactions, including kinetic (T_i), potential (V_{ij}), and nuclear-vibrational ($T_n V_n$) terms.

$$\theta_{phase} = 2 \arcsin\left(\sqrt{\psi_0 | M_{phase} | \psi_e} \right) \tag{92}$$

Eq. (92) represents the estimating unknown phase angles (θ_{phase}) of energy eigenstates (ψ_e) through interaction with measurement operator (M_{phase}) and ground state (ψ_0).

$$M_{opt}(x) = \alpha f_1(x) + \beta f_2(x) + \gamma f_3(x) \tag{93}$$

Eq. (93) represents the optimize material property (M_{opt}) by balancing multiple properties (f_1, f_2, f_3) with coefficients (α, β, γ) (see Figure 17).

Figure 17. Optimization of material properties using quantum simulation. The optimized material property (M_{opt}) is a combination of three individual properties (Property 1, Property 2, and Property 3) represented by distinct curves. The plot demonstrates the effectiveness of quantum simulation in achieving the desired material characteristics by showcasing the variation of the material property with respect to the input variable (x).

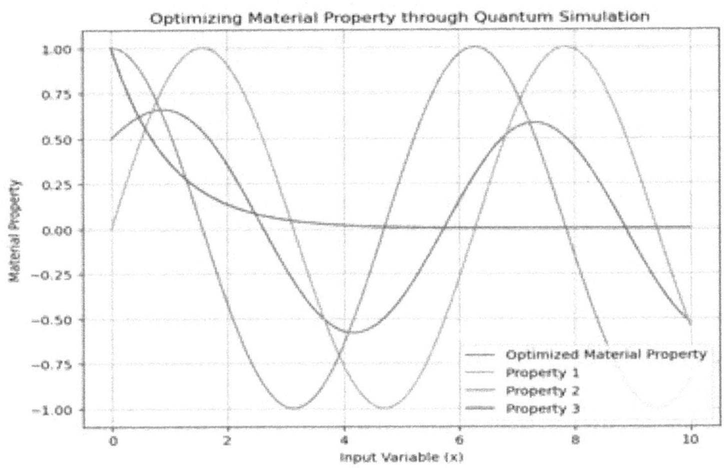

The molecular Hamiltonian demonstrates the complex quantum description of molecules. Quantum phase estimation demonstrates the ability to extract important information from complex energy landscapes. The emphasis on optimizing material properties highlights the multi-objective nature of material design tasks.

Quantum computing has potential applications in the aerospace and engineering sectors, particularly in solving complex problems such as aerodynamic simulations, structural optimization, and materials research. Improved computational capabilities could lead to the design of more fuel-efficient aircraft, optimized engineering solutions, and advancements in aerospace technology.

$$D_{drag} = C_D \cdot \rho \cdot \frac{v^2}{2} \tag{94}$$

Eq. (94) represents the classical model for drag (D_{drag}) with drag coefficient (C_D), air density (ρ), and velocity (v).

$$P_{transition}\left(\Phi_i, \Phi_j\right) = e^{-\beta H(\Phi_i)-\beta \Delta E(\Phi_i,\Phi_j)} \tag{95}$$

Eq. (95) represents the probability of transitioning between turbulent states (Φ_i, Φ_j) based on Hamiltonian (H) and energy difference (ΔE).

AI and ML algorithms can accelerate ML processes, enabling faster training of complex models and solving optimization problems. Quantum computing algorithms can address optimization challenges in telecommunications networks, leading to more efficient and secure telecommunications infrastructure.

$$T_{QVA}(N,\alpha) = O(N^{\frac{1}{\alpha}}) \tag{96}$$

Eq. (96) represents the asymptotic scaling of training time for a QVA with N parameters and Ansatz depth α, compared to exponential scaling for classical algorithms.

$$P_{success}(p,\Delta E) = 1 - e^{-\gamma p \Delta E} \tag{97}$$

Eq. (97) represents the probability of finding the ground state with QAOA, considering problem size (p), energy gap (ΔE), and algorithm parameter (γ).

$$L_{GAN}(G,D) = minV(D(x)) - maxE(G(z)) \tag{98}$$

Eq. (98) represents the minimax objective function for a generative adversarial network (GAN), balancing generator (G) and discriminator (D) performance (see Figure 18).

Figure 18. (a) Drag force (D_{drag}) is calculated based on the drag coefficient (C_D), air density (ρ), and velocity (v). The plot shows the non-linear relationship between velocity and drag force when using the classical drag model. This highlights the significant impact of velocity on drag force. (b) Asymptotic scaling of QVA training time, with varying numbers of parameters (N) and a set Ansatz depth (α) as an example. The graph shows how training time increases with the number of parameters, indicating the potential efficiency gains of quantum algorithms compared to classical ones. (c) The success probability of the QAOA for different problem sizes (p) is also shown. Success probability ($P_{success}$) is determined by problem size, energy gap (ΔE), and algorithm parameter (γ). The plot demonstrates how these factors affect the likelihood of finding the ground state using QAOA.

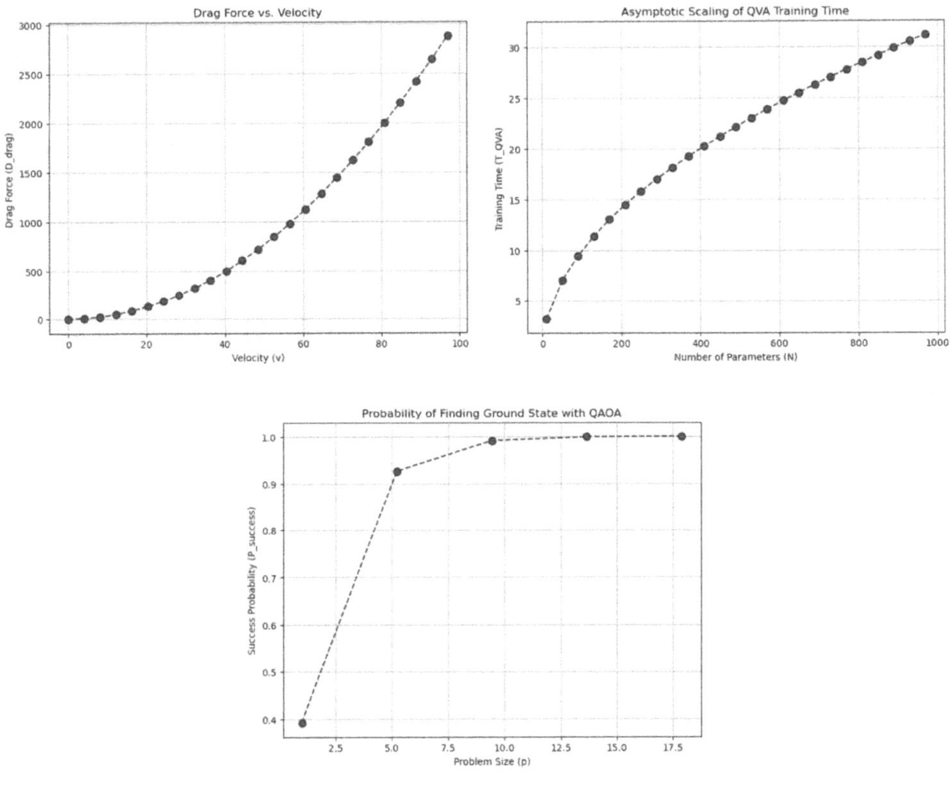

The training time of QVA highlights the potential for faster training by exploiting quantum parallelism. The success probability of QAOA showcases the ability to solve hard optimization problems relevant to ML tasks. The loss function of GAN emphasizes how quantum architectures could contribute to improved GAN learning dynamics.

Quantum computing can be used in environmental modelling and climate research to simulate complex environmental systems. This can contribute to climate research, weather prediction, and ecosystem modelling. The computational power of quantum computing may improve our understanding of climate dynamics, enabling more accurate predictions and informed decision-making in environmental management.

$$\Delta T(t) = f(G(t), A(t), O(t)) \tag{99}$$

Eq. (99) represents the $G(t)$ is ocean circulation, $A(t)$ is atmospheric dynamics, and $O(t)$ is external forcings like greenhouse gas concentrations.

$$MSE(f_q, f_c) = \sum_i (y_i - f_q(x_i))^2 - \sum_i (y_i - f_c(x_i))^2 \tag{100}$$

Eq. (100) represents the mean squared error difference between a quantum (f_q) and classical (f_c) climate prediction model.

Industries interested in exploring quantum applications must consider access to quantum computing resources. Cloud-based quantum computing services and collaborative initiatives have emerged to address this challenge. The potential impact of quantum computing on current cryptographic systems raises security and privacy concerns.

$$C_{cloud} = T_q \cdot C_q + T_c \cdot C_c + D \tag{101}$$

Eq. (101) represents the cost considering quantum time (T_q), classical time (T_c), per-unit costs (C_q, C_c), and distance-related overhead (D).

$$E_{network}(N, C_{comm}) = \frac{N^2}{C_{comm}} \tag{102}$$

Eq. (102) represents the network efficiency as a function of number of participants (N) and communication overhead (C_{comm}).

$$P_{break}(C_{alg}, Q_{resources}) = e^{-\frac{\alpha T_q}{Q_{resources}}} \tag{103}$$

Eq. (103) represents the probability of breaking a cryptosystem with classical algorithm (C_{alg}) given quantum resources ($Q_{resources}$) and a scaling factor (α) (see Figure 19).

Figure 19. (a) Network efficiency for quantum collaboration, (b) impact of quantum resources on cryptosystem security

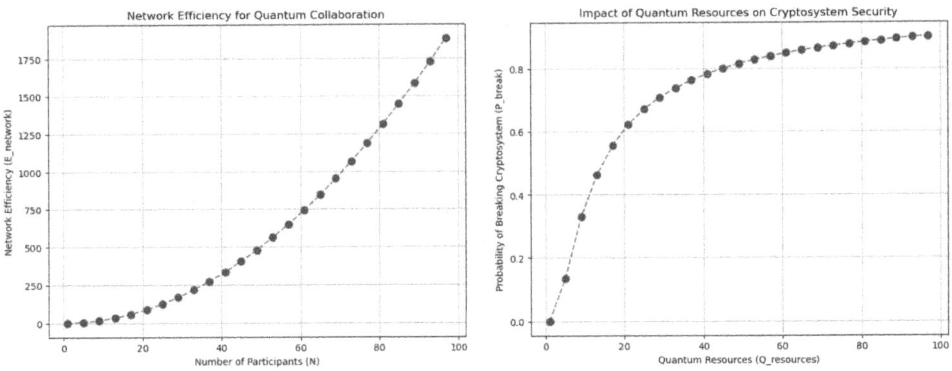

The cost model for cloud-based quantum computing emphasises the importance of efficient resource utilisation. The efficiency of collaborative research networks highlights the need to optimise communication within research communities. The challenge of post-quantum cryptography showcases the urgency of developing new cryptographic protocols that are resistant to quantum attacks.

The prospects for widespread adoption and impact on industries are significant as quantum computing technology matures. The future of quantum computing in various industries looks promising with the potential for achieving quantum advantage in specific domains, hybrid quantum-classical systems, quantum cloud computing services, quantum-safe cryptography, industry-specific quantum consortia, quantum-enhanced optimization, quantum ML advances, secure quantum communication networks, quantum-safe supply chain solutions, and quantum-enhanced environmental modelling.

CHALLENGES AND OPPORTUNITIES IN QUANTUM TECHNOLOGIES

The rapid evolution of quantum technologies has ushered in a new era of computational possibilities and scientific exploration. However, this transformative journey is not without its challenges, including qubit coherence, gate fidelity, scalability, and environmental impacts. Researchers and engineers are actively addressing these challenges to unlock the full potential of quantum computing.

$$T_c = e^{-\frac{\Delta E}{k_B T}} \tag{104}$$

Eq. (104) represents the exponential dependence of coherence time on energy gap (ΔE), Boltzmann constant (k_B), and temperature (T).

$$F_{gate} = 1 - p_{err} \tag{105}$$

Eq. (105) represents the probability of successful gate operation with error probability (p_{err}).

$$QC = QMA\left(p_{poly}, log\left(\frac{1}{\varepsilon}\right)\right) \tag{106}$$

Eq. (106) represents the identifying problems efficiently solved by quantum computers with polynomial time (p_{poly}) and constant error (ε).

Qubit coherence and decoherence are significant challenges in quantum computing. Qubits are highly susceptible to decoherence, which results in the loss of quantum information due to interactions with the external environment. Innovations in materials, qubit designs, and error correction techniques aim to extend coherence times. Additionally, gate fidelities and error rates are crucial as imperfections in quantum gates lead to errors during quantum operations, which can significantly impact the fidelity of computations. Advances in error correction techniques, gate engineering, and error-robust gate designs contribute to reducing error rates.

The scalability of quantum processors faces challenges related to qubit connectivity, gate fidelities, and resource requirements. Various QEC codes, such as surface codes and topological codes, offer scalable solutions. Hybrid quantum-classical architectures and innovative qubit coupling schemes aim to enhance qubit connectivity and overall scalability.

$$T_{quantum} \sim N^{\frac{6}{ln(2)}} \tag{107}$$

Eq. (107) represents the quantum runtime for factoring an N-bit integer compared to exponential classical runtime (see Figure 20(a)).

$$d_{code} = \min\{w_X | X \subseteq G and X is a generator set of the stabilizer group\} \tag{108}$$

Eq. (108) represents the minimum detectable and correctable error distance for a surface code.

$$p_{th} = \frac{\alpha}{1 + \beta t} \tag{109}$$

Eq. (109) represents the minimum error probability tolerable for fault-tolerant quantum computation with constant (α) and decay term (βt).

The Shor algorithm demonstrates the potential for quantum speedup in high-accuracy computations. The ability to detect and correct errors based on code structure is demonstrated by the surface code distance. The error-correcting code threshold highlights the critical limit for error rates in reliable quantum operations.

$$N_{qubits} \sim \alpha^{d-1} \tag{110}$$

Eq. (110) represents the exponential scaling of cluster state size with dimension (d) and constant factor (α).

$$L_{coupling} \sim e^{(-\gamma d)} \tag{111}$$

Eq. (111) represents the distance dependence of topological qubit coupling with decay rate (γ) and distance (d) (see Figure 20(b)).

Figure 20. (a) Exponential scaling of cluster state size with dimension. (b) Distance dependence of topological qubit coupling.

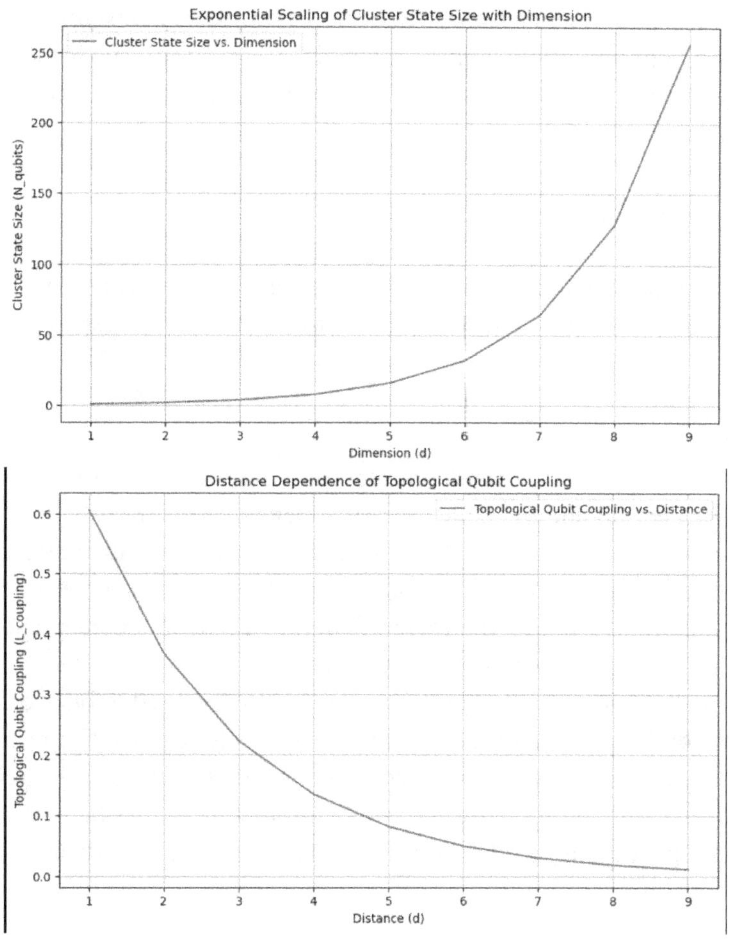

$$C_{hybrid} = T_q + \beta T_c \tag{112}$$

Eq. (112) represents the computational cost balancing quantum time (T_q) and classical time (T_c) with factor (β).

The scaling of cluster state demonstrates the possibility of exponential growth in qubit resources. The coupling of topological qubits demonstrates the benefits of distance-independent interactions in specific architectures. The cost of the hybrid algorithm emphasizes the balance between utilizing quantum capabilities and efficiently using classical processing power.

Standardizing quantum algorithms and applications is challenging due to the variability of quantum hardware. To address this, calibration techniques, adaptive algorithms, and hardware-agnostic approaches are being explored. The environmental impact of maintaining extremely low temperatures also presents challenges for large-scale quantum computing infrastructure. Innovations in cryogenic engineering and energy-efficient cooling systems aim to reduce the environmental footprint of quantum processors.

$$R_{algo}(H) = max_{\varphi} f_{algo}(H, \varphi) \tag{113}$$

Eq. (113) represents the performance of a hardware-agnostic algorithm across different hardware specifications (H) and optimization parameters (φ).

$$E_{cryo} = \alpha Q + \beta T + \gamma \tag{114}$$

Eq. (114) represents the modeling energy consumption with terms for qubit resources (Q), temperature (T), and constant factors (α, β, γ).

The importance of adaptable algorithms for diverse hardware platforms is highlighted by the hardware-agnostic algorithm performance. The energy required for cryogenic cooling highlights the trade-off between resource demands and energy efficiency when maintaining low temperatures.

Quantum software development is challenged by the need to bridge the gap between quantum hardware capabilities and software algorithms. Overcoming these challenges requires innovations in error correction codes, fault-tolerant architectures, and optimized error correction strategies. Collaborative efforts between academia, industry, and government research institutions are crucial for advancing quantum technologies.

$$F_{circuit}(T, p_{err}) = \alpha\left(e^{-\frac{T}{T_c}}\right) + \beta\left(1 - e^{-\frac{T}{T_c}}\right) p_{err}^{\gamma T} \tag{115}$$

Eq. (115) represents the modeling circuit fidelity considering error correction, execution time (T), error probability (p_{err}), coherence time (T_c), and factors (α, β, γ) (see Figure 21).

$$C_{software}(H, Q) = min \sum_i w_i D_i Q_i \tag{116}$$

Eq. (116) represents the optimizing software cost ($C_{software}$) with hardware specifications (H), qubit resource requirements (Q_i), weights (w_i), and distances (D_i).

Figure 21. Impact of error correction on circuit fidelity. Sample parameters include an initial error probability of 0.01, coherence time of 10 units, and factors (α, β, γ) determining the corrected and uncorrected circuit performances. The corrected circuit, represented by the blue curve, exhibits a nuanced dependence on execution time (T), while the uncorrected circuit, in orange, experiences an exponential decay with increasing time.

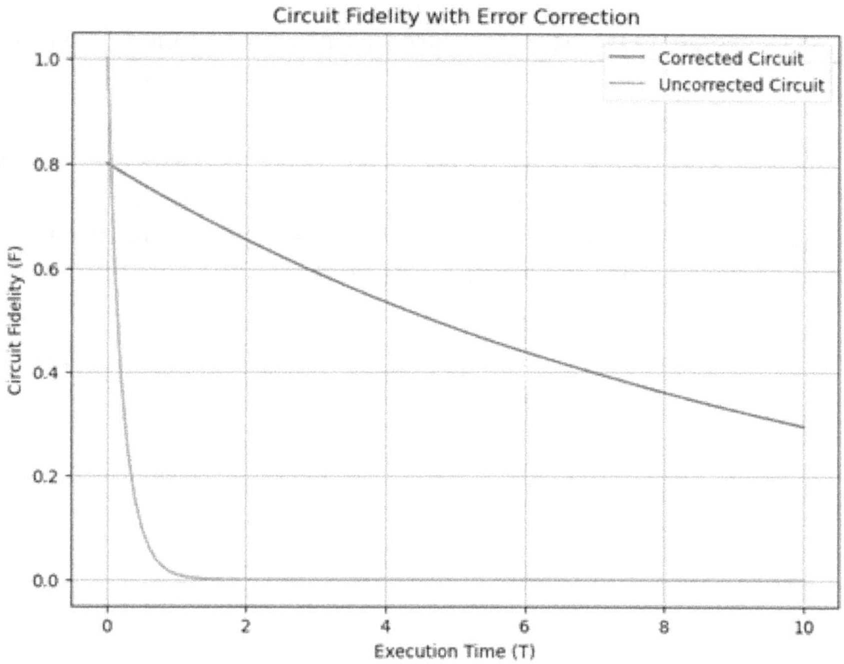

The critical error rate for reliable computations is emphasized by the fault-tolerant circuit threshold. The circuit fidelity demonstrates the dependence on error correction effectiveness and execution time. The need to balance hardware constraints and software resource requirements is highlighted by software optimization.

The rapid development of quantum technologies poses significant ethical challenges, including concerns about privacy, security, and responsible system development. Quantum information privacy is a major concern due to the unique capabilities of quantum systems, particularly in quantum communication. To address this, it is essential to employ quantum-resistant cryptographic protocols and quantum-safe encryption methods. The advent of powerful quantum algorithms poses security risks and threats that compromise widely used encryption schemes. To mitigate these risks, proactive efforts are necessary in developing and adopting quantum-resistant cryptographic protocols.

$$H_Q(\rho) = -\sum_i p_i log_2(p_i) \tag{117}$$

Eq. (117) represents the characterizing information uncertainty in a quantum state (ρ) with probabilities (p_i).

$$P_{intruder}(H_{eve}, QBER) = e^{-\frac{\beta H_{eve}}{QBER}} \tag{118}$$

Eq. (118) represents the estimating intruder's information gain (H_{eve}) with QBER and factor (β).

$$T_{decrypt}(N, \lambda) = O(N^{log(log(\lambda))}) \tag{119}$$

Eq. (119) represents the asymptotic scaling of decryption time for a protocol secure against adversaries with resources λ (see Figure 22).

Figure 22. Quantum entropy and the asymptotic scaling of the decryption time. The first subplot visualises the quantum state probabilities, showing the distribution of probabilities for different states. The second subplot shows the quantum entropy (H_Q) calculated from the probabilities. The third subplot shows the asymptotic scaling of the decryption time with the number of parameters (N). The green colour denotes the quantum state related components, providing insight into the quantum information uncertainty and the decryption time scaling.

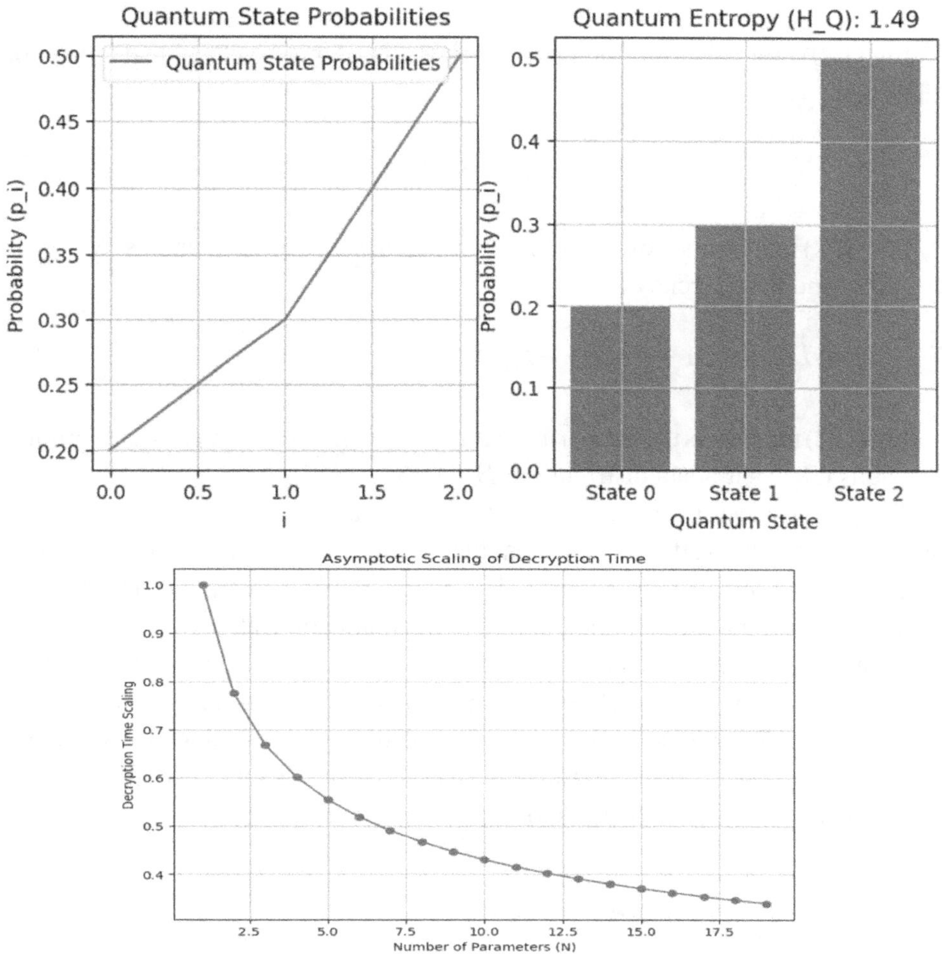

The quantum information entropy emphasises the potential vulnerability of quantum information. The impact of QBER on security is demonstrated by the intruder's information gain. The efficiency of post-quantum cryptography protocols emphasises the trade-off between security and computational cost.

The responsible development of quantum technologies requires navigating ethical considerations related to potential misuse, unintended consequences, and the dual-use nature of quantum capabilities. Establishing ethical guidelines, industry standards, and regulatory frameworks promotes responsible innovation and collaboration between researchers, industry stakeholders, and policymakers. This contributes to the ethical governance of quantum systems.

$$R_{dual} = P_{misuse} \cdot I_{impact} \tag{120}$$

Eq. (120) represents the combining probability of misuse (P_{misuse}) with potential impact (I_{impact}).

$$E_{engage} = \alpha N_{participants} + \beta diversity + \gamma transparency \tag{121}$$

Eq. (121) represents the evaluating engagement based on participants ($N_{participants}$), diversity, and transparency (γ).

$$F_{fairness}\left(D_{alg}, D_{data}\right) = 1 - \left|D_{alg}(x) - D_{data}(x)\right| \tag{122}$$

Eq. (122) represents the measuring algorithmic fairness by comparing algorithmic decisions (D_{alg}) with data distribution (D_{data}).

The importance of evaluating potential misuse and impact is highlighted in the risk assessment, while the multi-stakeholder engagement score emphasizes the need for diverse and transparent dialogue. The ethical AI framework metric provides a model for assessing algorithmic fairness in quantum applications.

Quantum communication surveillance raises concerns about potential surveillance and eavesdropping threats. However, advancements in quantum communication protocols, secure key distribution methods, and quantum-resistant encryption help mitigate these concerns. Integrating ethical considerations into the development of quantum algorithms promotes responsible decision-making.

$$P_{intercept} = \alpha\left(\frac{T_{intercept}}{T_{transmit}}\right) + \beta QBER + \gamma \tag{123}$$

Eq. (123) represents the estimating eavesdropping probability with intercept time ($T_{intercept}$), transmission time ($T_{transmit}$), QBER, and constants (α, β, γ).

$$H_{Eve}(\rho_{AB}) = H_{Eve}(\rho_A) + H_{Eve}(\rho_B) - H_{Eve}(\rho_{AB}) \tag{124}$$

Eq. (124) represents the quantifying information leakage to an eavesdropper (Eve) with individual and joint state entropies of Alice (A) and Bob (B).

$$T_{encrypt}(N, \kappa) = O(N^{\log^{\alpha} \kappa}) \tag{125}$$

Eq. (125) represents the asymptotic scaling of encryption time for an algorithm secure against adversaries with resources κ and factor α (see Figure 23).

Figure 23. (a) The left subplot shows how the probability of eavesdropping is estimated, taking into account intercept time ($T_{intercept}$), transmission time ($T_{transmit}$), and QBER. The curves represent different QBER values. (b) The middle subplot illustrates the amount of information that an eavesdropper (Eve) can obtain based on the individual and joint state entropies of Alice (A) and Bob (B). The plot demonstrates how the entropy of Eve for State A relates to information leakage. (c) The right subplot shows how encryption time scales asymptotically for an algorithm that is secure against adversaries with resources (κ). It explores how the encryption time scales with the number of parameters (N) and a scaling factor (α).

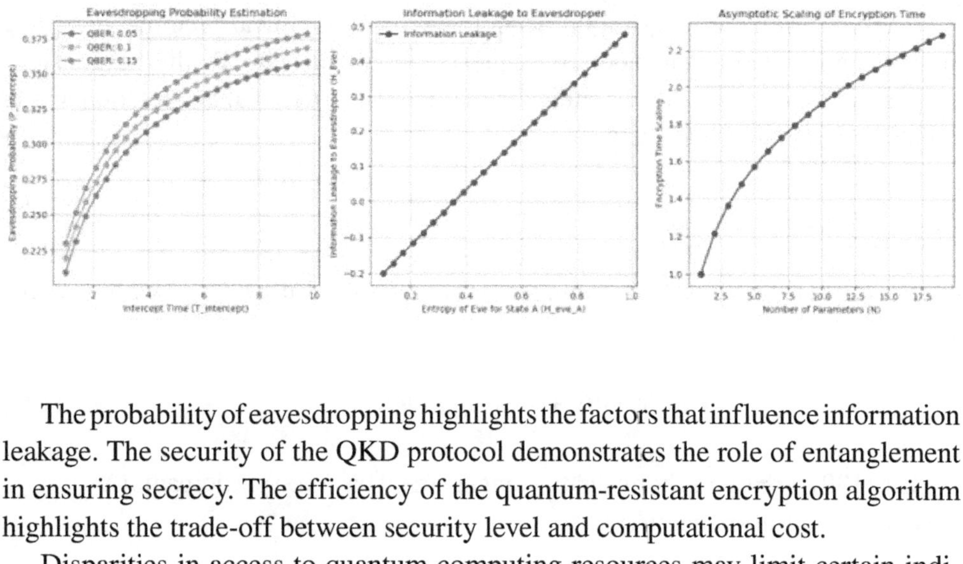

The probability of eavesdropping highlights the factors that influence information leakage. The security of the QKD protocol demonstrates the role of entanglement in ensuring secrecy. The efficiency of the quantum-resistant encryption algorithm highlights the trade-off between security level and computational cost.

Disparities in access to quantum computing resources may limit certain individuals or organizations from benefiting from quantum technologies. Initiatives promoting equitable access to quantum computing resources, educational programs, and collaborative research efforts can help address these disparities.

$$E_{access} = \alpha R_{resources} + \beta D_{education} + \gamma C_{collaboration} \tag{126}$$

Eq. (126) represents the measuring equitable access with resource availability ($R_{resources}$), educational opportunities ($D_{education}$), and collaborative participation ($C_{collaboration}$).

$$D_{workforce} = \sum_i p_i \cdot D_i \tag{127}$$

Eq. (127) represents the capturing diversity in the quantum workforce with group proportions (p_i) and respective diversity measures (D_i).

$$C_{contribution} = N_{developers} + C_{code} + R_{impact} \tag{128}$$

Eq. (128) represents the evaluating open-source contributions with active developers ($N_{developers}$), code complexity (C_{code}), and impact factor (R_{impact}).

The equitable access score emphasises the significance of resource availability, education, and collaboration. Meanwhile, the quantum workforce diversity index demonstrates the necessity of inclusive representation within the field. Finally, the open-source quantum software contribution metric encourages community participation and knowledge sharing.

The dual-use nature of quantum technologies presents ethical dilemmas regarding potential misuse and unintended consequences. To manage this duality responsibly, it is essential to establish clear ethical guidelines, international cooperation agreements, and regulatory frameworks.

$$R_{dual} = P_{misuse} \cdot I_{impact} \tag{129}$$

Eq. (129) represents the combining probability of misuse (P_{misuse}) with potential impact (I_{impact}).

$$E_{engage} = \alpha N_{participants} + \beta diversity + \gamma transparency \tag{130}$$

Eq. (130) represents the evaluating engagement based on participants ($N_{participants}$), diversity, and transparency (γ).

$$C_{agreement} = \alpha S_{objectives} + \beta E_{enforcement} + \gamma M_{monitoring} \tag{131}$$

Eq. (131) represents the assessing agreement effectiveness with shared objectives ($S_{objectives}$), enforcement mechanisms ($E_{enforcement}$), and monitoring strategies ($M_{monitoring}$).

The risk assessment highlights the significance of assessing both the potential for misuse and the impact. The score for multi-stakeholder engagement highlights the necessity for diverse and transparent dialogue. The effectiveness of international cooperation agreements is underscored by the importance of clear objectives, enforcement mechanisms, and monitoring strategies.

Additionally, environmental sustainability is a challenge due to the cryogenic conditions required for many quantum processors, which contribute to high energy consumption and potential environmental impacts. Innovations in energy-efficient quantum computing architectures, cryogenic engineering, and sustainable practices can reduce the environmental impact of quantum technologies. Addressing these ethical considerations requires collaboration among researchers, policymakers, industry leaders, ethicists, and the general public.

The quantum community is a global frontier where nations, institutions, and researchers collaborate to unlock the immense potential of quantum science. International collaboration in the quantum community is a driving force for progress, pushing the boundaries of knowledge and technological capabilities. The landscape of quantum technologies presents both challenges and opportunities. These include the need for cross-cultural collaboration, aligning goals and contributions of various global quantum initiatives, organizing and coordinating global quantum conferences and symposia, coordinating joint research projects across international borders, navigating geopolitical complexities, facilitating industry collaboration across borders, implementing cross-continental quantum testbeds, building quantum information networks, coordinating international funding mechanisms, addressing challenges related to quantum computing, and coordinating international efforts for quantum experiments in space.

International collaboration in the quantum community presents both challenges and opportunities. It highlights the interconnected nature of quantum research and technology development. Collaborative efforts on a global scale amplify the impact of individual contributions, accelerate the pace of discovery, and foster a sense of shared responsibility for addressing common challenges. International collaboration is crucial for unlocking the full potential of quantum technologies on a global scale. Building strong networks and nurturing partnerships are essential for achieving this goal.

International collaboration is crucial for unlocking the full potential of quantum technologies on a global scale. To unlock the full potential of quantum technologies on a global scale, the quantum community must address challenges such as cross-cultural collaboration, aligning goals and contributions of global quantum initiatives, managing funding mechanisms, and fostering collaboration.

SAFEGUARDING QUANTUM FUTURES: SECURITY IN THE QUANTUM REALM

The security of communication infrastructures is a crucial concern in the rapidly evolving landscape of quantum technologies. It is essential to develop and deploy quantum-safe communication protocols as traditional cryptographic methods, such as public-key encryption and digital signature schemes, are vulnerable to quantum attacks.

$$T_{Shor}(N) = O\left(2^{log^{\frac{2}{3}} N}\right) \tag{132}$$

Eq. (132) represents the quantum factoring runtime for an N-bit integer compared to exponential classical runtime.

$$Adv_{break}(C) \leq \varepsilon + \alpha Adv_{solve}(H) \tag{133}$$

Eq. (133) represents the relating security of cipher (C) to advantage of solving a hard problem (H) with constant (ε) and factor (α).

$$\sigma_{existential_{forgery}} = P_{forge}(\kappa) \tag{134}$$

Eq. (134) represents the probability of successfully forging a signature with security parameter (κ) (see Figure 24 and Figure 25).

Figure 24. (1) Quantum factoring runtime compared to classical factoring, (2) Conceptual relationship between cipher security and advantage of solving a hard problem, and (3) Placeholder plot for signature forgery probability with varying security parameters.

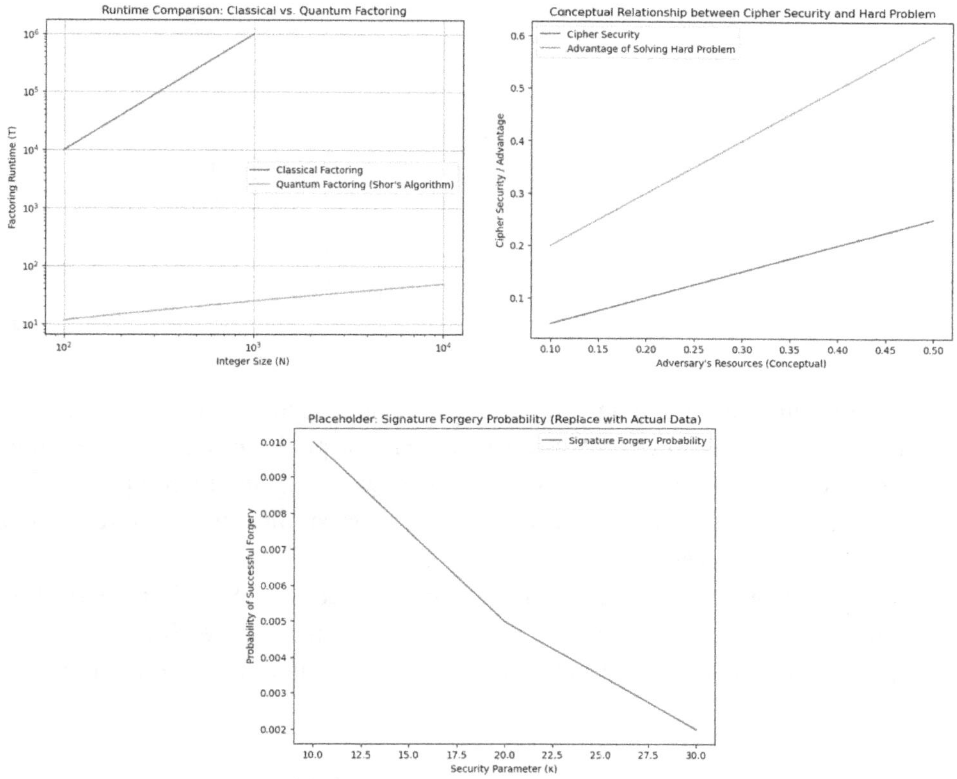

Figure 25. The figures illustrate quantum computing concepts: (1) Quantum factoring runtime vs. N-bit integer, (2) Security of cipher vs. advantage of solving a hard problem with varying cipher values, and (3) Forgery probability vs. security parameter κ.

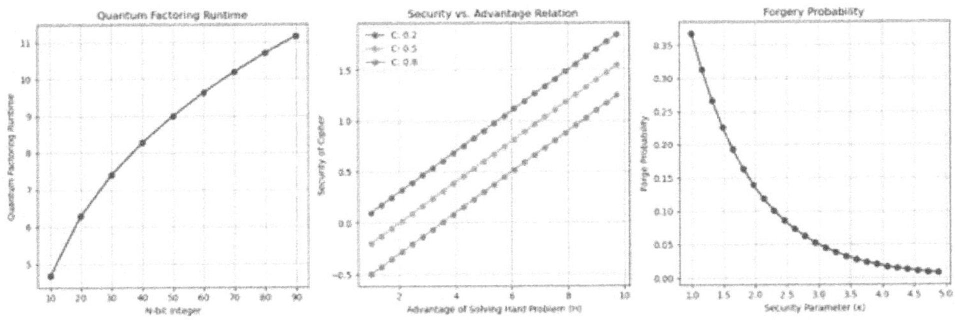

The Shor algorithm demonstrates the exponential speedup that quantum algorithms achieve for certain problems. The security of public-key encryption relies on hard problems that become solvable with powerful quantum computers. The vulnerability of digital signatures highlights the possibility of signature forgery resulting from weaknesses in traditional mechanisms.

Classical cryptography is vulnerable to quantum attacks, such as Shor's algorithm, when implemented on a sufficiently powerful quantum computer. To address this issue, researchers are actively developing and refining quantum-safe communication protocols.

$$T_{encapsulate}(\lambda) = O(\lambda^{log(log(\lambda))}) \tag{135}$$

Eq. (135) represents the asymptotic scaling of key encapsulation time for a secure protocol against adversaries with resources λ.

$$R_{2nd\_preimage} = P(find\,x_2 | h(x_1) = h(x_2)) \tag{136}$$

Eq. (136) represents the probability of finding a second preimage for a hash function (h) with distinct inputs (x_1, x_2).

$$L_{code}(\lambda) = O(\lambda^2) \tag{137}$$

Eq. (137) represents the code length for a multivariate cryptosystem secure against adversaries with resources λ (see Figure 26).

Figure 26. (a) Conceptual scaling of key encapsulation time for quantum-sage protocols. (b) Probability of finding a second preimage for different Hash functions. (c) Code length for multivariate cryptosystem with different security parameters.

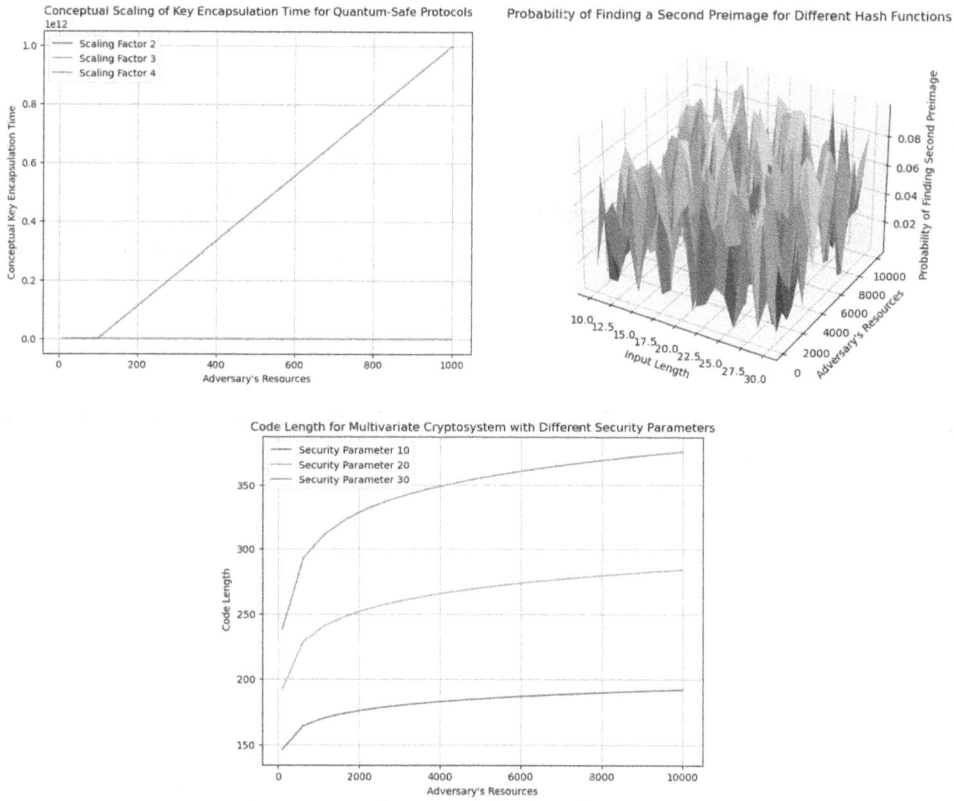

The lattice-based key encapsulation protocol efficiency shows potential for efficient and secure key exchange. Post-quantum hash function security emphasizes the importance of collision resistance in quantum algorithms. Multivariate cryptography code length highlights the trade-offs between security and key size in specific cryptosystems.

The landscape of post-quantum cryptography involves evaluating the security, efficiency, and compatibility of post-quantum cryptographic algorithms. Ongoing research and standardization efforts focus on identifying quantum-resistant cryptographic algorithms. Initiatives such as the NIST post-quantum cryptography standardization project play a pivotal role in evaluating and standardizing post-quantum cryptographic primitives.

$$C_{transition} = \alpha C_{infrastructure} + \beta T_{adoption} + \gamma M_{mitigation} \tag{138}$$

Eq. (138) represents the modeling transition cost with infrastructure upgrades ($C_{infrastructure}$), adoption time ($T_{adoption}$), and mitigation strategies ($M_{mitigation}$).

Lattice-based cryptography, hash-based signatures, code-based cryptography, multivariate polynomial cryptography, QKD, code-based quantum-secure protocols, hash-based quantum-secure protocols, and integration challenges offer opportunities for advancement.

$$T_{encapsulate}(\lambda) = O(d^{log(log(\lambda))}) \quad (139)$$

Eq. (139) represents the asymptotic scaling of key encapsulation time for a protocol secure against adversaries with resources λ, considering lattice dimension (d).

$$P_{forge}(\lambda) = e^{-\frac{\beta\lambda}{log(\lambda)}} \quad (140)$$

Eq. (140) represents the probability of successfully forging a signature with security parameter (λ) and factor (β).

$$L_{code}(\lambda) = O(\lambda^2 log(\lambda)) \quad (141)$$

Eq. (141) represents the code length for a secure code-based cryptosystem against adversaries with resources λ.

$$d_{ring}(\lambda) = O(\lambda) \quad (142)$$

Eq. (142) represents the ring dimension for a secure multivariate polynomial cryptosystem against adversaries with resources λ (see Figure 27 and Figure 28).

Figure 27. (a) Conceptual scaling of key encapsulation time for Lattice-based protocols. (b) Probability of signature forgery with different security parameters and factors.

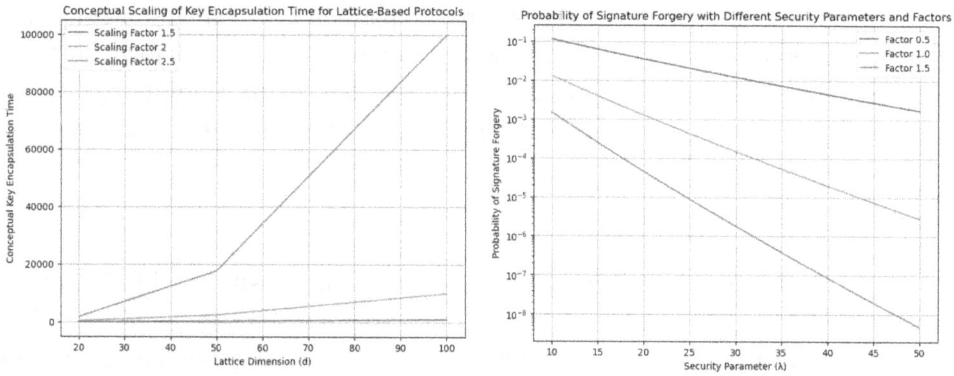

Figure 28. (a) Asymptotic scaling of key encapsulation time for a secure protocol against adversaries, with varying adversary resources (λ). Red markers highlight specific data points. (b) Probability of successfully forging a signature based on the security parameter (λ). Red markers indicate specific data points, providing insights into the forging probability. (c) Code length scaling for a secure code-based cryptosystem concerning adversary resources (λ). Red markers highlight key points on the plot. (d) Scaling of ring dimension for a secure multivariate polynomial cryptosystem against adversary resources (λ). Red markers highlight specific data points on the curve.

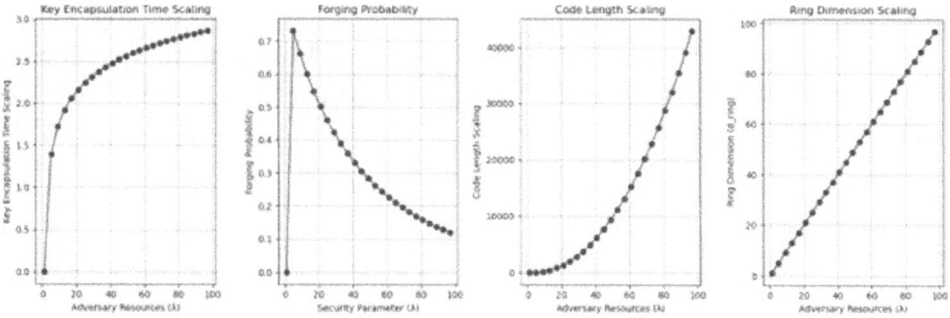

Lattice-based key encapsulation is an efficient method, but it depends on the lattice dimension. Hash-based signatures trade security for faster verification. Code-based cryptosystems offer compact key sizes, but decoding can be complex. Multivariate cryptography achieves good security with smaller rings.

Integration challenges include backward compatibility, migration strategies, and ensuring a smooth transition. Research and development efforts aim to tackle these challenges by proposing transition strategies, developing compatibility layers, and providing guidelines for organizations to gradually adopt quantum-safe protocols.

$$P_{smooth} = 1 - P_{disruption} \tag{143}$$

Eq. (143) represents the probability of a smooth transition with disruption probability ($P_{disruption}$).

Quantum-resistant virtual private networks (VPNs) and communication tools address the unique requirements of secure communication in the quantum era, including key exchange and authentication mechanisms. Research and industry collaborations aim to design and implement quantum-resistant VPNs and communication tools to provide secure and privacy-preserving communication solutions in anticipation of future quantum threats.

$$P_{secure} = 1 - P_{intercept}\left(T_{intercept}, QBER\right) - P_{compromise}\left(C_{devices}\right) \tag{144}$$

Eq. (144) represents the probability of secure communication considering eavesdropping probability ($P_{intercept}$) with intercept time ($T_{intercept}$) and QBER, and device compromise probability ($P_{compromise}$) based on device complexity ($C_{devices}$).

$$H_{Eve}(\rho_{AB}) = H_{Eve}(\rho_A) + H_{Eve}(\rho_B) - H_{Eve}(\rho_{AB}) - \mu \tag{145}$$

Eq. (145) represents the (quantifying information leakage to an eavesdropper (Eve) with individual and joint state entropies of parties (A, B) and mutual information (μ) (see Figure 29).

Figure 29. (a) Secure communication probability. (b) Conceptual impact of factors on secure communication probability.

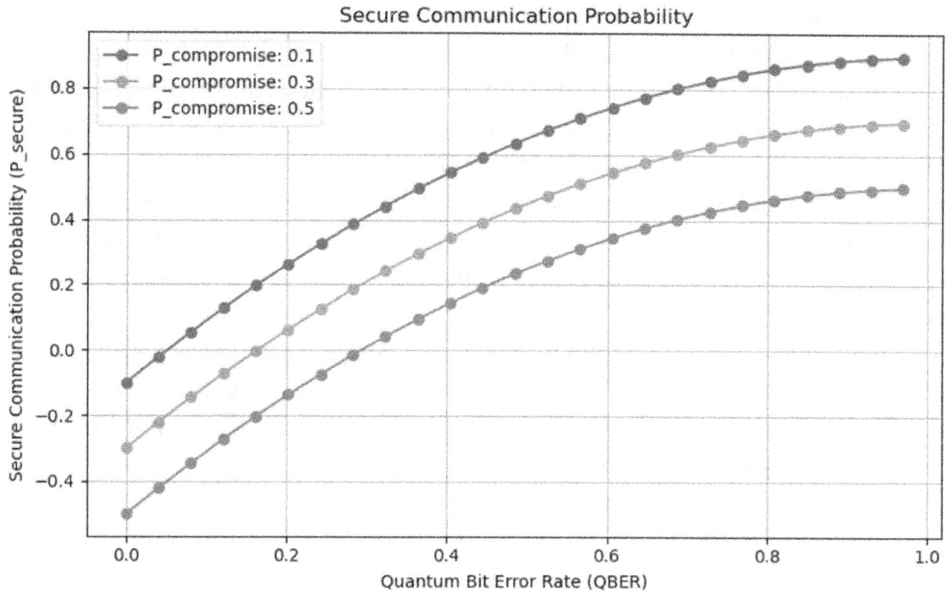

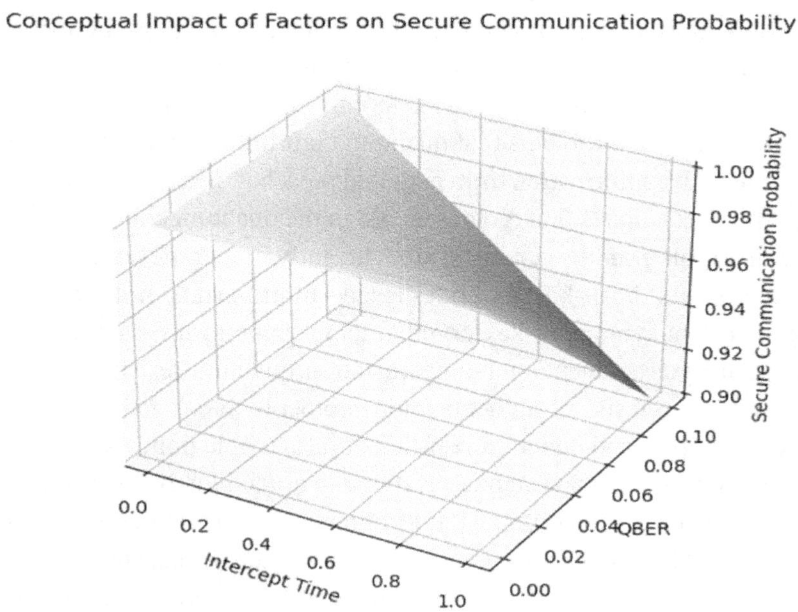

Quantum VPN security relies on minimising eavesdropping and device vulnerabilities. Quantum authentication protocols use entanglement for information security. Effective collaboration between research and industry requires shared expertise, diverse perspectives, and efficient coordination.

Large-scale deployments, collaboration with service providers, and adherence to regulatory considerations are necessary for real-world testing and deployment of quantum-resistant communication protocols. Industry collaborations, pilot deployments, and regulatory frameworks are contributing to the testing and deployment of quantum-resistant communication protocols at scale. The development and deployment of quantum-resistant communication protocols are critical pillars for ensuring the security and privacy of information in the quantum realm as quantum technologies advance.

Quantum-resistant cryptography is a crucial response to the threat posed by quantum computers, which have the ability to break widely-used cryptographic systems. The field aims to strengthen cybersecurity infrastructure against the transformative capabilities of quantum adversaries.

$$Adv_{break}(C) \leq \varepsilon + \alpha Adv_{solve}(H) \tag{146}$$

Eq. (146) represents the relating cipher security (C) to advantage of solving a hard problem (H) with constant (ε) and factor (α).

Quantum-resistant cryptography tackles challenges and opportunities in developing cryptographic techniques that can withstand the power of quantum attacks. Classical cryptographic systems face the looming threat of quantum attacks as they rely on mathematical problems that quantum algorithms can efficiently solve. To safeguard sensitive information from potential breaches, researchers aim to develop cryptographic techniques that remain secure in the quantum era.

Several cryptographic schemes are currently under development, including lattice-based, hash-based, code-based, multivariate polynomial, QKD, quantum-resistant symmetric-key cryptography, quantum-resistant hybrid cryptosystems, security assessments and standardization, quantum-resistant cryptographic libraries, quantum-resistant authentication protocols, blockchain and distributed ledger technologies, and quantum-resistant cryptographic policies.

Security assessments and standardization are crucial for evaluating the security of quantum-resistant cryptographic algorithms and achieving consensus on standardized protocols. Collaboration between industry, academia, and standardization organizations ensures the adoption of secure and interoperable quantum-resistant cryptographic solutions.

To address challenges related to usability, performance, and support for diverse platforms, quantum-resistant cryptographic libraries are being developed. Open-source initiatives contribute to the widespread adoption of these libraries across various domains.

$$P_{operations}(L) = \alpha T_{operation}(i) \cdot N_{operations}(i) \sum_i L_i \tag{147}$$

Eq. (147) represents the predicting overall library performance with operation times ($T_{operation}$) and counts ($N_{operations}$) across code sections (i) and library size (L).

It is essential to maintain the integrity of communication channels in the quantum era by implementing quantum-resistant authentication protocols. Research collaborations are exploring quantum-resistant solutions for blockchain and distributed ledger technologies to ensure the resilience of decentralized systems in the quantum realm.

$$P_{prove}(\kappa) = 1 - \varepsilon_{sound} - \varepsilon_{zero} - knowledge \tag{148}$$

Eq. (148) represents the probability of prover successfully convincing verifier without revealing secret information, with soundness error (ε_{sound}) and zero-knowledge error ($\varepsilon_{zero} - knowledge$), and security parameter (κ).

Developing policies and guidelines for the adoption of quantum-resistant cryptographic techniques involves addressing legal, regulatory, and compliance considerations. Harmonising global cryptographic policies ensures a consistent approach to quantum-resistant security. International collaboration on cryptographic policies contributes to the creation of a global framework for quantum-resistant security.

Identifying and addressing challenges specific to quantum-resistant cryptography is crucial for the advancement of quantum-safe cryptographic solutions. This includes the development of secure randomness sources, key management strategies, and resilience against quantum-inspired attacks. Collaborative efforts among researchers, industry professionals, and policymakers can facilitate the development and implementation of effective quantum-safe cryptographic solutions.

$$H_{random}(t) \geq log(N) \tag{149}$$

Eq. (149) represents the minimum information entropy (H_{random}) required for randomness source secure against adversaries with resources t, with N possible random outcomes.

$$P_{compromise}\left(C_{devices}\right) = e^{-\frac{\beta\lambda}{log(\lambda)}} \tag{150}$$

Eq. (150) represents the probability of key compromise with device complexity ($C_{devices}$) and security parameter (λ) with factor (β).

$$P_{attack}(\lambda) = \frac{\alpha T_{exploitation}}{T_{operation}} \tag{151}$$

Eq. (151) represents the probability of attack success based on exploitation time ($T_{exploitation}$) and operation time ($T_{operation}$) and factor (α) (see Figure 30).

Figure 30. (a) Minimum entropy for secure randomness source. (b) probability of key compromise with different device complexity and security parameters. (c) Probability of attack success with different exploitation and operation times (operation time: 0.0 to 1.0, Factor: 0.2 to 1.0).

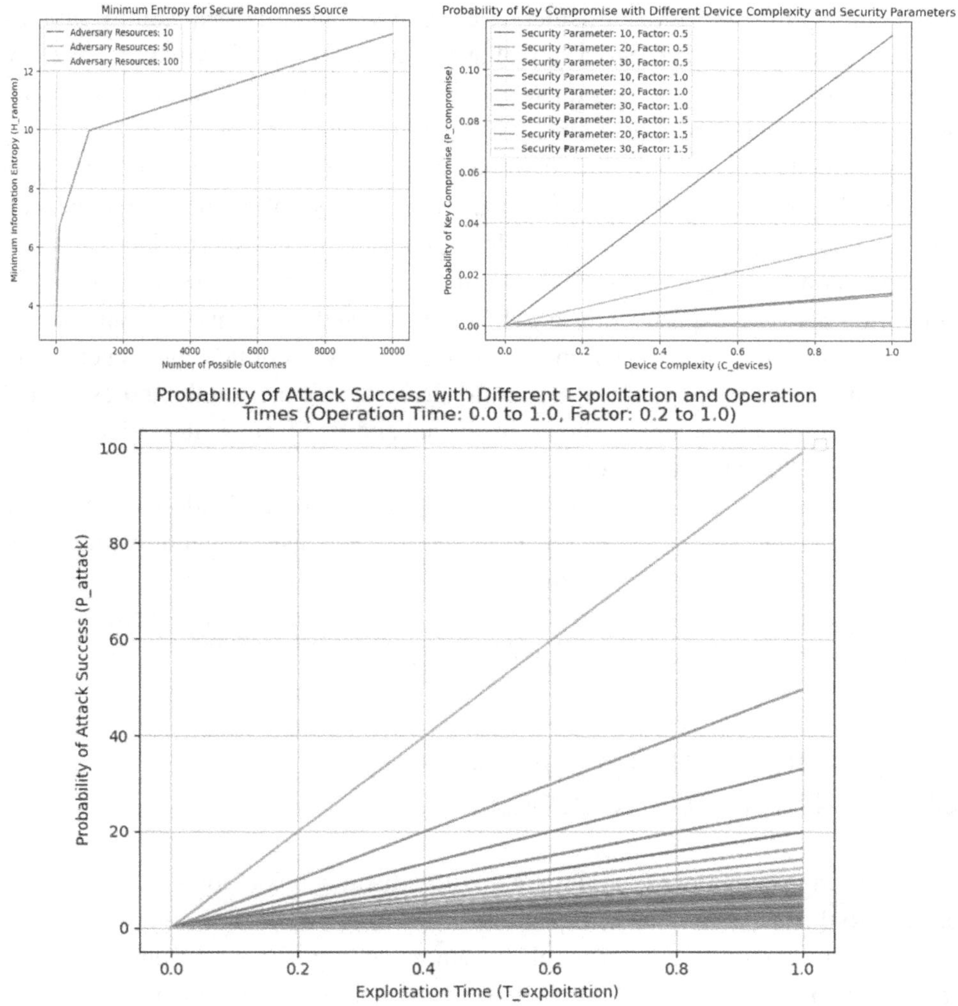

Educating stakeholders on the importance of quantum-resistant cryptography and the necessary steps for a smooth transition presents another opportunity. Collaborative efforts among researchers, industry experts, and standardization bodies contribute to the advancement of quantum-resistant cryptographic solutions.

The rapid development of quantum technologies presents complex challenges related to governance, ethics, and responsible development. To ensure ethical use, responsible innovation, and international cooperation, a robust regulatory framework is essential. This includes addressing emerging challenges in governance, ethical considerations in research and development, international collaboration in quantum governance, policies for quantum safety and security, standards for quantum computing, data privacy, and cryptography. Collaboration between policymakers, security experts, and the quantum community can lead to the development of policies that ensure the safe and secure deployment of quantum technologies.

Additionally, establishing international standards for quantum computing, addressing data privacy, and regulating quantum-safe cryptographic techniques can contribute to responsible governance. It is important to maintain a balanced and objective approach to these topics, avoiding biased or emotional language. Collaboration between experts, industry stakeholders, and policymakers can address regulatory challenges in quantum computing services, public engagement, and environmental impact. Technical terms should be used consistently and explained when first used. The language should be formal and free from grammatical errors, spelling mistakes, and punctuation errors.

Global quantum governance forums can facilitate communication and promote international cooperation on quantum governance. Responsible quantum innovation requires balancing innovation with ethical considerations and risk management. Future legislation should anticipate gaps in existing legal frameworks and adapt to the unique aspects of quantum technologies.

EXPLORING THE MORAL DIMENSIONS OF QUANTUM TECHNOLOGIES

Quantum research is a complex field with ethical considerations ranging from privacy to the responsible use of quantum computing power. The unique nature of quantum mechanics and its principles, such as superposition and entanglement, raise fundamental questions about the ethical implications of manipulating quantum states. Researchers and ethicists are in dialogue to develop ethical frameworks that respect the fundamental principles of quantum mechanics.

Privacy is a major concern in quantum computing due to the immense computational power of quantum computers. Researchers are collaborating to ensure that the advantages of quantum computing do not compromise individual and organizational privacy. The ethical implications of quantum computing's unparalleled processing power must be considered, especially in areas such as optimization, AI, and cryptography.

$$P_{secure} = 1 - P_{intercept}\left(T_{intercept}, QBER\right) - P_{compromise}\left(C_{devices}\right) \quad (152)$$

Eq. (152) represents the probability of secure communication considering eavesdropping probability ($P_{intercept}$) with intercept time ($T_{intercept}$) and QBER, and device compromise probability ($P_{compromise}$) based on device complexity ($C_{devices}$).

$$A_{amplification} = e^{\gamma N} \quad (153)$$

Eq. (153) represents the exponential increase in bias with amplification factor (γ) and number of qubits (N).

$$D_{circuit} = O(log(N)) \quad (154)$$

Eq. (154) represents the asymptotic scaling of circuit depth with number of qubits (N) for specific algorithms (see Figure 31).

Figure 31. (a) Exponential increase in bias with qubit number and amplification. (b) Conceptual impact of factors on secure communication probability.

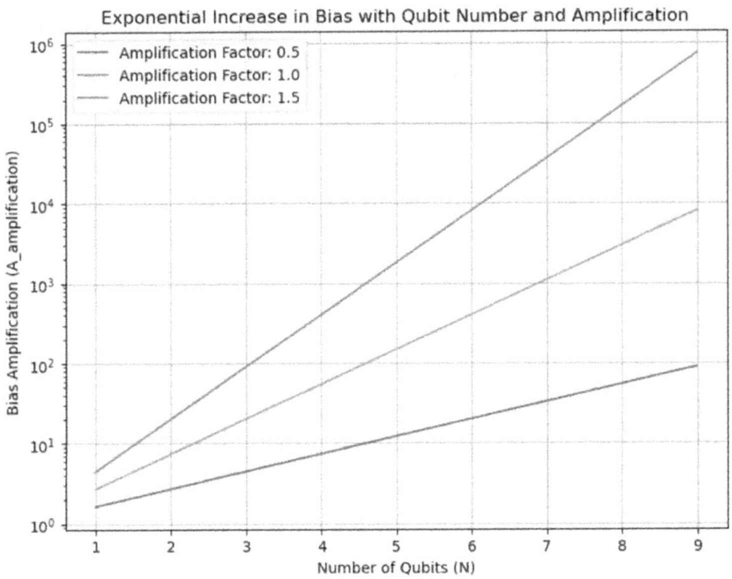

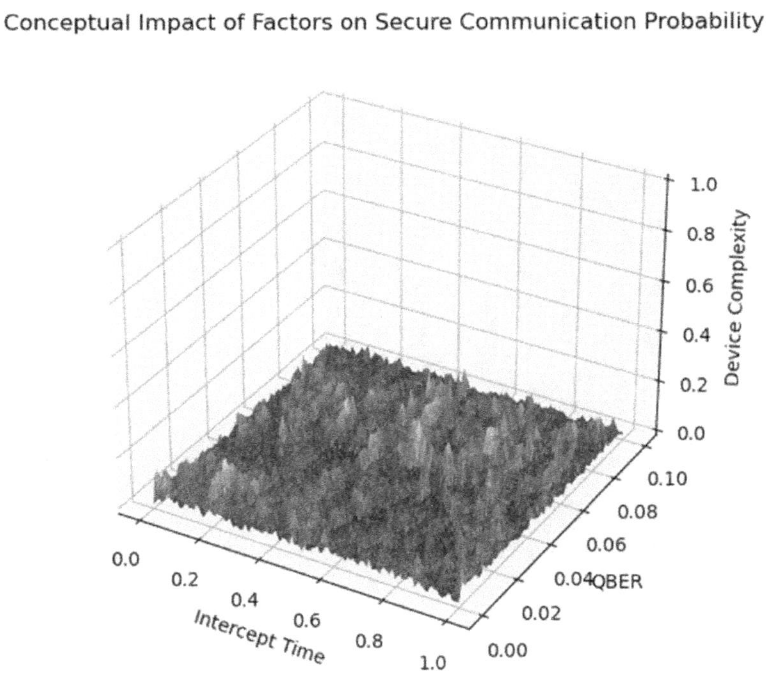

Quantum algorithms and bias raise concerns about algorithmic bias and potential unintended consequences. Ethical discussions focus on the development of energy-efficient quantum computing architectures and the promotion of sustainable practices in the manufacture and operation of quantum systems.

The boundaries of quantum simulation and ethical research are crucial, and guidelines are being developed to ensure responsible conduct of quantum simulation research. Access to quantum information and technologies raises ethical concerns about potential inequalities in access, contributing to a digital divide.

$$F_{simulation}(\tau) = e^{-\varepsilon\tau} \tag{155}$$

Eq. (155) represents the decrease in fidelity with time (τ) and error coefficient (ε).

$$P_{intercept}(QKD) \leq e^{-\frac{\beta\lambda}{log(\lambda)}} \tag{156}$$

Eq. (156) represents the probability of eavesdropping in quantum communication with security parameter (λ) and factor (β) (see Figure 32).

Figure 32. (a) Fidelity decrease with time for different coefficients. (b) Probability of eavesdropping with different security parameters and factors.

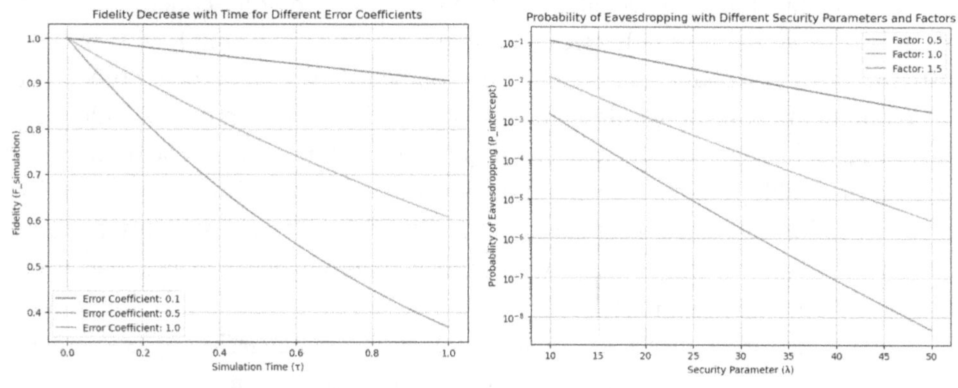

Quantum communication and secure information exchange also raise ethical considerations related to global surveillance, intelligence gathering, and the responsible use of secure communication channels. Ethical frameworks guide the development and deployment of quantum communication technologies. They emphasise transparency, accountability, and international cooperation to address security concerns while respecting the right to privacy.

Ethical considerations for funding quantum research include transparency, accountability, and responsible behaviour. Collaboration between researchers and institutions across international borders also raises ethical considerations related to cultural differences, legal frameworks, and the alignment of research values.

Public engagement in discussions about quantum research, ethical considerations, and the societal implications of advancing quantum technologies is essential.

Quantum research is a rapidly evolving field that is gaining increasing importance. It is characterized by the emergence of new technologies and applications that are transforming our understanding and interaction with the world. However, this field also presents challenges, particularly in terms of potential unintended consequences that could have significant social, economic, and geopolitical impacts. To manage these risks, researchers and policymakers must be aware of the potential consequences and take proactive measures.

Inclusivity and diversity are also important in the field, including removing barriers to entry, promoting diversity in research teams, and ensuring equal opportunities for participation.

Additionally, the ethical responsibilities of companies and industrial stakeholders in quantum research are crucial. These factors encompass environmental impact, ethical behaviour, and societal contribution. In conclusion, ethical considerations are crucial in quantum research to ensure the continued progress and advancement of knowledge in this rapidly evolving field.

The societal impact of quantum technologies is a growing concern, with the potential to revolutionise industries and shape human civilisation. However, being at the forefront of quantum research, development, and deployment comes with a profound ethical responsibility. Ethical considerations encompass a wide range of topics, including inclusive access to quantum technologies, ethical education in quantum fields, addressing economic disparities and job displacement, ethical communication of quantum advances, global health, environmental sustainability, sustainable development, ethical governance of quantum applications, educational equity, and cultural sensitivity.

In order to ensure equal access to quantum technologies, regardless of geographical location, socio-economic status or educational background, it is important to prioritize the development of ethical frameworks for quantum education programmes. These frameworks should prioritize responsible behaviour, transparency, and awareness of the societal impact of quantum technologies. Policies and initiatives should be developed to ensure that the economic benefits of quantum advances are shared inclusively, promoting economic empowerment and social justice.

Ethical communication of quantum advances is crucial, including transparency, accuracy, and avoidance of sensationalism. Quantum technologies can contribute to healthcare, drug discovery, and global health, while also addressing environmental impact, sustainable development, and educational equity. Collaboration between regulators, ethics committees, and industry stakeholders contributes to the responsible governance of quantum applications.

Quantum technologies have ethical implications that extend to human rights, social justice, global collaboration, ethical supply chains, ethical integration of AI, responsible business practices, consumer awareness, political advocacy, cultural exchange, media representation, disaster response, space exploration, access to healthcare, and international relations. Ethical considerations guide the development of quantum technologies to respect fundamental rights, address systemic inequalities, foster global collaboration, ensure fair labour practices, and promote ethical AI integration.

Industry stakeholders are guided by ethical frameworks in developing and maintaining ethical supply chains for quantum technologies, ensuring fair labour practices and responsible sourcing of materials. The integration of quantum technologies with AI requires ethical considerations such as fairness, transparency, and accountability. The quantum community also has an ethical responsibility to engage in policy discussions and advocate for ethical policies that govern the societal impact of quantum technologies.

Ethical representation in the media is crucial, avoiding sensationalism and ensuring accurate representation to the public. Collaboration between the quantum community and the media can help to inform the public about the societal implications of quantum technologies. Additionally, the quantum community has an ethical responsibility to use quantum technologies to contribute to disaster response and relief efforts, prioritising humanitarian objectives and adhering to ethical principles.

Finally, the quantum community has a responsibility to promote positive international relations, avoid geopolitical tensions, and contribute to global cooperation. By engaging in ethical reflection, collaborative initiatives, and global partnerships, the quantum community can navigate the complexities of societal impact and contribute positively to the collective future of humanity.

QUANTUM COMPUTING IN COLLABORATION

Quantum-enabled interdisciplinary research is an expanding field that combines quantum computing with various scientific domains. This collaboration involves researchers, scientists, and experts from various fields, including materials science, drug discovery, climate modelling, imaging, robotics, communications, secure information exchange, geophysics, astronomy, agriculture, social sciences, cognitive science, archaeology, entertainment, philosophy of mind, sports science, healthcare, and personalised medicine.

The potential of quantum computing to simulate molecular structures and chemical reactions is enabling innovative collaborations between quantum chemists and materials scientists. These collaborations are leading to the discovery of novel materials with tailored properties for applications in electronics, energy storage, and beyond. Collaborations in quantum biology are shedding light on quantum effects in biological processes, which could have implications for areas such as photosynthesis, enzyme catalysis, and sensory perception.

In drug discovery and pharmacology, quantum computing is collaborating with pharmaceutical researchers to revolutionise drug discovery and design through quantum pharmacology. Quantum algorithms can simulate molecular interactions, which helps identify new drug candidates and reduces the time and resources required for drug discovery. This technology has the potential to revolutionise various fields.

Quantum computing is also being used in climate modelling and finance to improve the accuracy and efficiency of models through quantum meteorology. Quantum ML and AI algorithms have the potential to deliver exponential speedups, which could transform the landscape of AI applications.

Quantum imaging and medical diagnostics could work with medical researchers and diagnosticians to explore quantum-enhanced imaging techniques. Quantum robotics and autonomous systems are contributing to the development of more efficient and adaptive autonomous systems. This is having an impact on areas such as self-driving vehicles, drone navigation, and industrial automation.

Quantum communication and secure information exchange can be achieved through quantum-safe communication protocols developed by cybersecurity experts. Collaborations in quantum sensing and geophysics are contributing to advances in geophysical exploration. Additionally, quantum algorithms for simulating celestial phenomena and analysing large astronomical datasets are contributing to a deeper understanding of the universe.

Quantum computing is being used in collaboration with healthcare professionals to develop personalised medicine and optimise healthcare. These collaborations contribute to the development of tailored treatment plans, drug discovery, and healthcare optimisation based on individual genetic and medical data. Collaborations in quantum linguistics and natural language processing are advancing language processing and translation. Collaborations between quantum philology and historical linguistics can aid in deciphering ancient scripts and preserving endangered languages. Collaborations between quantum consciousness researchers explore the links between quantum phenomena and consciousness, leading to interdisciplinary discussions on consciousness and cognitive processes.

The convergence of quantum computing and AI is a promising frontier for mutual advancement. Researchers and practitioners from both fields are collaborating to utilise the distinctive features of quantum systems to enhance ML, optimisation, and other aspects of AI. Collaborations in this area focus on quantum ML, including foundations, optimization algorithms, neural networks, reinforcement learning, support vector machines, clustering, generative models, natural language processing, robotics, hyperparameter tuning, transfer learning, anomaly detection, Bayesian networks, fairness and bias mitigation, robustness, explainability, autoML, and ensemble learning.

Quantum physicists and ML researchers are collaborating to establish the groundwork for quantum ML. They are using quantum algorithms to solve ML problems, and quantum optimisation algorithms to tackle complex problems in logistics, finance, and operations research. Additionally, quantum neural network models are being developed to leverage quantum entanglement and superposition for improved learning capabilities. Reinforcement learning algorithms are used to train intelligent agents. They exploit quantum parallelism to accelerate the learning process, which opens up avenues for more efficient training of autonomous systems.

Quantum computing experts are collaborating with generative modelling researchers to explore the potential of quantum computing in various fields, including generative modelling, natural language processing, reinforcement learning for robotics, hyper-parameter tuning, transfer learning, anomaly detection, Bayesian networks, fairness and bias mitigation, robustness in ML, explainability in AI, autoML and ensemble learning. These collaborations aim to unlock the full potential of quantum-enhanced AI and contribute to the development of more efficient and adaptive robotic systems.

Quantum physicists and ensemble learning researchers are collaborating to develop quantum algorithms that enhance ensemble learning techniques. The goal is to improve the diversity and accuracy of ensemble models, resulting in more robust predictions. Additionally, researchers are exploring the integration of quantum principles into robotics and AI algorithms to enhance their decision-making and learning capabilities. Quantum edge computing is being explored for AI, using algorithms that leverage quantum principles to perform tasks at the edge. This has the potential to reduce latency and improve efficiency. Additionally, quantum cognitive computing is being explored to simulate and enhance cognitive processes using quantum principles. There is also ongoing research into human-AI collaboration, with algorithms aimed at enhancing collaboration and fostering trust and transparency in human-robot interactions. Quantum AI is being used in drug discovery. Algorithms use quantum principles to simulate molecular interactions and optimize drug design processes. The collaboration between quantum physicists and AI researchers is expected to shape the next era of computational innovation.

BEYOND THE BINARY BY THE EVOLUTION OF QUANTUM COMMUNICATION

Quantum communication networks leverage quantum mechanics principles to ensure unprecedented levels of security and enable applications that were once deemed impossible with classical communication systems. These networks rely on quantum entanglement to form the backbone of QKD protocols, which use interconnected particles regardless of physical distance. Quantum teleportation enables the transfer of quantum states between particles separated by vast distances, providing a secure means of transmitting quantum information over quantum communication networks.

$$\text{Information Entropy}: H(\rho) = -\sum_i p_i \log_2(p_i) \tag{157}$$

$$\text{Mutual Information}: I(A:B) = H(A) - H(A|B) \tag{158}$$

Satellite-based quantum communication is another area where quantum communication networks can be utilised. Satellite-based systems, such as China's Micius satellite, demonstrate the feasibility of secure quantum communication on a global scale, overcoming terrestrial limitations. Quantum communication networks offer unparalleled security for financial transactions, safeguarding sensitive financial data from potential eavesdropping and interception.

$$TeleportationFidelity: F = tr(\rho_1 \rho_2) \tag{159}$$

The importance of quantum-safe cryptography cannot be overstated; as classical cryptographic systems are at risk of being compromised by quantum algorithms. Quantum communication networks enable the implementation of cryptographic protocols that are resilient to attacks from quantum adversaries, ensuring secure internet communication in the post-quantum era.

$$\text{Channel Capacity}: C = max I(A:B) - I(E:B) \tag{160}$$

These networks also have applications in healthcare, government, defence, smart cities, critical infrastructure, 5G networks, blockchain, cross-border diplomacy, education, environmental monitoring, IoT ecosystems, privacy preservation, and energy grid security.

Global standards and protocols are crucial for ensuring the interoperability of quantum communication networks. Quantum hubs and nodes facilitate the distribution of entangled particles and enable secure communication across the network.

As we approach the era of quantum computing, the need for post-quantum cryptography becomes imperative. Quantum communication networks play a crucial role in transitioning to quantum-safe cryptographic algorithms.

$$\text{Network Interoperability Index} = \frac{\sum Compatibility\_score_{ij} W_{ij}}{\sum W_{ij}} \tag{161}$$

Quantum internet protocols are crucial for developing a quantum communication landscape beyond classical communication. These protocols enable the transmission of quantum information across interconnected nodes, forming the foundation for a quantum internet. Quantum entanglement-based communication is the foundation for quantum communication networks, allowing secure and instantaneous transmission of quantum information between distant nodes. QKD protocols ensure secure key exchange between communicating parties. Quantum teleportation protocols enable the transfer of quantum states between particles separated by large distances.

$$\text{Bell Inequality Violation: } S = \sum a_i b_i - c_i d_i (\text{where} S > 2 \text{implies violation}) \tag{162}$$

Quantum repeaters and purification protocols overcome signal loss over long distances, extending the reach of quantum communication by maintaining the fidelity of transmitted quantum states. Quantum cryptographic protocols contribute to post-quantum security. Quantum teleportation-based quantum routing protocols direct the flow of quantum information. Quantum memory and storage protocols preserve the coherence of quantum information. Quantum clock synchronization protocols ensure coherent timing. QEC protocols enhance reliability. Quantum switching protocols optimize data routing (see Figure 33).

Figure 33. (a) Coverage comparison: terrestrial vs. satellite-based quantum communication. (b) Network interoperability visualization (Index: 0.70).

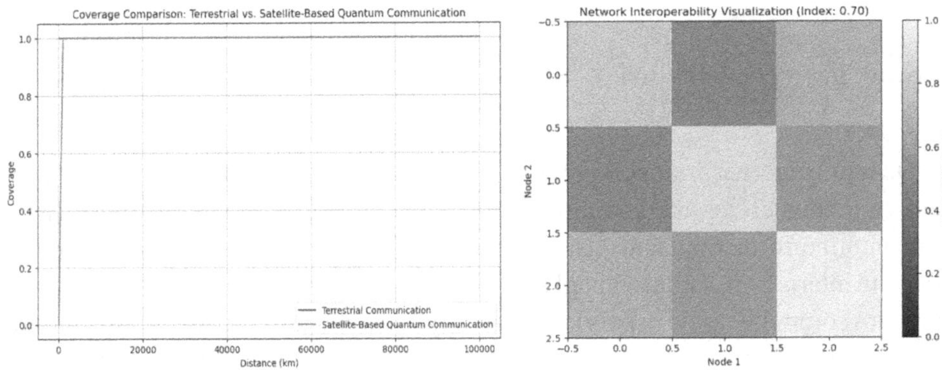

Quantum internet security architecture protocols ensure the security of quantum communication. Quantum multiplexing protocols improve communication efficiency by enabling simultaneous transmission of multiple quantum states through a shared communication channel. Quantum firewall protocols secure individual nodes within the quantum internet, while quantum resource management protocols optimize the utilization of quantum resources.

Quantum routing enables seamless integration of quantum communication with classical infrastructure in hybrid quantum-classical networks. Quantum network slicing protocols allow for tailored quantum communication services to be created. Quantum internet naming and addressing protocols provide unique identification and addressing of individual nodes within the quantum internet.

Quantum internet governance and standards protocols address the need for global coordination and standardization. Access protocols allow users to access quantum communication services. Integration with emerging technologies such as AI, blockchain, and edge computing enhances the capabilities and applications of the quantum internet.

QUANTUM COMPUTING'S IMPACT ON SOCIETY

Quantum literacy is essential for individuals and communities to navigate the transformative changes brought about by quantum technologies. It involves understanding quantum principles and their applications, nurturing quantum intelligence, and equipping individuals with the knowledge to comprehend, contribute to, and benefit from quantum advancements.

K-12 education is the foundation for quantum literacy, as it integrates fundamental quantum concepts into science and mathematics curricula. Higher education institutions play a crucial role in advancing quantum literacy through courses, degree programs, and research opportunities. Online platforms provide accessible learning resources, while continuing education programs cater to professionals seeking to enhance their quantum literacy.

Policymakers have a crucial role in shaping the regulatory landscape for quantum technologies. Quantum literacy programs tailored for policymakers ensure that informed decisions related to funding, regulations, and strategic planning are based on a comprehensive understanding of quantum principles. Public engagement initiatives aim to demystify quantum concepts for a broader audience, while advocacy groups promote quantum literacy as a community-led effort.

Quantum literacy initiatives prioritise inclusivity, bridging disciplinary gaps, and enabling innovation. International collaboration in quantum education enhances knowledge exchange and best practices. Outreach in developing regions empowers individuals with the knowledge and skills to participate in the quantum era. Citizen science programs bridge the gap between researchers and the public. Assessment and certification programs validate quantum knowledge and skills.

Quantum literacy programs focus on creative expression of quantum concepts, informed reporting, lifelong learning, cross-cultural understanding, and preparing future generations for a quantum-influenced world. Scientific outreach programs that incorporate quantum literacy inspire curiosity and interest in the younger generation. Industry collaboration aligns education with evolving workforce needs. Government support plays a crucial role in promoting quantum literacy. Policies that encourage funding, research, and collaboration in quantum education are essential.

The adoption of quantum technologies is expected to have a significant impact on the economy, industries, employment, security, and ethical considerations. Traditional industries may face disruption and evolution, while new industries may emerge. Small and medium enterprises (SMEs) will need to adapt to these changes, requiring tailored support, education, and resources.

The adoption of quantum technologies by governments and industries may position them as leaders in the quantum era, potentially influencing the global competitiveness of nations. The demand for skilled quantum workforce is rising. Equitable access to quantum technologies raises questions, and security challenges arise with quantum-resistant cryptographic techniques.

Ethical considerations are crucial in quantum computing, particularly in relation to data privacy and responsible use of computing power. Governance frameworks are necessary to manage and secure quantum data, addressing issues of ownership, privacy, and responsible data handling. Quantum technologies have the potential to revolutionise various fields, including healthcare, finance, energy, environmental science, smart cities, transportation, agriculture, education, telecommunications, social media, and cultural transformations.

Quantum technologies have the potential to enhance efficiency, productivity, innovation, and competitiveness across various sectors, including finance, logistics, manufacturing, and agriculture. Additionally, they can improve communication infrastructure, ensuring the privacy and integrity of digital communication channels.

In conclusion, the adoption of quantum technologies by society presents a unique opportunity to shape a future that is both technologically advanced and socially inclusive. Balancing economic growth, workforce development, ethical considerations, and access to quantum technologies is essential for fostering a future that is both technologically advanced and socially inclusive.

CONCLUSION

The chapter 'Embarking on Quantum Horizons' concludes by emphasising the revolutionary potential and ethical responsibilities associated with quantum technology. It stresses the importance of ethical frameworks to guide industry stakeholders in creating and managing supply chains, which in turn ensures fair labour standards and responsible material procurement. The integration of AI and quantum technologies requires the assessment of factors such as accountability, transparency, and justice. The ethical obligations for the quantum community include making beneficial contributions to disaster relief, ensuring ethical representation in the media, and promoting international collaboration.

Collaboration across scientific areas, particularly in finance, drug development, climate modelling, and related fields, is known to fuel innovation. The next phase of computational innovation is expected to be shaped by the mutually beneficial uses of AI and quantum computing.

The chapter discusses the importance of quantum communication networks and their applications in various fields such as smart cities, healthcare, governance, and defence. It advocates for the establishment of international standards and protocols, as well as the adoption of post-quantum encryption as quantum technologies advance.

The social implications of quantum technology on industry, education, and security are also explored. To ensure that individuals and communities are equipped to understand, contribute to, and benefit from quantum advancements, this chapter advocates for the promotion of quantum literacy. The potential impact of quantum adoption on the economy, industry, and society highlights the need for a comprehensive strategy that considers workforce development, ethical concerns, and equitable access.

In conclusion, the chapter highlights the importance of ethical governance, responsible innovation, and the development of a quantum-literate society for the peaceful integration of quantum technologies into our future.

REFERENCES

Aaronson, S., & Gottesman, D. (2002). Improved simulation of stabilizer circuits. *Physical Review A*, 66(3), 032305.

Aspuru-Huerva, A., Peruzzo, A., & Plenio, M. B. (2013). Searching for magic: Black-box optimization and the future of quantum computing. *Quantum Information Processing*, 12(1), 3–15.

Babbush, R., Hadley, C., & Rolls, D. (2013). Variational quantum eigensolver for the frequency spectrum. *Physical Review A*, 88(1), 012605.

Bennett, C. H., & Brassard, G. (1984). Quantum cryptography: Public key distribution and coin tossing. In *Secure multi-party computation* (pp. 17–62). Springer.

Biamonte, J., Wittek, P., Gyurif, N., & Faccin, P. (2017). Quantum machine learning. *Nature*, 549(7671), 465–474. 10.1038/nature2347428905917

Bravyi, S., & Kitaev, A. (2005). *Supercloning of entanglement.* arXiv preprint arXiv:quant-ph/0501050.

Deutsch, D., & Ekert, A. (2000). Quantum physics and society. *Nature*, 400(6744), 259–264.

Ekert, A. K. (1991). Quantum cryptography: Based on Bell's theorem. *Physical Review Letters*, 67(6), 661–663. 10.1103/PhysRevLett.67.66110044956

Farhi, E., Goldschmidt, M., & Gutmann, S. (2014). Quantum approximate optimization algorithm. arXiv preprint arXiv:1411.4440.

Fowler, A. G., Mariantoni, M., & Leung, D. W. (2012). Surface code quantum fault tolerance without topological recovery for all distances. *Physical Review. B*, 86(2), 024303.

Fowler, A. G., Mariantoni, M., Martinis, J. M., & Cleland, A. N. (2015). Surface code quantum hardware at the crossroads. *Physical Review A*, 92(5), 052329.

Giovannetti, V., Maccone, L., & Marinatto, L. (2011). Quantum metrology. *Physical Review. B*, 84(16), 161008.

Greenberger, D. M., Horne, M. A., & Shimony, A. (1983). Bell's theorem without inequalities. *American Journal of Physics*, 51(6), 437–440.

Griffiths, D. J., & Schroeter, D. F. (2018). *Introduction to quantum mechanics* (3rd ed.). Cambridge University Press. 10.1017/9781316995433

Grover, L. K. (1996). A fast quantum mechanical algorithm for database search. *Proceedings of the twenty-eighth annual ACM symposium on theory of computing* (pp. 212-223). ACM. 10.1145/237814.237866

Grover, L. K. (1997). A fast quantum mechanical algorithm for database search. In *Proceedings of the twenty-eighth annual ACM symposium on theory of computing* (pp. 212-226). ACM.

Hänsch, T. W., & Wallman, J. (2006). Fundamental limitations in setting the phase of a trapped ion. *Physical Review Letters*, 98(21), 210406.

Harty, T., McCaskey, M., & Shor, P. W. (2017). Stabilizing computations against qubit failures. In *Proceedings of the Annual Symposium on Foundations of Computer Science* (pp. 156-167). Academic Press.

Heisenberg, W. (1927). Über den unscharfen Begriff der Naturwissenschaftlichen Mengenbeziehung in der Kinematik und der Strahlungstheorie. *Zeitschrift für Physik*, 41(7-8), 445–467.

Horodecki, R., Horodecki, P., Horodecki, M., & Horodecki, K. (2009). Quantum entanglement. *Reviews of Modern Physics*, 81(2), 865–942. 10.1103/RevModPhys.81.865

Mabesoone, G. F. (2019). Quantum computing and algorithms. *Nature Physics*, 15(11), 1070–1074.

McClean, J. R., Romero, J., Babbush, R., & Aspuru-Huerva, A. (2016). The theory of variational quantum algorithms. arXiv preprint arXiv:1608.00238.

Nielsen, M. A., & Chuang, I. L. (2000). *Quantum Computation and Quantum Information*. Cambridge University Press.

Preskill, J. (2018). Quantum computing in the NISQ era and beyond. *Quantum : the Open Journal for Quantum Science*, 2, 79. 10.22331/q-2018-08-06-79

Scarani, V., Bechmann-Pasquinucci, H., Cerf, N. J., Dusek, M., Lütken, N., & Tittel, W. (2009). The security of practical quantum key distribution. *Reviews of Modern Physics*, 81(3), 1301–1350. 10.1103/RevModPhys.81.1301

Schuld, M., & Killback, R. (2019). An introduction to quantum machine learning. arXiv preprint arXiv:1904.05366.

Shor, P. W. (1997). Polynomial-time algorithms for prime factorization and discrete logarithms on a quantum computer. *SIAM Journal on Computing*, 26(5), 1464–1484. 10.1137/S0097539795293172

Sihare, S., & Khang, A. (2023). Effects of Quantum Technology on the Metaverse. In *Handbook of Research on AI-Based Technologies and Applications in the Era of the Metaverse* (pp. 174–203). IGI Global. 10.4018/978-1-6684-8851-5.ch009

Sihare, S. R. (2022a). Dynamic multi-party quantum key agreement protocol based on commutative encryption. *International Journal of Theoretical Physics*, 61(9), 242. 10.1007/s10773-022-05203-w

Sihare, S. R. (2022b). Multi-party quantum key agreement protocol for detection of collusive attacks in each sub-circle segment by headers. *International Journal of Theoretical Physics*, 61(7), 208. 10.1007/s10773-022-05184-w

Sihare, S. R. (2022c). Transformation of Classical to Quantum Image, Representation, Processing and Noise Mitigation. *International Journal of Image. Graphics and Signal Processing*, 12(5), 10. 10.5815/ijigsp.2022.05.02

Sihare, S. R. (2022d). Qubit and bit-based quantum hybrid secret key generation. *The European Physical Journal D*, 76(11), 222. 10.1140/epjd/s10053-022-00532-1

Sihare, S. R. (2023a). Potential of quantum computing to effectively comprehend the complexity of brain. *Applied Intelligence*, 53(22), 1–24. 10.1007/s10489-023-04857-1

Sihare, S. R. (2023b). Multi-Party Quantum Key Distribution Using Variational Quantum Eigensolvers. *Advanced Quantum Technologies*, 2300270.

Weinberg, S. (1992). *Dreams of a final theory: The search for the fundamental laws of nature*. Pantheon Books.

KEY TERMS AND DEFINITIONS

Ethical Frameworks: The creation and use of quantum technologies, encompassing ethical labour standards, ethical material procurement, and AI integration issues.

Global Standards and Protocols: The requirement for international protocols and standards to guarantee the safe transition to quantum-safe cryptographic algorithms and the interoperability of quantum communication networks.

Interdisciplinary Collaboration: Demonstrating how collaborative quantum research is and how quantum computing is interacting with other scientific fields including medication development, climate modelling, and healthcare.

Post-Quantum Cryptography: Admitting that quantum assaults can compromise traditional cryptography systems.

Quantum Literacy: The requirement that people and societies comprehend and manage the revolutionary shifts brought forth by quantum technology.

Quantum Technologies: Developments in quantum technologies, including transdisciplinary applications, quantum communication, and quantum computing.

Societal Impact: The possible effects of quantum technologies on education, industry, security, and society as a whole, as well as the moral dilemmas that come with their application.

Compilation of References

A universal two-bit gate for quantum computation. (1995). *Proc. R. Soc., 449*(1937), 679–683. 10.1098/rspa.1995.0066

Aaronson, S., & Kuperberg, G. (2007). Quantum versus classical proofs and advice. *Computational Complexity, 2007. CCC'07. Twenty-Second Annual IEEE Conference on.*

Aaronson, S., & Gottesman, D. (2002). Improved simulation of stabilizer circuits. *Physical Review A*, 66(3), 032305.

Abdollahi, A., & Pedram, M. (2006, March). Analysis and synthesis of quantum circuits by using quantum decision diagrams. [). IEEE.]. *Proceedings of the Design Automation & Test in Europe Conference*, 1, 1–6.

Achuthan, K., & Murali, S. S. (2015). Software Engineering in Intelligent Systems. In *Advances in Intelligent Systems and Computing* (Vol. 349). https://www.scopus.com/inward/record.url?eid=2-s2.0-84942759426&partnerID=tZOtx3y1

Adalı, E. (n.d.). *3. Uluslararası Bilgisayar Bilimleri ve Mühendisliği Konferansı = 3rd International Conference on Computer Science and Engineering : Sarajevo - Bosnia&Herzegovina, 20-23 Eylül (September) 2018*. IEEE.

Agarwal, M., Singh, A., Arjaria, S., Sinha, A., & Gupta, S. (2020). ToLeD: Tomato Leaf Disease Detection using Convolution Neural Network. *Procedia Computer Science*, 167, 293–301. 10.1016/j.procs.2020.03.225

Agrawal, D., Das, S., & El Abbadi, A. (2012). Data Management in the Cloud: Challenges and Opportunities. *Synthesis Lectures on Data Management*, 4(6), 1–138. 10.1007/978-3-031-01895-4

Aharonov, D., Kitaev, A., & Nisan, N. (1998, May). Quantum circuits with mixed states. In *Proceedings of the thirtieth annual ACM symposium on Theory of computing* (pp. 20-30).

Ahmadi, P., Muharam, F. M., Ahmad, K., Mansor, S., & Seman, I. A. (2017). Early detection of ganoderma basal stem rot of oil palms using artificial neural network spectral analysis. *Plant Disease*, 101(6), 1009–1016. 10.1094/PDIS-12-16-1699-RE30682927

Alagic, G., Alagic, G., Alperin-Sheriff, J., Apon, D., Cooper, D., Dang, Q., ... Smith-Tone, D. (2019). *Status report on the first round of the NIST post-quantum cryptography standardization process.* Academic Press.

Alattar, M. H., Al Farawn, A., & Ali, N. S. (2018). Anti-continuous collisions user-based unpredictable iterative password salted hash encryption. *International Journal of Internet Technology and Secured Transactions*, 8(4), 619–634. 10.1504/IJITST.2018.095944

Alazzam, A., & AlOmar, B. (2018). Using Average Uniform Algorithm to model educational data. *ITT 2017 - Information Technology Trends: Exploring Current Trends in Information Technology, Conference Proceedings,* 30–34. 10.1109/CTIT.2017.8259563

Alazzam, A., Yuzgec, E., & Lewis, H. W. (2013). A new optimization algorithm for non-convex problems. *IIE Annual Conference and Expo 2013*, 2784–2791.

Al-Mohammed, H. A., & Yaacoub, E. (2022). New Way to Generating and Simulation QKD. In *Proceedings of Sixth International Congress on Information and Communication Technology* (pp. 801--809). springer. 10.1007/978-981-16-1781-2_69

Al-Mohammed, H. A., Al-Ali, A., Yaacoub, E., Abualsaud, K., & Khattab, T. (2021). Detecting Attackers during Quantum Key Distribution in IoT Networks using Neural Networks. *GLOBECOM 2021-2021 IEEE Global Communications Conference.*

Al-Mohammed, H. A., Al-Ali, A., Yaacoub, E., Qidwai, U., Abualsaud, K., Rzewuski, S., & Flizikowski, A. (2021). Machine Learning Techniques for Detecting Attackers During Quantum Key Distribution in IoT Networks With Application to Railway Scenarios. *IEEE Access : Practical Innovations, Open Solutions*, 9, 136994–137004. 10.1109/ACCESS.2021.3117405

Al-Mohammed, H. A., & Yaacoub, E. (2021). On the Use of Quantum Communications for Securing IoT Devices in the 6G Era. *2021 IEEE International Conference on Communications Workshops, ICC Workshops 2021 - Proceedings*, 1–6. 10.1109/ICCWorkshops50388.2021.9473793

Amer, O., Krawec, W. O., & Wang, B. (2020). Efficient Routing for Quantum Key Distribution Networks. *Proceedings - IEEE International Conference on Quantum Computing and Engineering,* 137–147. 10.1109/QCE49297.2020.00027

Anderson, R. (2020). Security engineering: a guide to building dependable distributed systems. *John Wiley & Sons*. 10.1002/9781119644682

Aqsazafar. (2020). *SVM machine learning.* https://www.mltut.com/svm-machine-learning/

Asfaw, A., Corcoles, A., Bello, L., Ben-Haim, Y., Bozzo-Rey, M., Bravyi, S., Bronn, N., Capelluto, L., Vazquez, A. C., Ceroni, J., Chen, R., Frisch, A., Gambetta, J., Garion, S., Gil, L., De La Puente Gonzalez, S., Harkins, F., Imamichi, T., Kang, H., Karamlou, A. H., Loredo, R., McKay, D., Mezzacapo, A., Minev, Z., Movassagh, R., Nannicini, G., Nation, P., Phan, A., Pistoia, M., Rattew, A., Schaefer, J., Shabani, J., Smolin, J., Stenger, J., Temme, K., Tod, M., Wood, S., & Wootton, J. (2020). *Learn Quantum Computation Using Qiskit.* Qiskit-community.

Aspelmeyer, M., Jennewein, T., Pfennigbauer, M., Leeb, W. R., & Zeilinger, A. (2003). Long-distance quantum communication with entangled photons using satellites. *IEEE Journal of Selected Topics in Quantum Electronics*, 9(6), 1541–1551. 10.1109/JSTQE.2003.820918

Aspuru-Huerva, A., Peruzzo, A., & Plenio, M. B. (2013). Searching for magic: Black-box optimization and the future of quantum computing. *Quantum Information Processing*, 12(1), 3–15.

Awad, M., & Khanna, R. (2015). Efficient learning machines: Theories, concepts, and applications for engineers and system designers. *Efficient Learning Machines: Theories, Concepts, and Applications for Engineers and System Designers*, 1–248. 10.1007/978-1-4302-5990-9

Babar, Z., Chandra, D., Nguyen, H. V., Botsinis, P., Alanis, D., Ng, S. X., & Hanzo, L. (2018). Duality of quantum and classical error correction codes: Design principles and examples. *IEEE Communications Surveys and Tutorials*, 21(1), 970–1010. 10.1109/COMST.2018.2861361

Babber, K., & Singh, J. P. (2021). Quantum cryptography and security analysis. *Journal of Discrete Mathematical Sciences and Cryptography*. 10.1080/09720529.2019.1692452

Babbush, R., Hadley, C., & Rolls, D. (2013). Variational quantum eigensolver for the frequency spectrum. *Physical Review A*, 88(1), 012605.

Banerjee, S. (2019). *Open Quantum Systems: Dynamics of Nonclassical Evolution*. Springer. 10.1007/978-981-13-3182-4

Bansal, A., & Singhal, A. (n.d.). *Proceedings of the 7th International Conference Confluence 2017 on Cloud Computing, Data Science and Engineering : 12th-13th January 2017, Amity University, Noida, Uttar Pradesh, India*. Academic Press.

Bengtsson, I., & Życzkowski, K. (2006). *Geometry of Quantum States: An Introduction to Quantum Entanglement*. Cambridge University Press. 10.1017/CBO9780511535048

Benioff, P. (1980). The computer as a physical system: A microscopic quantum mechanical Hamiltonian model of computers as represented by Turing machines. *Journal of Statistical Physics*, 22(5), 563–591. 10.1007/BF01011339

Benioff, P. (1982). Quantum mechanical hamiltonian models of turing machines. *Journal of Statistical Physics*, 29(3), 515–546. 10.1007/BF01342185

Bennett, C. H. (1992). Quantum cryptography using any two nonorthogonal states. *Physical Review Letters*, 68(21), 3121–3124. 10.1103/PhysRevLett.68.312110045619

Bennett, C. H., Bernstein, E., Brassard, G., & Vazirani, U. (1997). Strengths and weaknesses of quantum computing. *SIAM Journal on Computing*, 26(5), 1510–1523. 10.1137/S0097539796300933

Bennett, C. H., & Brassard, G. (1984). Quantum cryptography: Public key distribution and coin tossing. In *Secure multi-party computation* (pp. 17–62). Springer.

Bennett, C. H., & Brassard, G. (1984). Quantum cryptography: public key distribution and coin tossing. *Proceedings of the International Conference on Computers, Systems & Signal Processing*, 175–179.

Bennett, C. H., & Brassard, G. (1989). Experimental quantum cryptography: the dawn of a new era for quantum cryptography: the experimental prototype is working]. *SIGACT News*, 20(4), 78–82. 10.1145/74074.74087

Bennett, C. H., & Brassard, G. (2014). Quantum cryptography: Public key distribution and coin tossing. *Theoretical Computer Science*, 560, 7–11. 10.1016/j.tcs.2014.05.025

Bennett, C. H., Brassard, G., Crépeau, C., Jozsa, R., Peres, A., & Wootters, W. K. (1993). Teleporting an unknown quantum state via dual classical and Einstein-Podolsky-Rosen channels. *Physical Review Letters*, 70(13), 1895–1899. 10.1103/PhysRevLett.70.189510053414

Bennett, C. H., & DiVincenzo, D. P. (2000). Quantum information and computation. *Nature*, 404(6775), 247–255. 10.1038/3500500110749200

Bennett, C. H., & Shor, P. W. (1998). Quantum information theory. *IEEE Transactions on Information Theory*, 44(6), 2724–2742. 10.1109/18.720553

Bennett, C., & Brassard, G. (1984). Quantum cryptography: Public key distribution and coin tossing. *Proceedings of IEEE International Conference on Computers, Systems, and Signal Processing*, (pp. 175-179). IEEE.

Bennett, C., Brassard, G., Popescu, S., Schumacher, B., Smolin, J., & Wootters, W. (1996). Purification of Noisy Entanglement and Faithful Teleportation via Noisy Channels. *Physical Review Letters*, 76(5), 722–725. 10.1103/PhysRevLett.76.72210061534

Bennett, C., & Wiesner, S. (1992). Communication via one- and two-particle operators on Einstein-Podolsky-Rosen states. *Physical Review Letters*, 69(20), 2881–2884. 10.1103/PhysRevLett.69.288110046665

Beresford, A. R., & Stajano, F. (2003). Location privacy in pervasive computing. *IEEE Pervasive Computing*, 2(1), 46–55. 10.1109/MPRV.2003.1186725

Berman, P., Arimondo, E., & Lin, C. (2012). *Advances in Atomic, Molecular, and Optical Physics*. Elsevier. Retrieved from https://www.elsevier.com/books/advances-in-atomic-molecular-and-optical-physics/berman/978-0-12-396482-3

Bernardino, H. S., & Barbosa, H. J. (2009). Artificial immune systems for optimization. *Studies in Computational Intelligence*, 193, 389–411. 10.1007/978-3-642-00267-0_14

Bernstein, D. J. (2009). Introduction to post-quantum cryptography. In *Post-quantum cryptography* (pp. 1–14). Springer Berlin Heidelberg. 10.1007/978-3-540-88702-7_1

Biamonte, J., Wittek, P., Pancotti, N., Rebentrost, P., Wiebe, N., & Lloyd, S. (2017). Quantum machine learning. *Nature*, 549(7671), 195–202. 10.1038/nature2347428905917

Black, P. E., Kuhn, D. R., & Williams, C. J. (2002). Quantum computing and communication. [). Elsevier.]. *Advances in Computers*, 56, 189–244.

Blum, C., & López-Ibáñez, M. (2007). Ant colony optimization. *Scholarpedia Journal*, 2(3), 1461. 10.4249/scholarpedia.1461

Bonaventure, O. (2011). *Computer Networking: Principles, Protocols and Practice*. Saylor foundation.

Branciard, C., Gisin, N., Kraus, B., & Scarani, V. (2005). Security of two quantum cryptography protocols using the same four qubit states. *Physical Review A*, 72(3), 032301. 10.1103/PhysRevA.72.032301

Brandão, F. G., & Vianna, R. O. (2004a). Robust semidefinite programming approach to the separability problem. *Physical Review A*, 70(6), 062309. 10.1103/PhysRevA.70.062309

Brandão, F. G., & Vianna, R. O. (2004b). Separable Multipartite Mixed States: Operational Asymptotically Necessary and Sufficient Conditions. *Physical Review Letters*, 93(22), 220503. 10.1103/PhysRevLett.93.22050315601074

Braunstein, S. L., & Loock, P. (2005). Quantum information with continuous variables. *Reviews of Modern Physics*, 77(2), 513–577. 10.1103/RevModPhys.77.513

Bravyi, S., & Kitaev, A. (2005). *Supercloning of entanglement*. arXiv preprint arXiv:quant-ph/0501050.

Broadbent, A., & Schaffner, C. (2016). Quantum cryptography beyond quantum key distribution. In *Designs, Codes, and Cryptography* (Vol. 78, Issue 1). Springer US. 10.1007/s10623-015-0157-4

Buhrman, H., Cleve, R., & Wigderson, A. (1998, May). Quantum vs. classical communication and computation. In *Proceedings of the thirtieth annual ACM symposium on Theory of computing* (pp. 63-68). 10.1145/276698.276713

Burgholzer, L., Raymond, R., & Wille, R. (2020, October). Verifying results of the IBM Qiskit quantum circuit compilation flow. In *2020 IEEE International Conference on Quantum Computing and Engineering (QCE)* (pp. 356-365). IEEE.

Busch, P., Heinonen, T., & Lahti, P. (2007). Heisenberg's uncertainty principle. *Physics Reports*, 452(6), 155–176. 10.1016/j.physrep.2007.05.006

Buttler, W. T., Hughes, R. J., Kwiat, P. G., Lamoreaux, S. K., Luther, G. G., Morgan, G. L., Nordholt, J. E., Peterson, C. G., & Simmons, C. M. (1998). Practical free-space quantum key distribution over 1 km. *Physical Review Letters*, 81(15), 3283–3286. 10.1103/PhysRevLett.81.3283

Cacciapuoti, A. S., Caleffi, M., Tafuri, F., Cataliotti, F. S., Gherardini, S., & Bianchi, G. (2020). Quantum Internet: Networking Challenges in Distributed Quantum Computing. *IEEE Network*, 34(1), 137–143. 10.1109/MNET.001.1900092

Caleffi, M. (2017). Optimal Routing for Quantum Networks. *IEEE Access : Practical Innovations, Open Solutions*, 5, 22299–22312. 10.1109/ACCESS.2017.2763325

Chin, E., Felt, A. P., Greenwood, K., & Wagner, D. (2011, June). Analyzing inter-application communication in Android. In *Proceedings of the 9th international conference on Mobile systems, applications, and services* (pp. 239-252). 10.1145/1999995.2000018

Chiueh, T. D., & Tsai, P. Y. (2008). *OFDM baseband receiver design for wireless communications*. John Wiley & Sons.

Chruściński, D., & Sarbicki, G. (2014). Entanglement witnesses: Construction, analysis and classification. *Journal of Physics. A, Mathematical and Theoretical*, 47(48), 483001. 10.1088/1751-8113/47/48/483001

Chun, W. H. K. (2008). *Control and freedom: Power and paranoia in the age of fiber optics*. MIT Press.

Chun-Yan, L., Hong-Yu, Z., Yan, W., & Fu-Guo, D. (2005). Secure quantum key distribution network with Bell states and local unitary operations. *Chinese Physics Letters*, 22(5), 1049–1052. 10.1088/0256-307X/22/5/006

Colbeck, R. (2009). Quantum and relativistic protocols for secure multi-party computation. *arXiv preprint arXiv:0911.3814*.

Collins, D., Gisin, N., & De Riedmatten, H. (2005). Quantum relays for long-distance quantum cryptography. *Journal of Modern Optics*, 52(5), 735–753. 10.1080/09500340412331283633

Cosma, G., Brown, D., Archer, M., Khan, M., & Graham Pockley, A. (2017). A survey on computational intelligence approaches for predictive modeling in prostate cancer. *Expert Systems with Applications*, 70, 1–19. 10.1016/j.eswa.2016.11.006

D, D., JG, L., S, D., W, M., A, C., & V, S. (2017). Cardiotocography versus intermittent auscultation of fetal heart on admission to labour ward for assessment of fetal wellbeing. *The Cochrane Database of Systematic Reviews, 1*(1). 10.1002/14651858.CD005122.pub5

Daemen, J., & Rijmen, V. (1999). *AES proposal: Rijndael*. Academic Press.

Dagmar, B., & Gerd, L. (2019). *Quantum Information: From Foundations to Quantum Technology Applications*. John Wiley & Sons.

Dastjerdi, Amir Vahid and Buyya, R. (2016). Fog Computing: Helping the Internet of Things Realize Its Potential. *Computer, 49*(8), 112–116. 10.1109/MC.2016.245

Dawy, Saad, Ghosh, Andrews, & Yaacoub. (2017). Toward massive machine type cellular communications. *IEEE Wireless Communications*, 120–128.

de Andrade, M. G., Dai, W., Guha, S., & Towsley, D. (2021). *A quantum walk control plane for distributed quantum computing in quantum networks*. Academic Press.

De Cola, T., Paolini, E., Liva, G., & Calzolari, G. P. (2011). Reliability options for data communications in the future deep-space missions. *Proceedings of the IEEE*, 99(11), 2056–2074. 10.1109/JPROC.2011.2159571

de Ronde, C. (2018). Quantum Superpositions and the Representation of Physical Reality Beyond Measurement Outcomes and Mathematical Structures. *Foundations of Science*, 23(4), 621–648. 10.1007/s10699-017-9541-z

Deutsch, D., & Ekert, A. (2000). Quantum physics and society. *Nature*, 400(6744), 259–264.

Di Candia, R., Yi itler, H., Paraoanu, G. S., & Jäntti, R. (2020). Two-way covert microwave quantum communication. In *arXiv*. 10.21203/rs.3.rs-93750/v1

Diaz, V. (2015). Backhauling with fibre? *Fibre Systems, 6*.

Dieks, D. G. B. J. (1982). Communication by EPR devices. *Physics Letters. [Part A]*, 92(6), 271–272. 10.1016/0375-9601(82)90084-6

Diffie, W., & Hellman, M. E. (2022). New directions in cryptography. In *Democratizing Cryptography: The Work of Whitfield Diffie and Martin Hellman* (pp. 365-390). 10.1145/3549993.3550007

Diffie, W., & Hellman, M. (1976). New directions in cryptography. *IEEE Transactions on Information Theory*, 22(6), 644–654. 10.1109/TIT.1976.1055638

Digital Evidence and Computer Crime - 2nd Edition. (n.d.). Retrieved September 7, 2021, from https://www.elsevier.com/books/digital-evidence-and-computer-crime/casey/978-0-08-047250-8

Djordjevic, I. (2012). *Quantum information processing and quantum error correction: an engineering approach*. Academic press.

Docs and Resources. (n.d.). *IBM Quantum Experience IBM*. Available: https://quantum-computing.ibm.com/docs/

Dodis, Y., Pointcheval, D., Ruhault, S., Vergniaud, D., & Wichs, D. (2013). Security analysis of pseudo-random number generators with input: /dev/random is not robust. *Proceedings of the ACM Conference on Computer and Communications Security*, 647–658. 10.1145/2508859.2516653

Doherty, A. C., Parrilo, P. A., & Spedalieri, F. M. (2004). Complete family of separability criteria. *Physical Review A*, 69(2), 022308. 10.1103/PhysRevA.69.022308

Du, K. L., & Swamy, M. N. S. (2016). Search and optimization by metaheuristics: Techniques and algorithms inspired by nature. *Search and Optimization by Metaheuristics: Techniques and Algorithms Inspired by Nature*, 1–434. 10.1007/978-3-319-41192-7

Dür , Vidal , Cirac, Linden, & Popescu. (2001). Entanglement Capabilities of Nonlocal Hamiltonians. *Physical Review Letters, 87*, 137901. Retrieved from https://link.aps.org/doi/10.1103/PhysRevLett.87.137901

Dzung, D., Naedele, M., Von Hoff, T. P., & Crevatin, M. (2005). Security for industrial communication systems. *Proceedings of the IEEE*, 93(6), 1152–1177. 10.1109/JPROC.2005.849714

Easttom, C. (2019). *Computer security fundamentals*. Pearson IT certification.

Einstein, A., Podolsky, B., & Rosen, N. (1937). Can Quantum-Mechanical Description of Physical Reality Be Considered Complete? *Physical Review*, 47(10), 777–780. 10.1103/PhysRev.47.777

Ekert, A. K. (1991). Quantum cryptography based on Bell's theorem. *Physical Review Letters*, 67(6), 661–663. 10.1103/PhysRevLett.67.66110044956

El Ashmawy, M. S. (2021). Error correction in quantum cryptography.

Elhoseny, M., Ramírez-González, G., Abu-Elnasr, O. M., Shawkat, S. A., Arunkumar, N., & Farouk, A. (2018). Secure Medical Data Transmission Model for IoT-Based Healthcare Systems. *IEEE Access : Practical Innovations, Open Solutions*, 6, 20596–20608. 10.1109/ACCESS.2018.2817615

Farhi, E., Goldschmidt, M., & Gutmann, S. (2014). Quantum approximate optimization algorithm. arXiv preprint arXiv:1411.4440.

Fernandez-Carames, T. M. (2019). From Pre-Quantum to Post-Quantum IoT Security: A Survey on Quantum-Resistant Cryptosystems for the Internet of Things. *IEEE Internet of Things Journal*, 7(7), 1–1. 10.1109/JIOT.2019.2958788

Ferrari, D., Tavernelli, I., & Amoretti, M. (2021). Deterministic algorithms for compiling quantum circuits with recurrent patterns. *Quantum Information Processing*, 20(6), 1–21.

Feynman, R. P. (1986). Quantum mechanical computers. *Foundations of Physics*, 16(6), 507–531. 10.1007/BF01886518

Fitzsimons, J. F. (2017). Private quantum computation: An introduction to blind quantum computing and related protocols. *NPJ Quantum Information*, 3(1), 1–10. 10.1038/s41534-017-0025-3

Fowler, A. G., Mariantoni, M., & Leung, D. W. (2012). Surface code quantum fault tolerance without topological recovery for all distances. *Physical Review. B*, 86(2), 024303.

Fowler, A. G., Mariantoni, M., Martinis, J. M., & Cleland, A. N. (2015). Surface code quantum hardware at the crossroads. *Physical Review A*, 92(5), 052329.

Freedman, S. J., & Clauser, J. F. (1972). Experimental Test of Local Hidden-Variable Theories. *Physical Review Letters*, 28(14), 938–941. 10.1103/PhysRevLett.28.938

Fuchs, C., & Giggenbach, D. (2009). Optical free-space communication on earth and in space regarding quantum cryptography aspects. *International Conference on Quantum Communication and Quantum Networking*. Springer.

Furusawa, A., Sørensen, J. L., Braunstein, S. L., Fuchs, C. A., Kimble, H. J., & Polzik, E. S. (1998). Unconditional quantum teleportation. *Science*, 282(5389), 706–709. 10.1126/science.282.5389.7069784123

G, J. M., & R., S. D. (2020). Automatic Tomato Plant Leaf Disease Classification using Multi-Kernel Support Vector Machine. *International Journal of Engineering and Advanced Technology*, 9(5), 560–565. 10.35940/ijeat.E9689.069520

Gill, S. S., Kumar, A., Singh, H., Singh, M., Kaur, K., Usman, M., & Buyya, R. (2020). Quantum Computing: A Taxonomy, Systematic Review and Future Directions. *Emerging Technologies*. DOI=arXiv:2010.15559 [cs.ET]

Gingrich, R. M., & Adami, C. (2002). Quantum entanglement of moving bodies. *Physical Review Letters*, 89(27), 2704021–2704024. 10.1103/PhysRevLett.89.27040212513186

Giovannetti, V., Maccone, L., & Marinatto, L. (2011). Quantum metrology. *Physical Review. B*, 84(16), 161008.

Gisin, N. (2018). Quantum communications. *Optics InfoBase Conference Papers: Vol. Part F112*. 10.1364/SPPCOM.2018.SpM3G.5

Gisin, N., & Thew, R. (2007). Quantum communication. *Nature Photonics*, 1(3), 165–171. 10.1038/nphoton.2007.22

Gottfried, K., & Yan, T. M. (2013). *Quantum mechanics: fundamentals*. Springer Science & Business Media.

Greenberger, D. M., Horne, M. A., & Shimony, A. (1983). Bell's theorem without inequalities. *American Journal of Physics*, 51(6), 437–440.

Griffiths, D. J. (2016). *Introduction to quantum mechanics*. Cambridge University Press.

Grover, L. K. (1996, July). A fast quantum mechanical algorithm for database search. In *Proceedings of the twenty-eighth annual ACM symposium on Theory of computing* (pp. 212-219). 10.1145/237814.237866

Grover, L. K. (1997). A fast quantum mechanical algorithm for database search. In *Proceedings of the twenty-eighth annual ACM symposium on theory of computing* (pp. 212-226). ACM.

Gu, Q., Sheng, L., Zhang, T., Lu, Y., Zhang, Z., Zheng, K., Hu, H., & Zhou, H. (2019). Early detection of tomato spotted wilt virus infection in tobacco using the hyperspectral imaging technique and machine learning algorithms. *Computers and Electronics in Agriculture*, 167, 105066. Advance online publication. 10.1016/j.compag.2019.105066

Gyongyosi, L., Bacsardi, L., & Imre, S. (2019). A survey on quantum key distribution. *Infocommunications Journal*, 11(2), 14–21. 10.36244/ICJ.2019.2.2

H.D., G. (2020). Machine Learning Approach towards Tomato Leaf Disease Classification. *International Journal of Advanced Trends in Computer Science and Engineering*, 9(1), 490–495. 10.30534/ijatcse/2020/67912020

Haghi Kashani, M., Madanipour, M., Nikravan, M., Asghari, P., & Mahdipour, E. (2021). A systematic review of IoT in healthcare: Applications, techniques, and trends. *Journal of Network and Computer Applications*, 192(July), 103164. 10.1016/j.jnca.2021.103164

Haitjema, M. (2007). *A Survey of the Prominent Quantum Key Distribution Protocols*. https://www.cse.wustl.edu/~jain/cse571-07/quantum.htm

Han, K. H., & Kim, J. H. (2000, July). Genetic quantum algorithm and its application to combinatorial optimization problem. In Proceedings of the 2000 congress on evolutionary computation. CEC00 (Cat. No. 00TH8512) (Vol. 2, pp. 1354-1360). IEEE

Hänsch, T. W., & Wallman, J. (2006). Fundamental limitations in setting the phase of a trapped ion. *Physical Review Letters*, 98(21), 210406.

Harty, T., McCaskey, M., & Shor, P. W. (2017). Stabilizing computations against qubit failures. In *Proceedings of the Annual Symposium on Foundations of Computer Science* (pp. 156-167). Academic Press.

Hashimoto, Y. (2021). Recent Developments in Multivariate Public Key Cryptosystems. In *International Symposium on Mathematics, Quantum Theory, and Cryptography*. Springer. 10.1007/978-981-15-5191-8_16

Hassija, V., Chamola, V., Saxena, V., Chanana, V., Parashari, P., Mumtaz, S., & Guizani, M. (2020). Present landscape of quantum computing. *IET Quantum Communication*, 1(2), 42–48.

Heathcote, A. (1990). Unbounded Operators and the Incompleteness of Quantum Mechanics. *Philosophy of Science*, 57(3), 523–534. 10.1086/289572

Heisenberg, W. (1927). Über den unscharfen Begriff der Naturwissenschaftlichen Mengenbeziehung in der Kinematik und der Strahlungstheorie. *Zeitschrift für Physik*, 41(7-8), 445–467.

Hensen, B., Bernien, H., Dréau, A. E., Reiserer, A., Kalb, N., Blok, M. S., Ruitenberg, J., Vermeulen, R. F. L., Schouten, R. N., Abellán, C., Amaya, W., Pruneri, V., Mitchell, M. W., Markham, M., Twitchen, D. J., Elkouss, D., Wehner, S., Taminiau, T. H., & Hanson, R. (2015). Loophole-free Bell inequality violation using electron spins separated by 1.3 kilometres. *Nature*, 526(7575), 682–686. 10.1038/nature1575926503041

Holevo, A. S. (2011). *Probablilistic and Statistical Aspects of Quantum Theory*. Springer Science & Business Media. 10.1007/978-88-7642-378-9

Horodecki, Horodecki, Horodecki, & Horodecki. (2009). Quantum entanglements. *Rev. Mod. Phys., 81*(2), 865-942. .10.1103/RevModPhys.81.865

Huang, Z., Joshi, S. K., Aktas, D., Lupo, C., Quintavalle, A. O., Venkatachalam, N., ... & Rarity, J. G. (2022). Experimental implementation of secure anonymous protocols on an eight-user quantum key distribution network. *NPJ Quantum Information, 8*(1), 25.

Huang, W., Wen, Q.-Y., Liu, B., Su, Q., & Gao, F. (2014). Cryptanalysis of a multi-party quantum key agreement protocol with single particles. *Quantum Information Processing*, 13(7), 1651–1657. 10.1007/s11128-014-0758-2

Hughes, R. J., Nordholt, J. E., Derkacs, D., & Peterson, C. G. (2002). Practical free-space quantum key distribution over 10 km in daylight and at night. *New Journal of Physics*, 4(1), 43. 10.1088/1367-2630/4/1/343

Hund, F. (1980). *History of quantum theory.* Academic Press.

Hwang, W. Y. (2003). Quantum key distribution with high loss: Toward global secure communication. *Physical Review Letters*, 91(5), 057901. 10.1103/PhysRevLett.91.05790112906634

Ignatovich, V. K. (2007). On EPR paradox, Bell's inequalities and experiments which prove nothing. *arXiv preprintquant-ph/0703192.*

Iwama, K., Kambayashi, Y., & Yamashita, S. (2002, June). Transformation rules for designing CNOT-based quantum circuits. In *Proceedings of the 39th annual Design Automation Conference* (pp. 419-424).

Jaeger, G. (2007). *Quantum information*. Springer.

Jeong, Y. C., Ji, S. W., Hong, C., Park, H. S., & Jang, J. (2020). Deterministic secure quantum communication on the BB84 system. *Entropy (Basel, Switzerland)*, 22(11), 1–13. 10.3390/e2211126833287036

Jiang, H., Wang, M., Liu, M., Yan, J., & Engineering, I. (2012). *A Quantum-inspired Ant-Based Routing algorithm forWSNs*. Academic Press.

Johnson, S. G., & Frigo, M. (2007). A modified split-radix FFT with fewer arithmetic operations. *IEEE Transactions on Signal Processing*, 55(1), 111–119. 10.1109/TSP.2006.882087

Kahn, D. (1968). *The Story of Secret Writing*. Weidenfeld & Nicolson.

Kaufman, C., Perlman, R., & Speciner, M. (2002). *Network Security* (2nd ed.). Prentice Hall.

Kaur, M., & Bhatia, R. (2019). Development of an improved tomato leaf disease detection and classification method. *2019 IEEE Conference on Information and Communication Technology, CICT 2019*. 10.1109/CICT48419.2019.9066230

Kelsey, J., Schneier, B., & Ferguson, N. (2000). Yarrow-160: Notes on the design and analysis of the yarrow cryptographic pseudorandom number generator. *Lecture Notes in Computer Science (Including Subseries Lecture Notes in Artificial Intelligence and Lecture Notes in Bioinformatics), 1758*, 13–33. 10.1007/3-540-46513-8_2

Kent, A. D., & Liebrock, L. M. (2011). Secure communication via shared knowledge and a salted hash in Ad-hoc environments. *Proceedings - International Computer Software and Applications Conference*, 122–127. 10.1109/COMPSACW.2011.30

Khan, M. M., Murphy, M., & Beige, A. (2009). High error-rate quantum key distribution for long-distance communication. *New Journal of Physics*, 11(6), 063043. 10.1088/1367-2630/11/6/063043

Khan, S., & Narvekar, M. (2020). Novel fusion of color balancing and superpixel based approach for detection of tomato plant diseases in natural complex environment. *Journal of King Saud University. Computer and Information Sciences*. Advance online publication. 10.1016/j.jksuci.2020.09.006

Kitaev, A., & Preskill, J. (2006). Topological Entanglement Entropy. *Physical Review Letters*, 96(11), 110404. 10.1103/PhysRevLett.96.11040416605802

Kotpalliwar, M. V., & Wajgi, R. (2015). Classification of attacks using support vector machine (SVM) on KDDCUP'99 IDS database. *Proceedings - 2015 5th International Conference on Communication Systems and Network Technologies, CSNT 2015*, 987–990. 10.1109/CSNT.2015.185

Kumar Sharma, D., Sreenivasa Chakravarthi, D., Ara Shaikh, A., Al Ayub Ahmed, A., Jaiswal, S., & Naved, M. (2021). The aspect of vast data management problem in healthcare sector and implementation of cloud computing technique. *Materials Today: Proceedings*, xxxx. Advance online publication. 10.1016/j.matpr.2021.07.388

Lanyon, B. P., Barbieri, M., Almeida, M. P., & White, A. G. (2008). Experimental quantum computing without entanglement. *Physical Review Letters*, 101(20), 200501. 10.1103/PhysRevLett.101.20050119113321

Li, C., Li, T., Liu, Y. X., & Cappellaro, P. (2021). Effective routing design for remote entanglement generation on quantum networks. *NPJ Quantum Information*, 7(1), 10. Advance online publication. 10.1038/s41534-020-00344-4

Li, J., Li, N., Li, L.-L., & Wang, T. (2016). One step quantum key distribution based on EPR entanglement. *Scientific Reports*, 6(1), 28767. 10.1038/srep2876727357865

Lindblad, G. (1999). A general no-cloning theorem. *Letters in Mathematical Physics*, 47(2), 189–196. 10.1023/A:1007581027660

Liu, J., & Gregory, T. Byrd, & Zhou, H. 2020. Quantum Circuits for Dynamic Runtime Assertions in Quantum Computation. In Proceedings of the Twenty-Fifth International Conference on Architectural Support for Programming Languages and Operating Systems (ASPLOS). Association for Computing Machinery, New York, NY, USA, 1017–1030. DOI:https://doi.org/10.1145/3373376.3378488

Liu, B., Gao, F., Huang, W., & Wen, Q. (2013). Multiparty quantum key agreement with single particles. *Quantum Information Processing*, 12(4), 1797–1805. 10.1007/s11128-012-0492-6

Liu, J., Bello, L., & Zhou, H. (2021, February). Relaxed peephole optimization: A novel compiler optimization for quantum circuits. In *2021 IEEE/ACM International Symposium on Code Generation and Optimization (CGO)* (pp. 301-314). IEEE.

Liu, W., Chen, J., Xu, Y., Tang, J., Tong, L., & Song, X. (2020). Quantum Algorithms and Experiment Implementations Based on IBM Q. *Computers, Materials & Continua*.

Liu, W., Huang, P., Peng, J., Fan, J., & Zeng, G. (2018). Integrating machine learning to achieve an automatic parameter prediction for practical continuous-variable quantum key distribution. *Physical Review. A*, 97(2), 1–9. 10.1103/PhysRevA.97.022316

Lo, H. K. (2001). Proof of unconditional security of six-state quantum key distribution scheme. *arXiv preprintquant-ph/0102138.*

Lo, H. K., & Chau, H. F. (1999). Unconditional security of quantum key distribution over arbitrarily long distances. *Science, 283*(5410), 2050-2056.

Lo, H. K., Ma, X., & Chen, K. (2005). Decoy state quantum key distribution. *Physical Review Letters*, 94(23), 230504. 10.1103/PhysRevLett.94.23050416090452

Lohachab, A., Lohachab, A., & Jangra, A. (2020). A comprehensive survey of prominent cryptographic aspects for securing communication in post-quantum IoT networks. *Internet of Things : Engineering Cyber Physical Human Systems*, 9(Mar), 100174. 10.1016/j.iot.2020.100174

Long. (2001). Grover algorithm with zero theoretical failure rate. *Physical Review A, 64*(2).

Mabesoone, G. F. (2019). Quantum computing and algorithms. *Nature Physics*, 15(11), 1070–1074.

Madhavi, R., Karri, R. R., Sankar, D. S., Nagesh, P., & Lakshminarayana, V. (2017). Nature inspired techniques to solve complex engineering problems. *Journal of Industrial Pollution Control*, 33(1), 1304–1311.

Madiwalar, S. C., & Wyawahare, M. V. (2017). Plant disease identification: A comparative study. *2017 International Conference on Data Management, Analytics and Innovation, ICDMAI 2017*, 13–18. 10.1109/ICDMAI.2017.8073478

Malathy, S., Santhiya, M., & Dhanaraj, R. K. (2022). Quantum Cryptographic Techniques. *Quantum Blockchain: An Emerging Cryptographic Paradigm*, 31-53.

Malina, L., Popelova, L., Dzurenda, P., Hajny, J., & Martinasek, Z. (2018). On Feasibility of Post-Quantum Cryptography on Small Devices. *IFAC-PapersOnLine*, 51(6), 462–467. 10.1016/j.ifacol.2018.07.104

Martınez, I. L. (2014). *Real world quantum cryptography* (Doctoral dissertation, University of Calgary).

Matsukevich, D. N., Maunz, P., Moehring, D. L., Olmschenk, S., & Monroe, C. (2008). Bell Inequality Violation with Two Remote Atomic Qubits. *Physical Review Letters*, 100(15), 150404. 10.1103/PhysRevLett.100.15040418518088

Mattle, K., Weinfurter, H., Kwiat, P. G., & Zeilinger, A. (1996). Dense Coding in Experimental Quantum Communication. *Physical Review Letters*, 76(25), 4656–4659. 10.1103/PhysRevLett.76.465610061348

Mavromatis, A., Ntavou, F., Salas, E. H., Kanellos, G. T., Nejabati, R., & Simeonidou, D. (2018). Experimental Demonstration of Quantum Key Distribution (QKD) for Energy-Efficient Software-Defined Internet of Things. *European Conference on Optical Communication, ECOC,* (1), 1–3. 10.1109/ECOC.2018.8535267

McClean, J. R., Romero, J., Babbush, R., & Aspuru-Huerva, A. (2016). The theory of variational quantum algorithms. arXiv preprint arXiv:1608.00238.

McEvoy, R., Curran, J., Cotter, P., & Murphy, C. (2007). *Fortuna: cryptographically secure pseudo-random number generation in software and hardware*. 10.1049/cp:20060479

Meunkaewjinda, A., Kumsawat, P., Attakitmongcol, K., & Srikaew, A. (2008). Grape leaf disease detection from color imagery using hybrid intelligent system. *5th International Conference on Electrical Engineering/Electronics, Computer, Telecommunications and Information Technology, ECTI-CON 2008, 1*, 513–516. 10.1109/ECTICON.2008.4600483

Mina, M., & Simion, E. (2020). A Scalable Simulation of the BB84 Protocol Involving Eavesdropping. *Cryptology EPrint Archive, 1*, 1–19. https://eprint.iacr.org/2020/1074

Mödersheim, S. (2009, March). Algebraic properties in Alice and Bob notation. In *2009 International Conference on Availability, Reliability and Security* (pp. 433-440). IEEE. 10.1109/ARES.2009.95

Mo, H., & Xu, L. (2011). Biogeography migration algorithm for traveling salesman problem. *International Journal of Intelligent Computing and Cybernetics*, 4(3), 311–330. 10.1108/17563781111160002

Mohajer, R., & Eslami, Z. (2017). Cryptanalysis of a multiparty quantum key agreement protocol based on commutative encryption. *Quantum Information Processing*, 16(8), 197. 10.1007/s11128-017-1647-2

Mokhtar, U., Ali, M. A. S., Hassenian, A. E., & Hefny, H. (2016). Tomato leaves diseases detection approach based on Support Vector Machines. *2015 11th International Computer Engineering Conference: Today Information Society What's Next? ICENCO 2015*, 246–250. 10.1109/ICENCO.2015.7416356

Mollazade, K., Omid, M., Akhlaghian Tab, F., Kalaj, Y. R., Mohtasebi, S. S., & Zude, M. (2013). Analysis of texture-based features for predicting mechanical properties of horticultural products by laser light backscattering imaging. *Computers and Electronics in Agriculture*, 98, 34–45. 10.1016/j.compag.2013.07.011

Mustafa, H. M., Badran, S. M., Al-Hamadi, A., & Al-Somani, T. F. (2011). On mathematical modeling of cooperative e-learning performance during face to face tutoring sessions (Ant Colony System approach). *2011 IEEE Global Engineering Education Conference, EDUCON 2011*, 338–346. 10.1109/EDUCON.2011.5773158

Nachman, B., Provasoli, D., de Jong, W. A., & Bauer, C. W. (2021). Quantum Algorithm for High Energy Physics Simulations. *Physical Review Letters*, 126(6), 062001.

Nahar, S., Pithawa, D., Bhardwaj, V., Rawat, R., Rawat, A., & Pachlasiya, K. (2023). Quantum Technology for Military Applications. *Quantum Computing in Cybersecurity*, 313-334.

Naikoo, J., Dutta, S., & Banerjee, S. (2019). Facets of quantum information under non-Markovian evolution. *Physical Review. A*, 99(4), 042128. 10.1103/PhysRevA.99.042128

Nielsen, M. A., & Chuang, I. L. (2001). Quantum computation and quantum information. *Physics Today*, 54(2), 60.

Nielsen, M. A., & Chuang, I. L. (2016). *Quantum Computation and Quantum Information.* Cambridge University Press.

Nithya, S., & Dhanuja, B., & AshaJerlin, M. (2016). Evolutionary algorithm – A review. *International Journal of Pharmacy and Technology*, 8(4), 25959–25966.

Noble, W. S. (2006). What is a support vector machine? *Nature Biotechnology*, 24(12), 1565–1567. 10.1038/nbt1206-156517160063

Noor-ul-Ain, W., Atta-ur-Rahman, M., Nadeem, M., & Abbasi, A. G. (2016). Quantum cryptography trends: a milestone in information security. In *Hybrid Intelligent Systems: 15th International Conference HIS 2015 on Hybrid Intelligent Systems, Seoul, South Korea, November 16-18, 201515* (pp. 25-39). Springer International Publishing. 10.1007/978-3-319-27221-4_3

Oppliger, R. (2021). *Cryptography 101: From Theory to Practice*. Artech House.

Orieux, A., & Diamanti, E. (2016). Recent advances on integrated quantum communications. *Journal of Optics*, 18(8), 083002. 10.1088/2040-8978/18/8/083002

Orzel, C. (2017). *How Do You Create Quantum?* Forbes. Retrieved from https://www.forbes.com/sites/chadorzel/2017/02/28/how-do-you-create-quantum-entanglement/?sh=308175441732

Osterloh, A., Amico, L., Falci, G., & Fazio, R. (2002). Scaling of entanglement close to a quantum phase transition. *Nature*, 608-610(6881), 608–610. Advance online publication. 10.1038/416608a11948343

Pan, D., Song, X. T., & Long, G. L. (2023). Free-space quantum secure direct communication: basics, progress, and outlook. *Advanced Devices & Instrumentation, 4*, 0004.

Pant, Y., & Bhadauria, H. S. (2017). Performance Study of Routing Protocols in Wireless Sensor Network. *Proceedings - 2016 8th International Conference on Computational Intelligence and Communication Networks, CICN 2016*, 134–138. 10.1109/CICN.2016.32

Pant, M., Krovi, H., Towsley, D., Tassiulas, L., Jiang, L., Basu, P., Englund, D., & Guha, S. (2019). Routing entanglement in the quantum internet. *NPJ Quantum Information*, 5(1), 1–9. 10.1038/s41534-019-0139-x

Papadimitriou, C. H. (2003). *Computational Complexity*. John Wiley and Sons Ltd.

Parakh, A. (2013). A probabilistic quantum key transfer protocol. *Security and Communication Networks*, 6(11), 1389–1395. 10.1002/sec.736

Patil, R. Y., & Devane, S. R. (2019). Network Forensic Investigation Protocol to Identify True Origin of Cyber Crime. *Journal of King Saud University. Computer and Information Sciences*. Advance online publication. 10.1016/j.jksuci.2019.11.016

Pemberton-Ross, P. J., & Kay, A. (2011). Perfect quantum routing in regular spin networks. *Physical Review Letters*, 106(2), 1–4. 10.1103/PhysRevLett.106.02050321405213

Peres, A. (1996). Separability Criterion for Density Matrices. *Physical Review Letters*, 77(8), 1413–1415. 10.1103/PhysRevLett.77.141310063072

Perry, R. T. (2006). The temple of quantum computing. *Riley Perry standard, Australia.* Available on: http://www.toqc.com/TOQCv1_1.pdf

Perseguers, S., Lewenstein, M., Acín, A., & Cirac, J. I. (2010). Quantum random networks. *Nature Physics*, 6(7), 539–543. 10.1038/nphys1665

Pfister, A. D., Salz, M., Hettrich, M., Poschinger, U. G., & Schmidt-Kaler, F. (2016). A quantum repeater node with trapped ions: A realistic case example. *Applied Physics. B, Lasers and Optics*, 122(4), 1–19. 10.1007/s00340-016-6362-7

Pirandola, S., Laurenza, R., Ottaviani, C., & Banchi, L. (2017). Fundamental limits of repeaterless quantum communications. *Nature Communications*, 8(1), 15043. Advance online publication. 10.1038/ncomms1504328443624

Pirker, A., & Dür, W. (2019). A quantum network stack and protocols for reliable entanglement-based networks. *New Journal of Physics*, 21(3), 033003. Advance online publication. 10.1088/1367-2630/ab05f7

Prasad, S., Kumar, P., Hazra, R., & Kumar, A. (2012). Plant leaf disease detection using Gabor wavelet transform. *Lecture Notes in Computer Science (Including Subseries Lecture Notes in Artificial Intelligence and Lecture Notes in Bioinformatics), 7677 LNCS*, 372–379. 10.1007/978-3-642-35380-2_44

Preskill, J. (2001). Lecture notes for ph219/cs219: Quantum information and computation.

Preskill, J. (2018). Quantum computing in the NISQ era and beyond. *Quantum : the Open Journal for Quantum Science*, 2, 79. 10.22331/q-2018-08-06-79

Quantum Computing and Communications - Google Books. (n.d.). Retrieved September 7, 2021, from https://books.google.co.in/books?hl=en&lr=&id=RQXpBwAAQBAJ&oi=fnd&pg=PR4&dq=quantum+computing+definitions&ots=BHTZvRoueA&sig=pr7ABq6SeIl0jd4siDiOZh2CxLE&redir_esc=y#v=onepage&q&f=false

Quantum computing and cybersecurity: How to capitalize on opportunities and sidestep risks | IBM. (n.d.). Retrieved September 8, 2021, from https://www.ibm.com/thought-leadership/institute-business-value/report/quantumsecurity# Ramentol10.1007/s10115-011-0465-6

Rarity, J. G., Tapster, P. R., Gorman, P. M., & Knight, P. (2002). Ground to satellite secure key exchange using quantum cryptography. *New Journal of Physics*, 4(1), 82. 10.1088/1367-2630/4/1/382

Rekaya, R., Robbins, K., Spangler, M., Smith, S., Hay, E. H., & Bertrand, K. (2013). *Ant Colony Algorithm with Applications in the Field of Genomics.* Ant Colony Optimization - Techniques and Applications. 10.5772/52051

Rivest, R. L., Shamir, A., & Adleman, L. (1978). A method for obtaining digital signatures and public-key cryptosystems. *Communications of the ACM*, 21(2), 120–126. 10.1145/359340.359342

Rosenberg, D., Harrington, J. W., Rice, P. R., Hiskett, P. A., Peterson, C. G., Hughes, R. J., Lita, A. E., Nam, S. W., & Nordholt, J. E. (2007). Long-distance decoy-state quantum key distribution in optical fiber. *Physical Review Letters*, 98(1), 010503. 10.1103/PhysRevLett.98.01050317358462

Saad, W., Bennis, M., & Chen, M. (2019). A vision of 6G wireless systems: Applications, trends, technologies, and open research problems. *ArXiv*, (June), 134–142.

Sardogan, M., Tuncer, A., & Ozen, Y. (2018). Plant Leaf Disease Detection and Classification Based on CNN with LVQ Algorithm. *UBMK 2018 - 3rd International Conference on Computer Science and Engineering*, 382–385. 10.1109/UBMK.2018.8566635

Sasaki, M. (2018). Quantum key distribution and its applications. *IEEE Security and Privacy*, 16(5), 42–48. 10.1109/MSP.2018.3761713

Scarani, V., Acín, A., Ribordy, G., & Gisin, N. (2004). Quantum cryptography protocols robust against photon number splitting attacks for weak laser pulse implementations. *Physical Review Letters*, 92(5), 057901. 10.1103/PhysRevLett.92.05790114995344

Scarani, V., Bechmann-Pasquinucci, H., Cerf, N. J., Dusek, M., Lütken, N., & Tittel, W. (2009). The security of practical quantum key distribution. *Reviews of Modern Physics*, 81(3), 1301–1350. 10.1103/RevModPhys.81.1301

Schlosshauer, M. (2019). Quantum decoherence. *Physics Reports*, 831, 1–57. 10.1016/j.physrep.2019.10.001

Schuld, M., & Killback, R. (2019). An introduction to quantum machine learning. arXiv preprint arXiv:1904.05366.

Sen, D. (2014). The uncertainty relations in quantum mechanics. *Current Science*, 107(2), 203–218. 10.18520/cs/v107/i2/203-218

Sethy, P. K., Negi, B., Behera, S. K., Barpanda, N. K., & Rath, A. K. (2017). An Image Processing Approach for Detection, Quantification, and Identification of Plant Leaf Diseases -A Review. *IACSIT International Journal of Engineering and Technology*, 9(2), 635–648. 10.21817/ijet/2017/v9i2/170902059

Shao, C., Li, Y., & Li, H. (2019). Quantum algorithm design: Techniques and applications. *Journal of Systems Science and Complexity*, 32(1), 375–452.

Shen, L. L., Bai, L., & Fairhurst, M. (2007). Gabor wavelets and General Discriminant Analysis for face identification and verification. *Image and Vision Computing*, 25(5), 553–563. 10.1016/j.imavis.2006.05.002

Shen, L., & Bai, L. (2006). A review on Gabor wavelets for face recognition. *Pattern Analysis & Applications*, 9(2–3), 273–292. 10.1007/s10044-006-0033-y

Shi, S., & Qian, C. (2019). *Modeling and Designing Routing Protocols in Quantum Networks*. Academic Press.

Shi, Y. (2004). Quantum entanglement in second-quantized condensed matter systems. *Journal of Physics. A, Mathematical and General*, 37(26), 6807–6822. 10.1088/0305-4470/37/26/014

Shor, P. W. (1995, October). Scheme for reducing decoherence in Quantum computer memory. *Physical Review A: General Physics*, 52(4), R2493–R2496. 10.1103/PhysRevA.52.R24939912632

Shor, P. W. (1997). Polynomial-time algorithms for prime factorization and discrete logarithms on a quantum computer. *SIAM Journal on Computing*, 26(5), 1464–1484. 10.1137/S0097539795293172

Shor, P. W. (1999). Polynomial-time algorithms for prime factorization and discrete logarithms on a quantum computer. *SIAM Review*, 41(2), 303–332. 10.1137/S0036144598347011

Shor, P. W., & Preskill, J. (2000). Simple proof of security of the BB84 quantum key distribution protocol. *Physical Review Letters*, 85(2), 441–444. 10.1103/PhysRevLett.85.44110991303

Sihare, S. R. (2022a). Dynamic multi-party quantum key agreement protocol based on commutative encryption. *International Journal of Theoretical Physics*, 61(9), 242. 10.1007/s10773-022-05203-w

Sihare, S. R. (2022b). Multi-party quantum key agreement protocol for detection of collusive attacks in each sub-circle segment by headers. *International Journal of Theoretical Physics*, 61(7), 208. 10.1007/s10773-022-05184-w

Sihare, S. R. (2022c). Transformation of Classical to Quantum Image, Representation, Processing and Noise Mitigation. *International Journal of Image. Graphics and Signal Processing*, 12(5), 10. 10.5815/ijigsp.2022.05.02

Sihare, S. R. (2022d). Qubit and bit-based quantum hybrid secret key generation. *The European Physical Journal D*, 76(11), 222. 10.1140/epjd/s10053-022-00532-1

Sihare, S. R. (2023a). Potential of quantum computing to effectively comprehend the complexity of brain. *Applied Intelligence*, 53(22), 1–24. 10.1007/s10489-023-04857-1

Sihare, S. R. (2023b). Multi-Party Quantum Key Distribution Using Variational Quantum Eigensolvers. *Advanced Quantum Technologies*, 2300270.

Sihare, S., & Khang, A. (2023). Effects of Quantum Technology on the Metaverse. In *Handbook of Research on AI-Based Technologies and Applications in the Era of the Metaverse* (pp. 174–203). IGI Global. 10.4018/978-1-6684-8851-5.ch009

Singh, H., Gupta, D. L., & Singh, A. (2014). Quantum Key Distribution Protocols: A Review. *IOSR Journal of Computer Engineering, 16*(2), 1–9. 10.9790/0661-162110109

Singh, D., & Singh, B. (2020). Investigating the impact of data normalization on classification performance. *Applied Soft Computing*, 97, 105524. Advance online publication. 10.1016/j.asoc.2019.105524

Singh, P. N., & Aarthi, S. (2021, February). Quantum Circuits–An Application in Qiskit-Python. In *2021 Third International Conference on Intelligent Communication Technologies and Virtual Mobile Networks (*ICICV) (pp. 661-667). IEEE

Singh, S. (2000). *The code book: the science of secrecy from ancient Egypt to quantum cryptography*. Anchor.

Singh, S. K., El Azzaoui, A., Salim, M. M., & Park, J. H. (2020). Quantum Communication Technology for Future ICT - Review. *Journal of Information Processing Systems*, 16(6), 1459–1478. 10.3745/JIPS.03.0154

Singh, T., Kumar, K., & Bedi, S. S. (2021). A review on artificial intelligence techniques for disease recognition in plants. *IOP Conference Series. Materials Science and Engineering*, 1022(1), 012032. Advance online publication. 10.1088/1757-899X/1022/1/012032

Sleator & Weinfurter. (1995). Realizable universal quantum logic gates. *Physical Review Letters, 74*(20).

Srivastava, A., & Chaudhury, P. (2021). Application of Nature-Inspired Swarm Optimization Algorithms in Artificial Neural Networks. *Proceedings - International Conference on Artificial Intelligence and Smart Systems, ICAIS 2021*, 1–6. 10.1109/ICAIS50930.2021.9395806

Srivastava, S., & Sahana, S. K. (2020). A survey on traffic optimization problem using biologically inspired techniques. *Natural Computing*, 19(4), 647–661. 10.1007/s11047-019-09731-z

Staff, S. (2020). *Quantum Cryptography & Encryption: What It Is & How It Works*. Sectigo. https://sectigo.com/resource-library/quantum-cryptography

Stallings, W. (2005). Cryptography and Network Security, 2005. Academic Press.

Stallings, W. (2003). *Cryptography and network security: Principles and practice*. Pearson Education India.

Steane, A. (1998). Quantum computing. *Reports on Progress in Physics*, 61(2), 117–173. 10.1088/0034-4885/61/2/002

Stinson, D. R. (2005). *Cryptography: Theory and practice*. CRC Press. 10.1201/9781420057133

Sujatha, B., Raju, S. V., & Rao, G. S. (2016). Proficient capability of QKD in Wi-Fi network system implementation. In *Communication and Electronics Systems (ICCES), International Conference on*. IEEE. 10.1109/CESYS.2016.7889981

Sun, Z., Huang, J., & Wang, P. (2016). Efficient multiparty quantum key agreement protocol based on commutative encryption. *Quantum Information Processing*, 15(5), 2101–2111. 10.1007/s11128-016-1253-8

Sun, Z., Yu, J., & Wang, P. (2016). Efficient multi-party quantum key agreement by cluster states. *Quantum Information Processing*, 15(1), 373–384. 10.1007/s11128-015-1155-1

Swathi & Rudra. (2022). An efficient approach for quantum entanglement purification. *International Journal of Quantum Information, 20*(4).

Swathi, M., & Rudra, B. (2021). Implementation of Reversible Logic Gates with Quantum Gates. *2021 IEEE 11th Annual Computing and Communication Workshop and Conference (CCWC)*. IEEE. 10.1109/CCWC51732.2021.9376060

Swathi, M., & Rudra, B. (2022). Novel Encoding method for Quantum Error Correction. *2022 IEEE 12th Annual Computing and Communication Workshop and Conference (CCWC)*. IEEE. 10.1109/CCWC54503.2022.9720880

Swathi, M., & Rudra, B. (2022). A Novel Approach for Asymmetric Quantum Error Correction With Syndrome Measurement. *IEEE Access : Practical Innovations, Open Solutions*, 10, 44669–44676. 10.1109/ACCESS.2022.3170039

Swathi, M., & Rudra, B. (2022). A Novel Architecture for Binary Code to Gray Code Converter Using Quantum Cellular Automata. In *Edge Analytics* (pp. 43–61). Springer. 10.1007/978-981-19-0019-8_4

Takeuchi & Yoshikawa. (2018). Minimum energy dissipation required for a logically irreversible operation. *Physical Review E, 97*(1).

Tanenbaum, A. S. (2011). *Computer Networks* (5th ed.). Prentice Hall.

Terhal, B. M. (2002). Detecting quantum entanglement. *Theoretical Computer Science*, 287(1), 313–335. 10.1016/S0304-3975(02)00139-1

Thapliyal, A. V. (1999). Multipartite pure-state entanglement. *Physical Review A*, 59(5), 3336–3342. 10.1103/PhysRevA.59.3336

Thapliyal, K., Pathak, A., & Banerjee, S. (2017). Quantum cryptography over non-Markovian channels. *Quantum Information Processing*, 16(5), 115. 10.1007/s11128-017-1567-1

Tokyo QKD Network. (2021). Tokyo QKD. http://www.tokyoqkd.jp/

Tomkos, I., Klonidis, D., Pikasis, E., & Theodoridis, S. (2020). Toward the 6G Network Era: Opportunities and Challenges. *IT Professional*, 22(1), 34–38. 10.1109/MITP.2019.2963491

Tsekeri, A., Amiridis, V., Louridas, A., Georgoussis, G., Freudenthaler, V., Metallinos, S., Doxastakis, G., Gasteiger, J., Siomos, N., Paschou, P., Georgiou, T., Tsaknakis, G., Evangelatos, C., & Binietoglou, I. (2021). Polarization lidar for detecting dust orientation: System design and calibration. *Atmospheric Measurement Techniques*, 14(12), 7453–7474. 10.5194/amt-14-7453-2021

Valivarthi, V. R. R. (2017). *Bell state measurements for quantum communication* (Doctoral dissertation, University of Calgary).

Van De Graaf, J. (1997). *Towards a formal definition of security for quantum protocols*. Université de Montréal.

Van Tilborg, H. C., & Jajodia, S. (Eds.). (2014). *Encyclopedia of cryptography and security*. Springer Science & Business Media.

Vidal, G., Latorre, J. I., Rico, E., & Kitaev, A. (2003). Entanglement in Quantum Critical Phenomena. *Physical Review Letters*, 90(22), 227902. 10.1103/PhysRevLett.90.22790212857342

Viswanathan, H., & Mogensen, P. E. (2020). Communications in the 6G Era. *IEEE Access : Practical Innovations, Open Solutions*, 8, 57063–57074. 10.1109/ACCESS.2020.2981745

Von Solms, R., & Van Niekerk, J. (2013). From information security to cyber security. *Computers & Security*, 38, 97–102. 10.1016/j.cose.2013.04.004

Wang, P., Sun, Z., & Sun, X. (2017). Multi-party quantum key agreement protocol secure against collusion attacks. *Quantum Information Processing*, 16(7), 170. 10.1007/s11128-017-1621-z

Weinberg, S. (1992). *Dreams of a final theory: The search for the fundamental laws of nature.* Pantheon Books.

Wei, T.-C., Nemoto, K., Goldbart, P. M., Kwiat, P. G., Munro, W. J., & Verstraete, F. (2003). Maximal entanglement versus entropy for mixed quantum states. *Physical Review A*, 67(2), 022110. 10.1103/PhysRevA.67.022110

Wiharto, N., Nashrullah, F. H., Suryani, E., Salamah, U., Prakisy, N. P. T., & Setyawan, S. (2021). Texture-Based Feature Extraction Using Gabor Filters to Detect Diseases of Tomato Leaves. *Revue d'Intelligence Artificielle*, 35(4), 331–339. 10.18280/ria.350408

Wilhite, S. E. (2012). Cloud computing? *HDA Now,* 12. 10.4018/IJCAC.2015040104

Wille, R., Van Meter, R., & Naveh, Y. (2019, March). IBM's Qiskit Tool Chain: Working with and Developing for Real Quantum Computers. In *2019 Design, Automation & Test in Europe Conference & Exhibition (DATE)* (pp. 1234-1240). IEEE.

Williams, C. P., & Williams, C. P. (2011). Quantum Cryptography. *Explorations in Quantum Computing*, 507-563.

Wootters, W. K., & Zurek, W. H. (1982). A single quantum cannot be cloned. *Nature*, 299(5886), 802–803. 10.1038/299802a0

Wootters, W. K., & Zurek, W. H. (2009). The no-cloning theorem. *Physics Today*, 62(2), 76–77. 10.1063/1.3086114

Yaacoub, E. (2021). Ad Hoc Networks Travel Hopping Enabled Resource Allocation (THEResA) and delay tolerant networking through the use of UAVs in railroad networks. *Ad Hoc Networks*, 122(June), 102628. 10.1016/j.adhoc.2021.102628

Yaacoub, E., & Alouini, M. S. (2020). A Key 6G Challenge and Opportunity—Connecting the Base of the Pyramid: A Survey on Rural Connectivity. *Proceedings of the IEEE, PP*, 1–50. 10.1109/JPROC.2020.2976703

Yanofsky, N. S., Mannucci, M. A., & Mannucci, M. A. (2008). Quantum Computing for Computer Scientists. Cambridge University Press. 10.1017/CBO9780511813887

Yildirim, T. (2010). *VOIP traffic classification in IPsec tunnelled networks* (Doctoral dissertation, RMIT University).

Yin, X., Ma, W., & Liu, W. (2013). Three-party quantum key agreement with two-photon entanglement. *International Journal of Theoretical Physics*, 52(11), 3915–3921. 10.1007/s10773-013-1702-4

Yusuf, I. I., Thomas, I. E., Spichkova, M., Androulakis, S., Meyer, G. R., Drumm, D. W., Opletal, G., Russo, S. P., Buckle, A. M., & Schmidt, H. W. (2015). Chiminey: Reliable Computing and Data Management Platform in the Cloud. *Proceedings - International Conference on Software Engineering. International Conference on Software Engineering*, 2, 677–680. 10.1109/ICSE.2015.221

Zbinden, H., Bechmann-Pasquinucci, H., Gisin, N., & Ribordy, G. (1998). Applied Physics B Lasers and Optics Quantum cryptography. *Applied Physics. B, Lasers and Optics*, 67(6), 743–748. 10.1007/s003400050574

Zhang, H., Ji, Z., Wang, H., & Wu, W. (2019). Survey on quantum information security. *China Communications*, 16(10), 1–36. 10.23919/JCC.2019.10.001

Zhang, Q., Goebel, A., Wagenknecht, C., Chen, Y.-A., Zhao, B., Yang, T., Mair, A., Schmiedmayer, J., & Pan, J.-W. (2006). Experimental quantum teleportation of a two-qubit composite system. *Nature Physics*, 2(10), 678–682. 10.1038/nphys417

Zhang, X., Xiang, H., & Xiang, T. (2018, December). An efficient quantum circuits optimizing scheme compared with qiskit (short paper). In *International Conference on Collaborative Computing: Networking, Applications and Worksharing* (pp. 467-476). Springer, Cham.

Zhao, L., Sakr, S., Liu, A., & Bouguettaya, A. (2014). Cloud data management. In *Cloud Data Management* (Vol. 9783319047). 10.1007/978-3-319-04765-2

Zhao, Y., & Qiao, C. (2021). *Quantum Transport Protocols for Distributed Quantum Computing*. Academic Press.

Zulehner, A., & Wille, R. (2019, January). Compiling SU (4) quantum circuits to IBM QX architectures. In *Proceedings of the 24th Asia and South Pacific Design Automation Conference* (pp. 185-190).

About the Contributors

Shyam R. Sihare completed his Ph.D. at Raksha Shakti University in Ahmedabad, India. He holds a Master's degree in Computer Science from Nagpur University, Nagpur, India, which he obtained in 2003. Additionally, he attained an M. Phil. in Computer Science from Madurai Kamraj University, Madurai, India. In 2011 and 2018, he successfully cleared the Professor Eligibility Test GSLET (Gujarat) and MS-SET (Maharashtra) in India, respectively. Furthermore, he completed his MCA from IGNOU in New Delhi, India, in 2011. Currently, Dr. Sihare serves as an Assistant Professor in Computer Science and Application at Dr. APJ Abdul Kalam Govt. College in Silvassa, Dadra & Nagar Haveli (UT), India. His research interests encompass a wide range of areas including Quantum Computing, Quantum Algorithms, Quantum Cryptography, and Classical Computer Algorithms.

* * *

Khalid Abualsaud (Senior Member, IEEE) is currently with the Computer Science and Engineering Department, Qatar University, Qatar. He has more than 25 years of professional expe- rience in information technology. He teaches courses in hardware and software systems. His research interests include health systems, wireless sensors for the IoT applications, cybersecurity, cloud computing, and computer network proto- cols. His researchwork has been presented in inter- national conferences and journals. He has participated actively in organizing several IEEE international conferences in Qatar, namely, ICIoT2020, IEEE WCNC'2016, PLM'2015, AICCSA'2014, RelMiCS'2011, and AICCSA'2008. He received several awards from different local and inter- national organizations. He is active in getting research funding from differ- ent sources, including Qatar National Research Foundation, the Supreme Committee for Delivery and Legacy (FIFA'2022), and some other orga- nizations in Qatar. He is also a LPI of NPRP 10-1205-160012 Research Project which achieved significant outcomes. He has served as a technical program committee (TPC) member and the chair for various reputable IEEE conferences. Recently, he served as a Guest Editor in Connected Healthcare Special Issue for IEEE Network. He is an Associate Editor of IET Quantum Communication journal.

Shruti Aggarwal is currently working as Associate Professor and batch head, in Department of Computer Science and Engineering at Chandigarh University. Previously, she has worked as Assistant Professor in Department of Computer Science at Sri Guru Granth Sahib World University, Fatehgarh Sahib, Punjab. She has done her doctoral degree in CSE from National Institute of Technology, Jalandhar in 2019. She completed her M.E. in C.S.E. from Panjab University, Chandigarh in 2011 and B.Tech. from Kurukshetra University in 2008. She has mentored many engineering graduates and guided more than 34 thesis in area of data mining and software engineering. During her more than decade of teaching experience, she has published more than 50 research works in highly reputed national and international journals and conferences. She has also worked with various industries in collaboration and is also member of various professional bodies and scientific societies.

Afnan Al-Ali received her Master of Science degree in Computer Engineering from the University of Basra, Basra, Iraq. She has finished her Ph.D. at Qatar University. Her research interests include machine learning, AI, computer vision, and object tracking. Additionally, she is interested in machine learning applications in healthcare.

Hasan Al-Mohammed received his bachelor's degree in Computer Engineering from Iraq University College, Basra, Iraq, in 2014, and his master's degree in Computing from Qatar University in June 2021. He is currently pursuing his Ph.D. in Quantum Optics at Qatar University. He has more than ten publications in international journals and conferences. His research interests include quantum radar, quantum computing, quantum communications, and security, as well as the Internet of Things (IoT) and sensor networks.

Afroj Jahan Badhon is pursuing her PhD Cyber Security domain in computer science and technology (CSE) from Chandigarh University, Punjab. She also working as an Instructor in Chandigarh University, Punjab. She has earned her Master's degree in computer science and engineering (CSE) from Jahangirnagar University, Savar, Bangladesh in 2019 in addition B.Sc. (C.S.E.) from Bangladesh university of Business and Technology, Dhaka, Bangladesh in 2017.

Vishal Bharti is currently working as Additional Director, in dept. of CSE. Previously he has worked as Professor and Head in Department CSE, School of Computing at DIT University, Dehradun. He has also worked as Professor and Head in AIT-CSE at Chandigarh University, Mohali, Punjab, Associate Prof. and; Head, Department of CSE at Dronacharya College of Engineering Gurgaon. He completed his Ph.D. in 2016. He did his M.Tech. and B.E. from Birla Institute of Technology, Mesra. Ranchi. He also holds Doctorate in Management Studies in addition to MBA in IT and E-MBA in S&M. He is having a mixed bag of experience of 14+ Years in both Academia and IT Industry He have five filed patents out of which two are in the process of Granting and Commercialization, Twenty-Seven copyrights and 13 Govt. Grants (SERB, DST, NSTEDB, EDI etc.) to his credit. He published more than Seventy-Eight research works at both National and; International domain like IEEE, Springer, Taylor and; Francis, Elsevir and was awarded with Best Faculty Award in 2010, Best Computer Teacher in 2019, Academic Leadership Award in 2019, Best Young HoD of the Year Award 2019. He is a member of more than 80+ International and National Societies/Conferences/Journals as Advisory Board.

Debotosh Bhattacharjee is working as a full professor in the Department of Computer Science and Engineering, Jadavpur University having sixteen years of post-PhD experience. His research interests pertain to the applications of machine learning techniques for Face Recognition, Gait Analysis, Hand Geometry Recognition, and Diagnostic Image Analysis. He has authored or coauthored more than 132 journals, 142 conference proceedings publications, 31 book chapters in Biometrics and Medical Image Processing. Two US patents have been granted on his works. Prof. Bhattacharjee has been granted sponsored projects by the Govt. of India funding agencies for a total amount of around Rs. 3 crores. For postdoctoral research, Dr. Bhattacharjee has visited different universities abroad like the University of Twente, The Netherlands; Instituto Superior Técnico, Lisbon, Portugal; University of Bologna, Italy; ITMO National Research University, St. Petersburg, Russia; University of Ljubljana, Slovenia; Northumbria University, Newcastle Upon Tyne, UK; Medical University Vienna, Austria and Heidelberg University, Germany. He is a Fellow of the West Bengal Academy of Science Technology, a life member of the Indian Society for Technical Education (ISTE, New Delhi), Indian Unit for Pattern Recognition and Artificial Intelligence (IUPRAI), and a senior member of IEEE (USA).

About the Contributors

Elias Yaacoub (Senior Member, IEEE) received the B.E. degree in electrical engineering from Lebanese University, in 2002, and the M.E. degree in computer and communications engi- neering and Ph.D. degree in electrical and com- puter engineering from the American University of Beirut (AUB), in 2005 and 2010, respectively. He worked as a Research Assistant with the Amer- ican University of Beirut, from 2004 to 2005, and Munich University of Technology, in Spring 2005. From 2005 to 2007, he worked as a Telecommunications Engineer with Dar Al-Handasah, Shair, and Partners. From November 2010 to December 2014, he worked as a Research Scientist/Research and Devel- opment Expert with Qatar Mobility Innovations Center (QMIC), where he led the Broadband Wireless Access Technology Team. Afterward, he joined the Strategic Decisions Group (SDG), where he worked as a Consultant, till February 2016. Then, he joined Arab Open University (AOU) as an Associate Professor and a Coordinator of the M.Sc. Program in information security and forensics. From February 2018 to August 2019, he worked as an Independent Researcher/a Consultant and he was also affiliated with AUB as a part-time Faculty Member. He has been an Associate Professor with the Computer Science and Engineering Department, Qatar University, since August 2019. His research interests include wireless communications, resource allocation in wireless networks, intercell interference mitigation techniques, antenna theory, sensor networks, and physical layer security.

Index

Symbols

A

B

D

E

H

I

M

N

P

Q

R

S

www.ingramcontent.com/pod-product-compliance
Lightning Source LLC
LaVergne TN
LVHW080633011026
849160LV00014B/1205
9781799895237